Biotechnology in Agriculture and Forestry

Edited by
T. Nagata (Managing Editor)
H. Lörz
J. M. Widholm

Volumes already published

Volume 1: Trees 1 (1986)
Volume 2: Crops 1 (1986)
Volume 3: Potato (1987)
Volume 4: Medicinal and Aromatic Plants 1 (1988)
Volume 5: Trees 11 (1989)
Volume 6: Crops 11 (1988)
Volume 7: Medicinal and Aromatic Plants 11 (1989)
Volume 8: Plant Protoplasts and Genetic Engineering 1 (1989)
Volume 9: Plant Protoplasts and Genetic Engineering 11 (1989)
Volume 10: Legumes and Oilseed Crops 1 (1990)
Volume 11: Somaclonal Variation in Crop Improvement 1 (1990)
Volume 12: Haploids in Crop Improvement 1 (1990)
Volume 13: Wheat (1990)
Volume 14: Rice (1991)
Volume 15: Medicinal and Aromatic Plants Ill (1991)
Volume 16: Trees 111 (1991)
Volume 17: High-Tech and Micropropagation 1 (1991)
Volume 18: High-Tech and Micropropagation 11 (1992)
Volume 19: High-Tech and Micropropagation Ill (1992)
Volume 20: High-Tech and Micropropagation IV (1992)
Volume 21: Medicinal and Aromatic Plants IV (1993)
Volume 22: Plant Protoplasts and Genetic Engineering 111 (1993)
Volume 23: Plant Protoplasts and Genetic Engineering IV (1993)
Volume 24: Medicinal and Aromatic Plants V (1993)
Volume 25: Maize (1994)
Volume 26: Medicinal and Aromatic Plants VI (1994)
Volume 27: Somatic Hybridization in Crop Improvement 1 (1994)
Volume 28: Medicinal and Aromatic Plants VII (1994)
Volume 29: Plant Protoplasts and Genetic Engineering V (1994)
Volume 30: Somatic Embryogenesis and Synthetic Seed 1 (1995)
Volume 31: Somatic Embryogenesis and Synthetic Seed 11 (1995)
Volume 32: Cryopreservation of Plant Germplasm 1 (1995)
Volume 33: Medicinal and Aromatic Plants VIII (1995)
Volume 34: Plant Protoplasts and Genetic Engineering VI (1995)
Volume 35: Trees IV (1996)
Volume 36: Somaclonal Variation in Crop Improvement 11 (1996)
Volume 37: Medicinal and Aromatic Plants IX (1996)
Volume 38: Plant Protoplasts and Genetic Engineering VII (1996)
Volume 39: High-Tech and Microprogation V (1997)
Volume 40: High-Tech and Microprogation VI (1997)
Volume 41: Medicinal and Aromatic Plants X (1998)
Volume 42: Cotton (1998)
Volume 43: Medicinal and Aromatic Plants XI (1999)
Volume 44: Transgenic Trees (1999)
Volume 45: Transgenic Medicinal Plants (1999)
Volume 46: Transgenic Crops 1 (1999)
Volume 47: Transgenic Crops 11 (2001)
Volume 48: Transgenic Crops Ill (2001)
Volume 49: Somatic Hybridization in Crop Improvement 11 (2001)
Volume 50: Cryopreservation of Plant Germplasm 11 (2002)
Volume 51: Medicinal and Aromatic Plants XII (2002)
Volume 52: Brassicas and Legumes: From Genome Structure to Breeding (2003)
Volume 53: Tobacco BY-2 Cells (2004)
Volume 54: *Brassica* (2004)
Volume 55: Molecular Marker Systems in Plant Breeding and Crop Improvement (2004)
Volume 56: Haploids in Crop Improvement II (2005)

Volumes in preparation

Tropical Crops I
Tropical Crops II
Plant Metabolomics

Biotechnology in Agriculture and Forestry 56

Haploids in Crop Improvement II

Edited by
C.E. Palmer, W.A. Keller, and K.J. Kasha

With 22 Figures and 12 Tables

Springer

Series Editors

Professor Dr. Toshiyuki Nagata
University of Tokyo
Gradulate School of Science
Department of Biological Sciences
7-3-1 Hongo, Bunkyo-ku
Tokyo 113-0033, Japan

Professor Dr. Horst Lörz
Universität Hamburg
Biozentrum Klein Flottbek
Zentrum für Angewandte Molekularbiologie
der Pflanzen (AMP II)
Ohnhorststraße 18
22609 Hamburg, Germany

Professor Dr. Jack Widholm
University of Illinois
285A E.R. Madigan Laboratory
Department of Crop Sciences
1201 W. Gregory
Urbana, IL 61801, USA

Volume Editors

Dr. C.E. Don Palmer
Dr. Wilfred A. Keller
NRC – Plant Biotechnology Institute
110 Gymnasium Place
Saskatoon, Saskatchewan
Canada S7N 0W9

Professor Dr. Ken J. Kasha
Departmenr of Plant Agriculture
University of Guelph
Guelph, Ontario
Canada N1G 2W1

ISSN 0934-943X
ISBN 978-3-642-06073-1 e-ISBN 978-3-540-26889-5

Springer is a part of Springer Science+Business Media
springeronline.com

Printed in Germany

Editor: Dr. Dieter Czeschlik, Heidelberg, Germany
Desk editor: Dr. Andrea Schlitzberger, Heidelberg, Germany
Cover Design: Design & Production, Heidelberg

Printed on acid-free paper 31/3150-WI – 5 4 3 2 1 0

Dedicated to the memory
of Professor Dr. Georg Melchers

Preface

On planning this monograph, our intent was to examine first the current status of knowledge of the fundamental aspects of gametophyte-to-sporophyte development and, second, the haploidy progress in representative species where it is being used for plant improvement. Consequently, the monograph is divided arbitrarily into two sections.

The first section deals with the molecular, cytological and biochemical aspects of haploid embryogenesis. In this case, microspore embryogenesis is emphasized as this system still represents the primary route to haploid and doubled haploid embryo development in most species. Here, the authors have presented an up-to-date review of the regulation of microspore embryo induction and development.

The second section is devoted to the utilization of haploids in the improvement of specific crop species. Here, we have grouped them into families containing commercially important crops. Although the Fabaceae, Euphorbiaceae and Malvaceae families contain commercially important species, these were not included because of the scarcity of literature on the induction, development and use of haploids in these species. Nevertheless, as we gain more basic understanding of the induction and regulation of haploid embryogenesis, the use of this technology will be of great value in the improvement of these and other species.

The chapter on utilization of haploid cells and embryos (Chap. I.8) addresses their potential use in gene transformation, mutation, selection and artificial seed technology. Microspore-derived embryos offer a convenient system for studies of storage product accumulation and metabolism. In Chapter I.6 the use of such embryos and haploid cell cultures for storage lipid and protein metabolism is examined. The treatment is restricted to *Brassica* as there have been no reports of such studies with other species.

For the chapter on miscellaneous species (Chap. II.5), the intent was to provide coverage of those crop species that could not be conveniently included in the major families. Here, the authors have covered five families with emphasis on the use of gynogenesis for doubled haploid production. This method is quite successful in the Liliaceae and Cucurbitaceae and may be useful even in cases where androgenesis is applicable as novel genetic recombinations may be uncovered. At the National Research Council of Canada Plant Biotechnology Institute in Saskatoon, research is ongoing, aimed at the potential application of doubled haploids to the improvement of commercially important members of the Umbelliferae, Labiatae and Caryophyllaceae. The results of those investigations will help in understanding species differences in haploid embryogenic response. Doubled haploid technology is of significant value in gene mapping and identification of quantitative trait loci (QTL), both of which are important for crop development.

With the current emphasis on a bioeconomy, renewable resources and sustainable development, existing crop species and emerging ones may have to be manipulated to produce biological molecules of commercial interest. There will be a need to improve crop adaptation to biotic and abiotic insults. Haploid technology is likely to be a valuable component of any strategy aimed at these improvements.

Bringing this material together as an overview should stimulate interest and the development of new concepts and mechanisms that will lead to further improvements and utilization of these very important haploid systems.

The editors greatly appreciate the cooperation of all the authors who contributed to this monograph, and we hope we have succeeded in highlighting the advances made in haploid embryo development and its potential uses. The editors wish to acknowledge the excellent technical assistance of Keith Pahl and Marie Mykytyshyn in editing this monograph. This work was performed as part of Genome Prairie's Enhancing Canola through Genomics project.

C.E. Palmer, W.A. Keller, and K.J. Kasha October 2004

Contents

List of Contributors

AIONESEI, T.
Max F. Perutz Laboratories, University Departments at the Vienna Biocenter, Institute of Microbiology and Genetics, Dr. Bohrgasse 9/4, 1030 Vienna, Austria

ANDERSEN, S.B.
Department of Agricultural Sciences, The Royal Veterinary and Agricultural University, Thorvaldsensvej 40, 1871 Frederiksberg C, Denmark

BOUTILIER, K.
Plant Research International. P.O. Box 16, 6700 AA Wageningen, The Netherlands

CLÉMENT, C.
Université de Reims Champagne Ardenne, UFR Sciences, Biologie et Physiologie Végétales, BP 1039, 51687 Reims Cedex 2, France

DEVAUX, P.
Florimond Desprez, Biotechnology Laboratory, 3 rue Florimond Desprez, P.O.B. 41, 59242 Cappelle en Pévèle, France

FIERS, M.
Plant Research International. P.O. Box 16, 6700 AA Wageningen, The Netherlands

FRIEDT, W.
Institute of Crop Science and Plant Breeding I, Justus-Liebig-University of Giessen, Heinrich-Buff-Ring 26–32, 35392 Giessen, Germany

FUKUOKA, H.
National Institute of Vegetable and Tea Science, Mie 514-2392, Japan

HEBERLE-BORS, E.
Max F. Perutz Laboratories, University Departments at the Vienna Biocenter, Institute of Microbiology and Genetics, Dr. Bohrgasse 9/4, 1030 Vienna, Austria

HORN, R.
Department of Genetics and Biochemistry, Clemson University,
100 Jordan Hall, Box 340324, Clemson, South Carolina 29634-0324, USA

JAKŠE, M.
Agronomy Department, Biotechnical Faculty, University in Ljubljana,
Jamnikarjeva 101, 1000 Ljubljana, Slovenia

JUHÁSZ, A.G.
Vegetable Crops Research Institute, Budapest, 1775 Pf 95, Hungary

KASHA, K.J.
Department of Plant Agriculture, University of Guelph, Guelph, Ontario,
N1G 2W1, Canada

KELLER, W.A.
NRC – Plant Biotechnology Institute, 110 Gymnasium Place, Saskatoon,
Saskatchewan, S7N 0W9, Canada

LIU, C.-M.
Plant Research International. P.O. Box 16, 6700 AA Wageningen,
The Netherlands

NICHTERLEIN, K.
Joint FAO/IAEA Division of Nuclear Techniques in Food and Agriculture,
Vienna, Austria. Present address: FAO, Regional Office for Europe,
Viale delle Terme di Caracalla, Rome, Italy

PALMER, C.E.
NRC – Plant Biotechnology Institute, 110 Gymnasium Place, Saskatoon,
Saskatchewan, S7N 0W9, Canada

PICKERING, R.
New Zealand Institute for Crop and Food Research Limited,
Private Bag 4704, Christchurch, New Zealand

SANGWAN, R.S.
Université de Picardie Jules Verne, Androgenèse et Biotechnologies, 33,
rue Saint-Leu, 80039 Amiens, France

SANGWAN-NORREEL, B.
Université de Picardie Jules Verne, Androgenèse et Biotechnologies, 33,
rue Saint-Leu, 80039 Amiens, France

TAI, G.C.C.
Potato Research Centre, Agriculture and Agri-Food Canada, P.O. Box 20280, Fredericton, New Brunswick, E3B 4Z7, Canada

TAKAHATA, Y.
Faculty of Agriculture, Iwate University, Morioka 020-8550, Japan

TASHPULATOV, A.S.
Max F. Perutz Laboratories, University Departments at the Vienna Biocenter, Institute of Microbiology and Genetics, Dr. Bohrgasse 9/4, 1030 Vienna, Austria

TOURAEV, A.
Max F. Perutz Laboratories, University Departments at the Vienna Biocenter, Institute of Microbiology and Genetics, Dr. Bohrgasse 9/4, 1030 Vienna, Austria

VAN DER GEEST, A. (Lonneke) H.M.
Plant Research International. P.O. Box 16, 6700 AA Wageningen, The Netherlands

WAKUI, K.
Junior College, Tokyo University of Agriculture, Setagaya 156-8502, Japan

WESELAKE, R.J.
Department of Chemistry and Biochemistry, University of Lethbridge, 4401 University Dr, Lethbridge, Alberta, T1K 3M4, Canada. Present address: Department of Agriculture, Food and Nutritional Science, 410 Agriculture/Forestry Centre, University of Alberta, Edmonton, Alberta, T6G 2P5, Canada

ZARHLOUL, M.K.
Institute of Crop Science and Plant Breeding I, Justus-Liebig-University of Giessen, Heinrich-Buff-Ring 26–32, 35392 Giessen, Germany

ZORINIANTS, S.
Max F. Perutz Laboratories, University Departments at the Vienna Biocenter, Institute of Microbiology and Genetics, Dr. Bohrgasse 9/4, 1030 Vienna, Austria

Section I Molecular, Cytological, and Biochemical Aspects of Haploid Embryogenesis

I.1 Overview of Haploidy

C.E. Don Palmer and Wilfred A. Keller[1]

1 Introduction

The term haploid sporophyte is generally used to designate sporophytes having the gametic chromosome number, and although the first haploids in flowering plants were identified over 80 years ago (Belling and Blakeslee 1922), it was not until Guha and Maheshwari (1964) reported the first in vitro culture anther-derived haploids from *Datura* that their potential for crop improvement was seriously contemplated as the value of quickly achieving homozygous lines was recognized. Since then, haploids and doubled haploids have been reported in a vast number of species and several cultivars have been developed using doubled haploids (Maluszynski et al. 2003a; Thomas et al. 2003). Consequently, the use of germplasm collection, classical plant breeding methods and genetic engineering for crop improvement can be supplemented with doubled haploid technology (Baenzinger 1996; Khush and Virmani 1996; Gepts 2002). Here we present a brief overview of the occurrence and experimental induction of haploidy.

2 Natural Occurrence of Haploids

Naturally occurring haploids have been reported in a number of species including tobacco, rice and maize (see review by Harlow et al. 1996). In *Brassica* homozygous diploid lines from naturally occurring haploids were reported by Thompson (1974). In barley, *Hordeum vulgare*, the *hap* initiator gene controls haploidy and spontaneous haploids were recovered at high frequency from barley (Hagberg and Hagberg 1980). In maize, the indeterminate gametophyte gene (*ig*) results in a monoploid embryo either from the sperm cell or from the egg cell (Kermicle 1969). Although doubled haploids can be recovered from such spontaneous haploids, the frequencies are too low for breeding purposes.

[1] NRC – Plant Biotechnology Institute, 110 Gymnasium Place, Saskatoon, Saskatchewan, S7N 0W9, Canada

Biotechnology in Agriculture and Forestry, Vol. 56
Haploids in Crop Improvement II (ed. by C.E. Palmer, W.A. Keller, and K.J. Kasha)

Haploid-inducing lines of *Zea mays* have been used to produce haploids by development of the unfertilized egg cell (Eder and Chalyk 2002) at frequencies of up to 8.0% and one line of maize showed a high frequency of haploidy (Coe 1959). The phenomenon of semigamy where disturbances during fertilization result in embryo formation from the egg cell without participation of the sperm cell occurs naturally in cotton (Turcotte and Feaster 1974). In this process both egg and sperm nuclei divide independently and may produce a chimera from which haploids can be isolated. In some cases, the haploid embryo is produced from either the egg cell or the sperm cell.

3 Induction of Haploidy

With the recognition of the importance of doubled haploids in plant breeding, extensive efforts were made to induce haploid embryogenesis and increase the frequency at which doubled haploids could be recovered. There are now four methods generally applicable to the production of haploids in plants at frequencies useful for a breeding program, and a recent monograph detailed the protocols applicable to haploid and doubled haploid production in a number of species (Maluszynski et al. 2003b). These methods are:

1. Androgenesis, where cultured anthers or isolated microspores undergo embryogenesis/organogenesis directly or through intermediate callus.
2. Gynogenesis, where cultured unfertilized isolated ovules, ovaries of flower buds, develop embryos from cells of the embryo sac.
3. Wide hybridization crosses followed by chromosome elimination from one parent of a cross, usually the pollinating parent.
4. Parthenogenesis, where there is development of an embryo by pseudogamy, semigamy or apogamy.

3.1 Androgenesis

Under the appropriate culture conditions, responsive microspores undergo cell division and organize embryos (sporophytes) rather than gametophytes. A number of factors influence embryogenic response of cultured anthers and microspores and these have been extensively reviewed (Ferrie et al. 1995; Sopory and Munshi 1996; Wang et al. 2000; Touraev et al. 2001). For some species and genotypes these requirements may be more stringent than for others.

Culture response is influenced by genotype and donor plant growth conditions. The nuclear stage of the microspore is a key factor for embryogenic response and the mid-uninucleate stage to the early binucleate stage is the responsive stage in most cases. There is a requirement for elevated temperature and/or nutritional or osmotic stress during the initial stages of culture

(Touraev et al. 2001). The type and levels of carbohydrates or polymers also influence embryogenic outcome (Ilic-Grubor et al. 1998; Touraev et al. 2001). In some cases, there may be a requirement for temperature or osmotic pretreatment of anthers, isolated microspores or flower buds to ensure success. Even when all these conditions are met, some species and genotypes may not respond to culture. During microspore embryogenesis all stages characteristic of zygotic embryo development are evident. After 21–28 days in culture, fully developed embryos are recognizable. Because of the potential to produce large numbers of embryos, the use of isolated microspore culture is increasing in importance in both monocots and dicots. However, of 193 species reported in 1996, anther culture was used in 105 species (Maluszynski et al. 1996).

3.2 Gynogenesis

This is an alternative route to haploid embryogenesis where under the appropriate culture conditions the unfertilized egg cell of the embryo sac develops into an embryo by yet unknown mechanisms. Other cells of the embryo sac, antipodals or synergids may produce the embryo (Mukhambetzhanov 1997). The culture response is still genotype dependent (Alan et al. 2003; Bohanec et al. 2003) and culture media composition and stage of embryo sac development are important considerations for successful culture (Keller and Korzun 1996; Sita 1996). Depending on the species, unfertilized ovule, ovary or flower bud can be cultured. In some members of the Chenopodiaceae, Lilliaceae and Cucurbitaceae, gynogenesis is the main route to doubled haploid production. Even where anther or microspore culture is successful, gynogenetic haploids have been produced, e.g. in barley, maize, rice and wheat (Sita 1996). Embryogenic frequency is low in many cases, but relatively high frequencies have been reported in other cases (Martinez et al. 2000; Alan et al. 2003). Genetic stability of the doubled haploids and the absence of albinism are attributes of this method (Touraev et al. 2001).

3.3 Wide Hybridization Crosses

This method of haploid production, through the elimination of all the chromosomes of the pollinating parent of a wide cross, is sometimes referred to as the bulbosum method as it came into prominence with the recovery of haploid *Hordeum vulgare* in a cross with *Hordeum bulbosum* as the pollinating parent (Kasha and Kao 1970). After fertilization, there is usually endosperm failure and the embryo must be rescued and cultured in vitro. Doubled haploids are recovered by treating either the embryo or the plantlet with colchicine. This approach is now widely used in cereals, especially wheat, where maize pollination yielded wheat haploids with high efficiency (Laurie and Bennett 1988; Kisana et al. 1993; Mujeeb-Kazi and Riera-Lizarazu 1996).

In a number of cereal species these crosses result in haploid recovery (see Khush and Virmani 1996). Other species of Poaceae can be used as pollinators (Falk and Kasha 1983), and with maize there are genotypic differences affecting the efficiency of pollination (Verma et al. 1999). The advantages of this method are genotype independence, drastic reduction in albinism and absence of gametoclonal variation. Doubled haploid lines produced by this method compared favorably with those produced by anther culture and by single seed descent (Guzy-Wrobelska and Szarejko 2003). However, instances of reduced fertility and pollinating parent chromosome retention have been reported (Riera-Lizarazu et al. 1996).

3.4 Parthenogenesis

With this method of haploid induction, the egg cell of the embryo sac usually develops into an embryo without the active involvement of the sperm nucleus. This process is referred to as pseudogamy; where the embryo develops from any haploid cell of the embryo sac other than the egg cell the process is called apogamy.

A distinction can be made between this form of parthenogenesis and embryo development by chromosome elimination and gynogenesis. In the latter two processes, there is usually endosperm failure and the embryo must be rescued for continued development in vitro. In the former, there is endosperm development and embryo maturation occurs in vivo.

Parthenogenesis can be induced by pollination with inactivated pollen or a variety of chemical treatments (Khush and Virmani 1996; Sestili and Ficcadeni 1996). The genes controlling haploidy such as indeterminate gametophyte (*ig*) in maize (Kermicle 1969) and the haploid initiator gene of barley (Hagberg and Hagberg 1980) induce embryo development by parthenogenesis. The frequency of parthenogenetic haploids is usually too low for plant breeding purposes (Khush and Virmani 1996). However, in potato (*Solanum tuberosum* L.) the use of special pollinator species induced haploidy at high enough frequency to be of value in breeding programs (Hutten et al. 1994; Peloquin et al. 1996; Straadt and Rasmussen 2003). Even though anther culture is reasonably successful as in potato (Rokka 2003), parthenogenesis through superior pollinators is still preferred to anther culture (Peloquin et al. 1996).

4 Conclusion

From the foregoing discussion, it can be concluded that the choice of method for haploid and doubled haploid production will depend on the species, genotype, efficiency of the generation, genetic stability of the doubled haploids and the ease of application of the method. Doubled haploids can occur spontaneously, but in most cases chromosome doubling of haploids is required to restore fertility. This is achieved by the use of antimicrotubule agents. In many cases androgenesis can be efficient especially in responsive genotypes where isolated microspore culture is used. Apart from doubled haploid production, this method is convenient for mutagenesis, transformation, basic research studies and other uses relevant to crop improvement. Gynogenetic haploid production is efficient in a few species and has been employed in cases where androgenesis and other methods prove intractable. Chromosome elimination technique is widely applicable in monocots even though in many cases androgenesis is efficient. The advantages are greater genetic stability and absence of albinism.

Parthenogenetic haploids have been detected in nature at low frequency and are inducible by pollination with inactivated pollen. The use of special pollinator species has allowed the recovery of haploid embryos, arising from pseudogamy, at high efficiencies in members of the Solanaceae.

References

Alan AR, Mutschler MA, Brants A, Cobb E, Earle ED (2003) Production of gynogenic plants from hybrids of *Allium apa* L. and *A. roylei Stearn*. Plant Sci 165:1201–1211

Baenzinger PS (1996) Reflections on doubled haploids in plant breeding. In: Jain SM, Sopory SK, Veilleux RE (eds) In vitro haploid production in higher plants, vol 1. Kluwer, Dordrecht, pp 35–48

Belling J, Blakeslee AF (1922) The assortment of chromosome in triploid *Daturas*. Am Nat 56:339–346

Bohanec B, Jakse M, Havey MJ (2003) Genetic analyses of gynogenetic haploid production in onion. J Am Hortic Sci 128:571–574

Coe EH (1959) A line of maize with high haploid frequency. Am Nat 93:381–382

Eder J, Chalyk S (2002) In vitro haploid induction in maize. Theor Appl Genet 104:703–708

Falk DE, Kasha KJ (1983) Genetic studies of the crossability of hexaploid wheat with rye and *Hordeum bulbosum*. Theor Appl Genet 64:303–307

Ferrie AMR, Palmer CE, Keller WA (1995) Haploid embryogenesis. In: Thorpe TA (ed) In vitro embryogenesis in plants, vol 20. Kluwer, Dordrecht, pp 309–344

Gepts P (2002) A comparison between crop domestication, classical plant breeding and genetic engineering. Crop Sci 42:1780–1790

Guha S, Maheshwari SC (1964) In vitro production of embryos from anthers of *Datura*. Nature 204:497

Guzy-Wrobelska J, Szarejko I (2003) Molecular and agronomic evaluation of wheat doubled haploid lines obtained through maize pollination and anther culture methods. Plant Breed 122:305–313

Hagberg A, Hagberg G (1980) High frequency of spontaneous haploids in the progeny of an induced mutation barley. Hereditas 93:341–343

Harlow C, Hamza S, Chupeau Y, Pelletier G (1996) Conditional lethal markers: spontaneous haploid selection in plants. In: Jain SM, Sopory SK, Veilleux RE (eds) In vitro haploid production in higher plants, vol 1. Kluwer, Dordrecht, pp 297–315

Hutten RCB, Scholberg EJMM, Huigen DJ, Hermsen JG, Jocobsen E (1994) Analysis of dihaploid induction and production ability and seed parent×pollinator interaction in potato. Euphytica 72:61–64

Ilic-Grubor K, Attree SM, Fowke LC (1998) Induction with polyethylene glycol (PEG) as osmoticum in a low sucrose medium. Plant Cell Rep 17:329–333

Kasha KJ, Kao KN (1970) High frequency haploid production in barley (*Hordeum vulgare* L.). Nature 225:874–876

Keller ERJ, Korzun L (1996) Ovary and ovule culture for haploid production. In: Jain SM, Sopory SK, Veilleux RE (eds) In vitro haploid production in higher plants, vol 1. Kluwer, Dordrecht, pp 217–236

Kermicle JL (1969) Androgenesis conditioned by a mutation in maize. Science 166:1422–1424

Khush GS, Virmani S (1996) Haploids in plant breeding. In: Jain SM, Sopory SK, Veilleux RE (eds) In vitro haploid production in higher plants, vol 1. Kluwer, Dordrecht, pp 11–33

Kisana NS, Nkongolo KK, Quick JS, Johnson DL (1993) Production of doubled haploids by anther culture and wheat × maize method in a wheat breeding program. Plant Breed 110:96–102

Laurie DA, Bennett MD (1988) The production of haploid wheat plants from wheat × maize crosses. Theor Appl Genet 76:393–397

Maluszynski M, Kasha KJ, Forster BP, Szarejko I (2003a) Doubled haploid production in crop plants, a manual. Kluwer, Dordrecht

Maluszynski M, Kasha KJ, Szarejko I (2003b) Published doubled haploid protocols in plant species. In: Maluszynski M, Kasha KJ, Forster BP, Szarejko I (eds) Doubled haploid production in crop plants, a manual. Kluwer, Dordrecht, pp 309–335

Maluszynski M, Szarejko I, Sigurbjornsson B (1996) Haploid and mutation techniques. In: Jain SM, Sopory SK, Veilleux RE (eds) In vitro haploid production in higher plants, vol 1. Kluwer, Dordrecht, pp 67–73

Martinez LE, Agüero CB, Lopez ME, Galmarini CR (2000) Improvement of in vitro gynogenesis induction in onion (*Allium apa* L.) using polyamines. Plant Sci 156:221–226

Mujeeb-Kazi A, Riera-Lizarazu O (1996) Polyhaploid production in the Triticeae by sexual hybridization. In: Jain SM, Sopory SK, Veilleux RE (eds) In vitro haploid production in higher plants, vol 1. Kluwer, Dordrecht, pp 275–296

Mukhambetzhanov SK (1997) Culture of nonfertilized female gametophytes in vitro. Plant Cell Tissue Organ Cult 48:111–119

Peloquin SJ, Gabert AC, Ortiz R (1996) Nature of 'pollinator' effect in potato (*Solanum tuberosum* L.) haploid production. Ann Bot 77:539–542

Riera-Lizarazu O, Rines HW, Phillips RL (1996) Cytological and molecular characterization of oat × maize partial hybrids. Theor Appl Genet 93:123–135

Rokka VM (2003) Anther culture through direct embryogenesis in a genetically diverse range of potato (*Solanum*) species and their interspecific and intergeneric hybrids. In: Maluszynski M, Kasha KJ, Forster BP, Szarejko I (eds) Doubled haploid production in crop plants, a manual. Kluwer, Dordrecht, pp 235–240

Sestili S, Ficcadeni N (1996) Irradiated pollen for haploid production. In: Jain SM, Sopory SK, Veilleux RE (eds) In vitro haploid production in higher plants, vol 1. Kluwer, Dordrecht, pp 263–274

Sita GL (1996) Gynogenic haploids in vitro. In: Jain SM, Sopory SK, Veilleux RE (eds) In vitro haploid production in higher plants, vol 5. Kluwer, Dordrecht, pp 175–193

Sopory S, Munshi M (1996) Anther culture. In: Jain SM, Sopory SK, Veilleux RE (eds) In vitro haploid production in higher plants, vol 1. Kluwer, Dordrecht, pp 145–176

Straadt IK, Rasmussen OS (2003) AFLP analysis of *Solanum phureja* DNA introgression into potato dihaploid. Plant Breed 122:352–356

Thomas WTB, Forster BP, Gertsson B (2003) Doubled haploids in breeding. In: Maluszynski M, Kasha KJ, Forster BP, Szarejko I (eds) Doubled haploid production in crop plants, a manual. Kluwer, Dordrecht, pp 337–349

Thompson KF (1974) Homozygous diploid lines from naturally occurring haploids. In: Proc 4th Int Rapskongr, Giessen, Germany, pp 119–124

Touraev A, Pfosser M, Heberle-Bors E (2001) The microspore: a haploid multipurpose cell. Adv Bot Res 35:53–109

Turcotte EL, Feaster CV (1974) Methods of producing haploids: semigamy production of cotton haploids. In: Kasha KJ (ed) Haploids in higher plants – advances and potential. University of Guelph, Guelph, pp 53–64

Verma V, Baines MS, Mangat GS, Nanda GS, Gosal SS, Singh K (1999) Maize genotypes show striking differences for induction and regeneration of haploid wheat embryos in wheat × maize system. Crop Sci 49:1722–1727

Wang M, van Bergen S, van Duijn B (2000) Insights into a key developmental switch and its importance for efficient plant breeding. Plant Physiol 124:523–530

I.2 Pathways to Microspore Embryogenesis

TATIANA AIONESEI, ALISHER TOURAEV, and ERWIN HEBERLE-BORS[1]

1 Introduction

Plant embryo formation comprises two processes that are of fundamental interest to developmental biologists: first, the establishment of the precise spatial organization of the component cells derived from a single, fertilized egg cell – pattern formation; and second, the generation of cellular diversity within the developing embryo – cytodifferentiation (Lindsey and Topping 1993). Both processes are tightly coordinated to create a recognizable morphological structure and are integrated by the apical–basal polarity established very early in the embryo and maintained to create the species-specific 'gestalt' of the mature plant.

Embryogenesis is the initial phase of the life cycle, taking place within the confines of the ovule and later the seed, and comprises the events during which a unicellular zygote undergoes morphological and cellular changes resulting in the formation of a mature zygotic embryo. These events can be subdivided into (1) establishment of the basic body plan, including the meristems, (2) differentiation of the primary plant tissue types, (3) generation of specialized storage organs essential for seed germination, and (4) features enabling the new individual to remain dormant until conditions are favorable for germination and post-embryogenic development (Goldberg et al. 1994).

Plants display a remarkable potential for cellular totipotency and it appears that any differentiated plant cell that retains its nucleus has the ability to revert to the embryogenic condition and regenerate an entire plant (Reynolds 1997). Embryogenesis in higher plants not only occurs after the fusion of egg and sperm cell in the ovule but also can develop naturally from unfertilized ovules (Koltunov 1993) or on the surface of leaves as in *Malaxis* (Taylor 1967), and it can be induced by experimental manipulation. Somatic embryo development was first observed independently by Steward et al. (1958) and Reinert (1959) in in vitro cultured carrot cells. Somatic embryo formation usually requires first a treatment of diploid cells with plant hormones, mostly auxin, in specific culture conditions and, later, auxin withdrawal to allow embryogenesis to continue (Toonen et al. 1994).

[1] Max F. Perutz Laboratories, University Departments at the Vienna Biocenter, Institute of Microbiology and Genetics, Dr. Bohrgasse 9/4, 1030 Vienna, Austria

Biotechnology in Agriculture and Forestry, Vol. 56
Haploids in Crop Improvement II (ed. by C.E. Palmer, W.A. Keller, and K.J. Kasha)

Male reproductive cells were first demonstrated to be totipotent in *Datura innoxia* by Guha and Maheshwari (1964). Embryogenesis from microspores inside cultured anthers or in isolated microspore cultures is induced by defined stress treatments, which convert a cell destined to produce a male gametophyte, i.e. a pollen grain, into an embryogenic cell which will develop into a sporophyte, i.e. an embryo (Touraev et al. 1997).

Microspore culture has several advantages which make this experimental system attractive to study plant embryogenesis: its main advantage is that it starts from single, defined haploid cells which are available in large amounts as a synchronous population and proceed to form embryos in a highly efficient manner. In tobacco, rapeseed, wheat and barley microspore embryogenesis is induced directly from freshly isolated, single cells, by a specific stress treatment (starvation or heat shock); hormonal treatments are not required to induce or maintain embryogenesis (Heberle-Bors 1999).

The focus of this chapter is to give an overview of the mechanisms underlying the early events of microspore embryogenesis. In particular, the developmental switch from gametophytic to sporophytic development, the pathways of division of embryogenic microspores and apical–basal polarity formation during embryogenesis will be dealt with.

2 Embryogenic Induction of Microspores

Stress pretreatment of microspores, either prior to microspore isolation in vivo or in vitro after isolation of the microspores, has been shown to redirect microspores, destined to develop into pollen grains, from gametophytic to a default sporophytic development. The inductive stress pretreatment has been used on donor plants (Heberle-Bors 1983), on severed buds or spikes (Gaillard et al. 1991), on excised anthers (Hoekstra et al. 1992) or on isolated microspores (Touraev et al. 1996a). Four categories of stress can be distinguished: heat shock in rapeseed, wheat and tobacco; cold shock in maize, wheat, barley, rice and other species; carbohydrate and nitrogen starvation in tobacco, wheat, rice and barley; and, finally, colchicine treatment in rapeseed (see Touraev et al. 2001 for a review).

In *Nicotiana tabacum*, embryogenic cells were originally obtained from isolated mid-bicellular pollen grains cultured under starvation conditions (Kyo and Harada 1986). Starvation in a medium without sugar but with nitrogen, followed by culture in a rich medium, produced a high number of embryos (Benito-Moreno et al. 1988). A heat-shock treatment was not effective at this stage of development, but was effective to induce embryogenesis at an earlier stage when unicellular microspores were used. Up to 75% of the initial population of unicellular microspores were induced to divide when they were starved in a sucrose-free medium for 6 days at 33 °C and then cultured in a sucrose-containing medium at 25 °C (Touraev et al. 1996a).

Many laboratories focused on *Brassica napus* because this species is an important crop species and because the whole process from microspore isolation to seedling formation requires only 2 weeks compared to 6 weeks in tobacco and other species. Microspores in the late unicellular or early bicellular stages are competent for embryo formation. Gametophytic development is maintained at 18 °C and can be changed to sporophytic development by elevating the temperature to 32 °C (Custer et al. 1994). About 8 h of in vitro culture at 32 °C is sufficient to induce sporophytic development.

In *Triticum aestivum*, culture of excised anthers under starvation and heat shock conditions (33 °C, 4 days) induced the formation of embryogenic microspores at high frequency (Touraev et al. 1996b). After isolation of the embryogenic microspores from the stressed anthers, up to 70 % of the cultured wheat microspores were induced to divide, and embryos usually developed by direct embryogenesis from single defined embryogenic cells (Touraev et al. 1996b).

3 Cellular Changes and Cell Cycle Events During Induction of Embryogenesis

The ultrastructure of microspore embryogenesis has been studied in detail in several species; *Nicotiana tabacum* (Dunwell and Sunderland 1974a,b), *Brassica napus* (Zaki and Dickinson 1990), *Datura inoxia* (Sunderland et al. 1974), *Hyoscyamus niger* (Raghavan 1978). Taken together, these studies give a good overview of the subcellular changes that occur during the conversion of a gametophytically programmed cell to an embryo founder cell: (1) fragmentation of the vacuole by formation of cytoplasmic strands from the perinuclear to the subcortical cytoplasm, (2) movement of the nucleus to the center of the microspore resulting in a central phragmosome, (3) increase in size of the cell, (4) formation of a new cell wall below the exine, (5) size reduction of the nucleolus, (6) compaction of chromatin, (7) structural simplification of the plastids, (8) size reduction of the starch grains, and (9) no marked structural change of mitochondria (Dunwell and Sunderland 1974a,b; Sangwan and Camefort 1984; Zaki and Dickinson 1990; Garrido et al. 1995; Touraev et al. 1996a; see also Chap. I.4, this Vol.).

3.1 Subcellular Changes During Embryogenic Induction

3.1.1 Nicotiana tabacum

Subcellular changes associated with embryogenic induction have been studied by a number of research groups, and contradictory observations exist depending upon whether the observations were followed in bicellular pollen grains or unicellular microspores of cultured anthers or microspore cultures. In bicellular pollen grains inside cultured anthers, a major change during the developmental switch from gametophytic to sporophytic development was the regression of the cytoplasmic organization in the vegetative cell (Dunwell and Sunderland 1974a,b, 1975; Rashid et al. 1981, 1982). Organelles appeared in complete disarray and in various states of disintegration and are associated with the disappearance of ribosomes and the appearance of zones of multivesiculate bodies resembling lysosomes. Mitochondria were condensed and slightly reduced in number, while plastids were strongly reduced in number and in structure and lacked starch grains seen in the pollen grains before culture. When the vegetative cell began to divide, the newly formed cells returned to the ultrastructural profile of metabolically active cells, i.e. increase in the number of ribosomes, lipid centers, starch-containing plastids and a change in mitochondrial morphology. The sequential disappearance of cytoplasmic organelles during embryogenic induction followed by the repopulation of the cell by a new set of organelles has led to the view that embryogenic pollen grains are dormant structures which remain inactive but can be activated to form embryos or tissues (Rashid et al. 1982).

When immature pollen grains of tobacco in direct culture were analyzed they were found to undergo regression of its cytoplasmic contents, similar to what had been described earlier in cultured anthers, to produce a large vacuole, dilatation of the generative cell wall, loss of nuclear pores in the vegetative nucleus, marked chromatin condensation, a decrease in size of the nucleolus and dedifferentiation of the plastids (Garrido et al. 1995). Furthermore, isolated late-unicellular microspores and bicellular pollen of tobacco were found to swell, the nucleus to move to a more central position and cytoplasm strands to form that pass through the vacuole and connect the perinuclear cytoplasm with the subcortical cytoplasm (Touraev et al. 1996a). Similar events have been observed in microspores of rapeseed (Zaki and Dickinson 1991), wheat (Touraev et al. 1996b) and rice (Raina and Irfan 1998).

In another study, proplastids were found to be specific to pollen grains of androgenic species, while differentiated plastids or amyloplasts were thought to be characteristic of recalcitrant species (Sangwan and Sangwan-Norreel 1987b). During in vitro embryogenesis of androgenic species, proplastids of the microspores were found to transform into amyloplasts and then into chloroplasts, while the amyloplasts detected in microspores of non-androgenic species retained their original morphology and the number of starch grains per plastid increased in vivo and in in vitro cultures.

3.1.2 *Brassica napus*

In the early phase of embryogenesis from heat-stressed isolated microspores, major changes in the cytoplasmic organization of the microspores take place which trigger the division of the embryogenic pollen grains. The nucleus moves from a peripheral to a central position, the cell develops a thick fibrillar wall, situated immediately adjacent to the intine of embryogenic pollen, starch synthesis commences in the plastids, the ribosome population increases, and cytoplasmic aggregates of an unidentified globular material appear (Zaki and Dickinson 1990). The cytoplasmic granules have been speculatively considered as heat-shock granules that probably shield mRNAs from damage (Telmer et al. 1995).

3.2 Cell Cycle Events and Cytoskeletal Changes During Embryogenic Induction

A simple fact testifies to the close link between microspore embryogenesis, the cytoskeleton and the cell cycle: *Brassica* microspores cultured under non-inducible temperatures can be induced to develop into embryos by adding colchicine to the culture medium (Zhao et al. 1996). Apart from the notion that it may exert a kind of chemical stress or that disruption of microtubules as such trigger totipotency (Zhao et al. 1996), colchicine's mechanism of action on embryogenic induction in microspores is, however, not yet understood. Another – the originally intended – effect of colchicine is in doubling the chromosome number of microspore embryos. It disrupts spindle formation in mitosis and nuclear reconstitution and lack of cytokinesis results in chromosome doubling. Amongst the different antimicrotubular cell cycle drugs colchicine is the most effective (Rao and Suprasanna 1996) and has been used not only on small plants (Guo and Pulli 2000) but also on microspores at the start of culture (Barnabas et al. 1999). The discovery that colchicine is a trigger for microspore embryogenesis evidently preceded its use on microspores as a diploidizing agent.

Even apart from the effect of colchicine, the interrelationship of microspore embryogenesis and the cell cycle goes deep. In tobacco and *Brassica* embryogenesis can be induced from the G1 phase of unicellular microspores until the mid-bicellular stage of young pollen grains (Touraev et al. 2001). Originally, the later stages of pollen grains were considered non-responsive to embryogenesis because of the starch grains they contain (Sangwan and Sangwan-Norreel 1987a). However, *Brassica* pollen grains, which already contain starch grains, can be induced toward embryogenesis by a precisely timed strong heat shock (Binarova et al. 1997). The fact that in the best studied systems, such as tobacco and *Brassica*, embryogenesis can be induced in microspores and in pollen grains at different phases of the cell cycle is at odds with the general assertion, derived from experimentation in other species, that the developmental window for embryogenic induction is rather narrow, restricted to unicellular microspores in most species.

In normal pollen development, first pollen mitosis is an asymmetrical division resulting in a small generative cell and a large vegetative cell (early bicellular pollen). The generative cell undergoes a rapid cell cycle immediately after first pollen mitosis and is arrested in the G2 phase, while the vegetative cell does not undergo a further cell cycle but is arrested in G1(G0). Embryogenic induction by nitrogen-carbohydrate starvation of tobacco immature pollen leads to the derepression of the G1 arrest in the cell cycle of the vegetative cell (Zarsky et al. 1992). In the generative nucleus DNA replication is completed during embryogenic induction and is followed by DNA replication in the vegetative cell, resulting in a pollen grain in which both cells are arrested in the G2 phase of the cell cycle. When tobacco microspores were isolated in the G1 phase and stressed by starvation and heat treatments, they underwent DNA replication as well during stress and again arrested in the G2 phase of the cell cycle (A. Touraev and E. Heberle-Bors, unpubl. observ.). In both G1 microspores and early bicellular pollen grains, the G2 arrest was overcome after transfer of the stressed microspores to a rich medium at normal temperature. Microspores isolated in the G2 phase underwent mitosis during the stress treatment and underwent a cell cycle arrest (A. Touraev and E. Heberle-Bors, unpubl. observ.). In those cases in which a symmetrical division occurred both cells seemed to be arrested in the G1 phase of the cell cycle, while after an asymmetric division the vegetative cell was arrested in G1 while the generative cell underwent its natural process of DNA replication and G2 arrest (A. Touraev and E. Heberle-Bors, unpubl. observ.).

In *Brassica napus*, isolated microspores and young bicellular pollen grains were not arrested by the inductive stress, but derepression of the developmentally programmed cell-cycle arrest of the vegetative cell in immature pollen grains and the initiation of sustained cell divisions are common to both embryogenic induction in *Brassica* and tobacco (Binarova et al. 1993).

Reorganization of the cytoskeleton is a key event in the induction of embryogenesis (Zaki and Dickinson 1990, 1991). In heat-stressed *Brassica* the first structural change seen after the transfer of in vitro cultured microspores to an elevated temperature is, along with the movement of the nucleus to the center inside the microspore, the appearance of a preprophase band (PPB) of microtubules, which does not form during the first haploid mitosis in normal pollen development (Simmonds and Keller 1999). The appearance of PPBs in heat-induced microspores marks sporophytic development, and its integrity is critical to the development of the first consolidated cell wall (Simmonds and Keller 1999).

4 Direct Embryogenesis Versus Indirect Plant Formation

It is one of the central dogmas of plant tissue culture that the hormones auxin and cytokinin are the decisive factors governing regeneration. In fact, the original anther culture experiments on *Datura* involved the inclusion of auxin and cytokinin in the medium (Guha and Maheshwari 1964). Papers are still published today in which anthers are plated on culture media in which variations of type, concentration and combination of hormones are presented as essential experimental factors for success in doubled haploid formation (Assani et al. 2003; Chaturvedi et al. 2003). Also, the concept of initiating somatic embryogenesis by an auxin treatment, followed by auxin withdrawal going back to Reinert (1959), has been applied. Only slowly the idea, derived from the results of microspore culture and many cytological investigations, gained ground that the hormones play a secondary role at best in microspore embryogenesis. In the most efficient microspore culture systems today hormones are not included in the medium, neither for induction nor for regeneration, and the general experience is that their addition is mostly harmful, leading to less plants rather than more or to plants with low quality (Touraev et al. 1996a).

An anther simply cannot be compared with any other explant in plant tissue culture. Somatic embryogenesis or organogenesis invariably proceed from surface cells of an explant, while in microspore embryogenesis it is not the somatic tissues in contact with the medium but the microspores inside the anther which produce embryos. Furthermore, unlike in somatic explants where the cells are connected to each other via their cell walls and plasmodesmata, the microspores or young pollen grains are highly individualized cells with specialized cell walls and without symplastic contact with the surrounding anther wall tissues. Even in cereals where the microspores are initially in symplastic contact with the tapetum, severance of the symplastic ties is one of the first events in cultured anthers, accompanied by parallel changes in exine properties, seen under the light microscope as a red shift in light reflection (Hoekstra et al. 1992). It is this separation of the microspore from the culture medium via the anther wall and the resulting deprivation of nutrients – a starvation stress, which is simulated in ab initio microspore cultures – that triggers embryogenesis (Heberle-Bors 1989).

Many microspores whether in cultured anthers or isolated in culture do not survive the stress treatment. The surviving ones do not follow a simple and single path of development after they have been transferred to non-stress conditions. Again, many do not survive the change and simply die. A minority of the surviving microspores may proceed with normal pollen development and either germinate with longer or shorter pollen tubes or develop into giant pollen grains (Touraev et al. 1996a). Others undergo a symmetrical division and continue to divide. In all systems studied so far, the majority of dividing microspores do not develop into embryos but stop dividing and abort. Only a minority of multicellular structures continue to divide to eventually form an embryo. Furthermore, even after formation of an embryo, abortion of the embryo may happen. Still in the torpedo-shaped stage abortion may take place, leading to failure to develop into a seedling (Fig. 1). The latter is particularly frequent in species like apple (Höfer et al. 1999). In *Brassica* such embryos can be rescued by an ABA treatment (Hansen 1994).

It is here that the effect of added hormones seems to come in. In hormone-containing media the multicellular structures seem to be able to undergo fur-

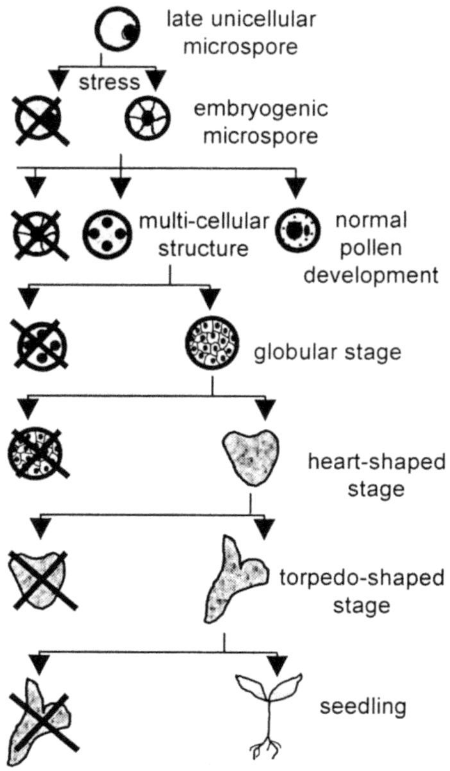

Fig. 1. Possible fates of microspores in vitro

ther growth and develop into calli (Sangwan and Sangwan-Norreel 1987a). However, no experimental data support this reasonable claim. In hormone-free media in contrast, it seems as if embryogenic microspores are endowed with varying amounts of growth factors, carried over from the anther or induced by the stress treatment, which trigger cell divisions but eventually run out and cause the multicellular structure to stop dividing. (It must be remembered at this point that genetic individuality cannot explain these differences. Although microspores are the products of meiosis, in the experimental systems used (tobacco, *Brassica*, barley, wheat) the cultured microspores are genetically identical because the donor plants are inbreds.)

Growth can only proceed when the multicellular structures undergo embryogenesis. Hormone autotrophy is a hallmark of embryogenesis. Higher plant embryos require the setting up of a dual and opposite apical–basal gradient of auxin and cytokinin. A failure to do so results either in abortion or in organogenesis of either shoots or roots (Strabala et al. 1996; Coenen and Lomax 1997). In anther or microspore cultures on hormone-free media, therefore, a strong selection pressure exists which allows further growth only to true embryos. Direct embryogenesis is therefore the rule. This is true, amongst others, for anther and microspore cultures of tobacco (Touraev et al. 1996a), *Brassica* (Custer et al. 1994), barley (Hoekstra et al. 1992) and wheat (Touraev et al. 1996b).

However, callus formation has been observed in hormone-free media of cereals, possibly as a result of cross-feeding of hormones by the surrounding embryos or by conditioning of the medium through ovaries (J'Aiti et al. 1999). These calli may regenerate plantlets by secondary embryogenesis or organogenesis (Sangwan and Sangwan-Norreel 1987a; Ferrie et al. 1995). The quality of these embryos, however, is often reduced. Although often diploid, others are of higher ploidy or aneuploid and, most importantly in cereals, albinos (Löschenberger et al. 1995). Other types of somaclonal variation may also exist but have not been analyzed. Embryos formed through direct embryogenesis, on the other hand, are more frequently green and haploid (Schumann 1990; Löschenberger et al. 1995).

Addition of hormones to anther culture media has sometimes led to an increase in plant formation (Nitsch and Nitsch 1969; Germana and Chiancone 2003). The hormones used were usually auxin and auxin analogs. In these attempts, callus was often and seemingly formed from somatic anther tissues, with or without simultaneous formation of microspore embryos. The subsequent regeneration of both clonal and recombinant diploid plants required the use of genetic markers to distinguish the two (Pal 1983; Gomez et al. 2001).

5 Division Pathways of Embryogenic Microspores

Several theories concerning the fate of microspores after being reprogrammed towards embryogenesis have been postulated over the years. The early development of embryogenic microspore can be traced from Feulgen squashes and light and electron microscopic observations from fixed materials. Using these methods, the division pattern of the daughter cells of embryogenic microspores have been followed in many species; *Datura*, *Hyoscyamus*, *Brassica*, tobacco, wheat and barley (Sunderland and Wicks 1971; Sunderland et al. 1974; Raghavan 1978; Sun 1978; Fan et al. 1988).

Meiosis in diploid microsporocytes within the pollen sacs results in the formation of microspores which leads finally to the formation of a large number of male gametophytes – pollen grains. In the majority of angiosperms, post-meiotic pollen development involves only a single asymmetrical mitotic cell division, first pollen mitosis, which separates microsporogenesis and gametophyte development or gametogenesis (Twell and Howden 1998). The asymmetrical division produces two unequal daughter cells, the vegetative and generative cells that have different structures and developmental fates. The generative cell divides again later into the sperm cells, second pollen mitosis, either during pollen development or, still later, after pollination, in the pollen tube. No correlation exists between the ability to undergo microspore embryogenesis in vitro and the timing of second pollen mitosis.

Taking into consideration the fate of the two cells resulting from the first cell division in microspore cultures, Sunderland and Evans (1980) described three major pathways. (To clarify terminology, the term microspore culture is used here for simplicity as a generic term, including true microspores and immature pollen grains.) In the A pathway, the vegetative cell of a bicellular pollen grain resulting from a regular asymmetrical division subsequently divides and forms eventually an embryo. The A pathway does not distinguish whether the asymmetric division took place prior or during culture. The generative cell sometimes divides into two sperm cells but usually dies. In the B pathway, unicellular microspores divide symmetrically to form two cells with two large nuclei, which resemble vegetative nuclei. The C pathway involves the fusion of the daughter nuclei produced by the first pollen mitosis, usually after symmetrical division. By inference the C pathway can occur only in cultured microspores because it requires an incomplete separation of the two daughter cells (Sunderland and Wicks 1971; Sunderland 1973; Sunderland et al. 1974). The latter pathway is of particular significance as it leads to the spontaneous formation of doubled haploid plants, making artificial chromosome doubling unnecessary.

Later, based on the assumption that a vegetative cell and a generative cell are the result of the first haploid mitosis, Raghavan (1997) considered three general routes to multicellularity of bicellular pollen to be important: (1) repeated divisions of the vegetative cell; (2) repeated divisions of the generative cell; and (3) repeated divisions of both the vegetative and generative cells.

Based on the classifications by Sunderland and Evans (1980) and by Raghavan (1997), five pathways are proposed here, termed A to E (see Fig. 2).

5.1 Repeated Division of the Vegetative Cell (A Pathway)

The origin of an embryo from the vegetative cell of a pollen grain was first observed in tobacco anther cultures (Sunderland and Wicks 1971). After about 6 days of anther culture, the vegetative cell has lost its morphogenetic individuality and has undergone cellular changes, which lead to a symmetrical division of the vegetative cell and culminate eventually in the appearance of an embryo which breaks through the anther wall and turns into a green seedling (Sunderland and Wicks 1971). This type of division pattern was designated as the A pathway (Sunderland 1973) and is predominant when young bicellular pollen grains are cultured in vitro. The generative cell either dies or sometime divides into two sperm cells, which then die as well. In tobacco anther cultures a further subdivision of the A pathway into A-1, A-2, A-3 has been proposed with respect to the occurrence of an intermediate free nuclear or cellular phase during the division of the vegetative cell and also of different

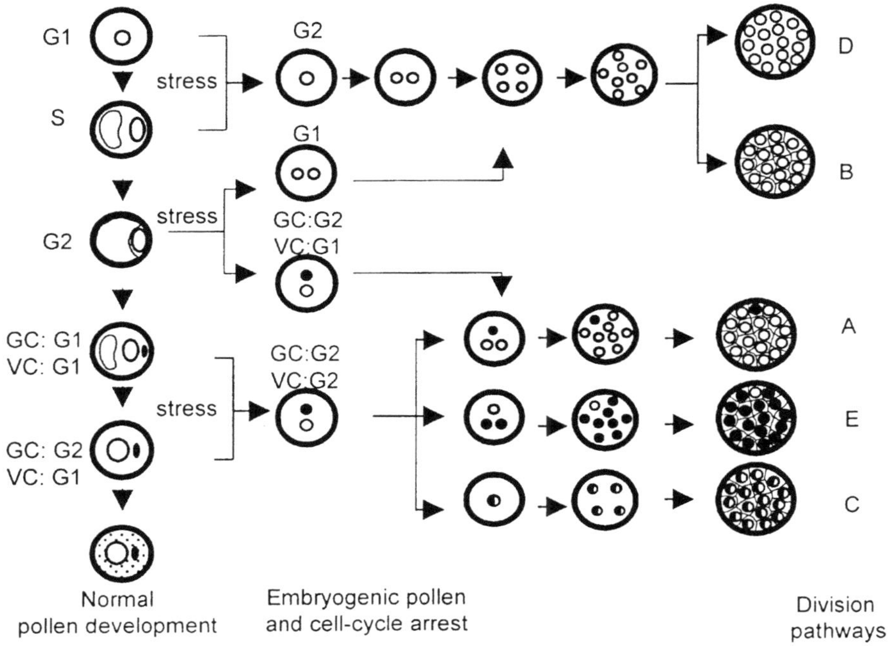

Fig. 2. Different fates of nuclei during early stages of microspore embryogenesis. In *pathways A* and *E*, *black circles* represent nuclei of the generative cell or its descendants; *white circles* indicate nuclei of the vegetative cell or its descendants. In the *C pathway*, *halfblack/halfwhite circles* represent nuclear fusions between the two nuclei. In the *B* and *D pathways*, *circles* indicate symmetrical nuclei resulting from first haploid mitosis and their division products. *G1*, *S* and *G2* indicate phases of the cell cycle. *GC* Generative cell; *VC* vegetative cell. (Modified after Raghavan 1986)

fates of the generative cell in the final products formed (Misoo et al. 1979). However, it is not certain whether these distinctions are widely applicable and what their functional significance with respect to the further development of these structures is.

Embryogenesis resulting from division of the vegetative nucleus has been observed with small differences also in other species: *Brassica napus* (Fan et al. 1988), wheat (Reynolds 1993), *Datura inoxia* (Sunderland et al. 1974), *Datura metel* (Iyer and Raina 1972), *Solanum surattense*, *Luffa cylindrica* (Sinha et al. 1978). Division of the vegetative nucleus is, however, not always followed by the formation of embryos. In some case, callus formation has been reported after division of the vegetative nucleus in several species, such as barley (Sun 1978), maize (Miao et al. 1978), wheat (Wang et al. 1973), rice (Chen 1977) and rye (Sun 1978).

5.2 Repeated Divisions of the Generative or Both, the Vegetative and Generative, Cells (E Pathway)

In normal gametophytic pollen development, the generative cell contains relatively few organelles and stored metabolites compared with the rich cytoplasm of the vegetative cell, and it completes a further mitotic division to form the two sperm cells. When a pollen grain is induced to pursue the sporophytic pathway the generative cell has been reported to be capable of dividing independently to produce an embryo or a callus (Raghavan 1976, 1978). Evidence for an origin of the embryo from the generative cell of a pollen grain was obtained from anther cultures of *Hyoscyamus niger*. In this species, the vegetative cell does not divide, or will undergo only a few divisions, and the division products of the vegetative cell form a suspensor-like structure on the organogenic part of the embryos derived from the generative cell (Raghavan 1976, 1978). The formation of an embryo from the generative cell in the presence of a passive or less active vegetative cell has been observed in several species, such as barley (Sunderland et al. 1979), maize (Guo et al. 1978), wheat (Pan et al. 1983) and rice (Qu and Chen 1984). When embryos were formed by repeated division of both vegetative and generative cells the embryos formed were chimeras at the cellular level containing cells with less (vegetative) or more (generative) compact chromatin (Sun 1978; Sunderland et al. 1979), although one has to bear in mind that the two cells are daughter cells of the same mother cell. This pathway, whether it involves exclusive divisions of the generative cell or divisions of both the generative and vegetative cells, was designated as the E pathway (Sun 1978). The significance of this pathway for the formation of embryos versus calli or the quality of the resulting plantlets are not known. In *Hyoscyamus niger*, at least, the resulting plantlets appeared to be true to type, apart from their haploid chromosome number (Raghavan 1979).

5.3 Symmetrical and Repeated Divisions of the Microspore (B and D Pathways)

When unicellular microspores are cultured in vitro to induce embryogenesis, a symmetrical division of the microspore nucleus may generate two identical or similarly sized cells within the confines of the microspore wall, which divide repeatedly to form an embryo. In a subtype of the B pathway the microspore nucleus divides into two vegetative-type nuclei without subsequent cytokinesis. This may result in a syncytium of up to 30 nuclei, which may or may not cellularize at various numbers of nuclei. Non-cellularizing syncytia die. This type of division pattern was designated as the B pathway (Sunderland and Wicks 1971).

In *Brassica napus* the formation of haploid embryos normally starts with a symmetrical division of the microspores compared with an asymmetric mitosis characteristic of normal development (Zaki and Dickinson 1990; Telmer et al. 1993). To determine whether the symmetry of pollen mitosis I is the key to subsequent sporophytic development, *Brassica* microspores were supplied with the antimicrotubular drug colchicine in lower concentration than those required for diploidization. The treatment led to an increase in the proportion of cultured microspores with symmetrical division (Zaki and Dickinson 1991; Zhao et al. 1996).

In tobacco, both the A and B pathways have been observed (Sunderland and Wicks 1971). Similar to *Brassica*, low levels of colchicine are effective in inducing symmetrical divisions of the tobacco microspore nucleus, producing two similar-sized daughter cells (Eady et al. 1995). The B pathway was originally reported to be relatively rare (Sunderland and Wicks 1971), while experiments with microspore cultures indicate that it may be more frequent (Zaki and Dickinson 1991). In wheat, cell tracking experiments showed that the first division of stressed microspores was always symmetrical and was followed by cytokinesis leading to embryo formation (Indrianto et al. 2001).

Whether a symmetrical division of the microspore plays a decisive role in the developmental fate of the microspore is still under question. Colchicine treatment of isolated tobacco microspores under maturation condition revealed that microspore continued to develop in a pollen-like manner (Eady et al. 1995). Similarly, the treatment of isolated microspores with an anther extract also induced a symmetrical division, and under maturation condition produced pollen grains with two similar-sized cells from which invariably only one cell produced a pollen tube after transfer to germination medium (Touraev et al. 1995). On the other hand, addition of lithium to a pollen maturation medium resulted in symmetrical divisions of tobacco microspores followed by a few cell divisions (Zonia and Tupy 1995). If the plane of division plays a decisive role in further sporophytic development of microspores, those that divide symmetrically should not develop in a gametophytic-like manner (Touraev et al. 2001). In any case it is clear that in both microspores and bicellular pollen grains it is a symmetrical division that initiates sporophytic development (callus or embryo).

A deviation of the B pathway, known as the D pathway, has been described in cultured anthers of wheat (Zhu et al. 1978; Pan et al. 1983). In this case, the two identical nuclei divide repeatedly to generate a cluster of free nuclei, but it is unclear whether the embryos or calli produced developed from the syncytia seen earlier in the cultures.

5.4 Nuclear Fusion (C Pathway)

The C pathway involves the fusion between two haploid vegetative cell nuclei or between one vegetative cell nucleus and a haploid generative cell nucleus which may be in the G1 or the G2 phase of the cell cycle. This type of embryogenic event predominates in anther cultures of *Datura innoxia* where the combination of endoreduplication and nuclear fusion leads to chromosome doubling and higher ploidy levels. Nuclear fusion occurs when first pollen mitosis is not completed by cytokinesis and the two nuclei synchronously entering into the next division under a common metaphase plate and spindle, resulting in two nuclei, each with more than one set of chromosomes (Sunderland et al. 1974). When one or both nuclei have undergone DNA replication prior to nuclear fusion, ploidy levels higher than diploid can be obtained in the embryos (Sunderland et al. 1974). Nuclear fusion is the main mechanism for spontaneous chromosome doubling of barley microspores following mannitol and cold pretreatment (Kasha et al. 2001). This appears to result from the disruption of microtubules, leading to a failure of phragmoplast formation, enabling nuclear fusion to occur between the two closely associated nuclei.

The C pathway has been invoked to explain the occurrence not only of triploid plants in *Datura*, but also of heterozygous transformants after bombardment of microspores with foreign DNA and regeneration of diploid plants in barley microspore cultures (Yao et al. 1997).

Light microscopy analysis of microspore cultures has revealed many insights into the early events of microspore embryogenesis and several pathways of division have been proposed that can lead to embryo formation (Fig. 2). These pathways often vary with the species and, moreover, are influenced by the pretreatment methods (Kasha et al. 2001). Their functional significance with respect to further development of the multicellular structures (callus versus embryo formation, ploidy, albino formation) is still not well understood.

6 Identification of the Developmental Fate of Microspores by Cell Tracking

Unfortunately, in all the pathways described above, the developmental fate of the pollen grains was not followed throughout the culture. Thus, it is not possible to say with certainty what the consequences of the different pathways are for the later development of the multicellular structures. Attempts to monitor the early events in individual microspores and microspore embryos have been made in barley (Bolik and Koop 1991; Kumlehn and Lörz 1999). Single microspores of barley were selected after initial preculture of anthers floating on a liquid medium and analyzed for their development after transferring them into microdroplets for single-cell cultures (Bolik and Koop 1991). Two types of microspores were distinguished. Type A had a rich cytoplasm of granular appearance; a vacuole was not visible, and the nucleus was located close to the cell wall. In the other type of microspores (type B), the nucleus was located in a cytoplasmic pocket in the center of the cell with less cytoplasm, which was smooth in appearance and arranged in a thin peripheral layer with visible cytoplasmic strands traversing the large vacuole. The first type of cell was identified as embryogenic microspores; in contrast, the second type was regarded as non-embryogenic because, under the conditions employed, divisions were almost never observed in these types of cell (Bolik and Koop 1991).

By using cell tracking of the entire process of embryogenesis, the development of isolated wheat microspores undergoing in vitro embryogenesis has been followed (Indrianto et al. 2001). Microspores isolated either directly from freshly excised anthers or from anthers in cut and cold-treated spikes or from in vitro cultured and starvation and heat-stressed anthers were immobilized in Sea Plaque agarose, and – apart from dead microspores – three types of microspores at the start of the microspore culture were identified and were followed up to the formation of embryos which were visible to the naked eye (Indrianto et al. 2001).

Type 1 microspores had a large vacuole and the nucleus was situated close to the microspore wall, usually opposite the single germ pore. This type was identical to non-stressed late unicellular microspores present in anthers in vivo. In type 2 and 3 microspores, the big central vacuole was fragmented and cytoplasmic strands passed through the vacuole and connected a cytoplasmic pocket around the nucleus with the subcortical cytoplasm. In type 2 microspores the cytoplasmic pocket with the nucleus was situated close to the microspore wall opposite the germ pore, while in type 3 microspores the cytoplasmic pocket with the nucleus was in the center of the microspore (Indrianto et al. 2001).

Tracking revealed that all three type of microspores produced multicellular structures, although at different frequencies. Multicellular structures were formed at high frequency from type 3 microspores, while type 2 and 1 microspores divided only rarely. Moreover, microspores that had not been subjected to any prior stress treatment also divided and formed multicellular structures at low frequency. Most importantly, type 1 microspores always converted first into type 2 and then into type 3 microspores, be it in cultures made from pretreated anthers or from anthers dissected from freshly cut tillers (Fig. 3). These type 3 microspores were very similar to the second type of microspores described by Bolik and Koop (1991) in barley. Similarly, type 2 microspores converted first into type 3, before, after a few days in a rich medium, they absorbed their vacuole and differentiated into cytoplasm-rich and starch-accumulating cells which were morphologically very similar to the first type of barley microspores identified as the embryogenic type by Bolik and Koop (1991).

The superficially conflicting conclusions in the two tracking studies can easily be harmonized when one assumes that the microspores which Bolik and Koop (1991) were tracking were already induced by the previous, short anther culture pretreatment, and that under the conditions of their droplet culture only the microspores with already filled cytoplasm were able to divide. Types 1, 2 and 3, according to Indrianto et al. (2001), and type A

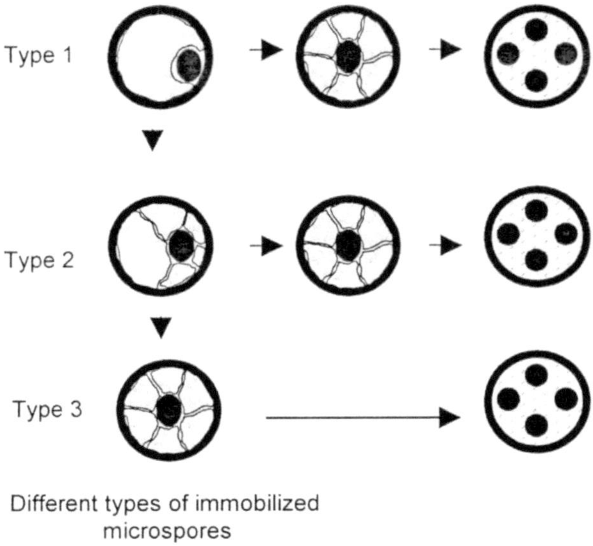

Fig. 3. Development of different types of immobilized wheat microspore classified at the start of culture according to their cytological features. Conversion of *types 1* and *2* into *type 3* microspores and further development of microspores. (After Indrianto et al. 2001)

according to Bolik and Koop (1991), represent stages of a continuous process which can fail for a number of reasons, leading to death (Fig. 3).

In the light of these cell-tracking experiments, we can conclude that theoretically each cultured microspore can be reprogrammed to divide and produce an embryo and that the star-like wheat microspore of type 3, identified by cell tracking as an embryogenic cell, is required for further embryogenic development.

7 Apical–Basal Polarity Formation of Embryogenic Microspores

In *Arabidopsis* and other species the first signs of apical–basal polarity are evident already in the egg cell. The zygote undergoes an asymmetrical transverse division to generate two daughter cells that are unequal in size and follow distinct developmental pathways. The small apical daughter cell generates most of the embryo, while the basal cell produces the very basal end of the embryo and the entire suspensor (Mayer and Jürgens 1998). During microspore embryogenesis, the establishment of polarity precedes rupture of the microspore wall and determines both rupture site and orientation of the body axis. Moreover, polarity seems to be inherited from the gametophyte cells which become structurally polarized as the nucleus moves laterally towards the pollen wall in the late microspore (Hause et al. 1994; Ilic-Grubor et al. 1998). This hypothesis was proven by tracking the developmental pattern of wheat microspore embryos (Indrianto et al. 2001).

A comparative morphological study of microspore-derived and zygotic embryos of *Brassica napus* L. was conducted by using scanning electron microscopy (Ilic-Grubor et al. 1998). Their observations indicated that a precise cell division pattern was absent during development of microspore-derived embryos and that early embryo development from microspores and young bicellular pollen grains had an irregular division pattern but that the final cotyledon stage of embryo was similar to its zygotic counterparts. In contrast, a precise cell division pattern exists during zygotic embryogenesis of *Brassica napus*, resulting in a linear cotyledon stage embryo with three spatial domains: the apical domain (cotyledons, shoot apex), central domain (hypocotyl) and basal domain (root apex; Yeung et al. 1996). The authors concluded that the strictly determined cell divisions occurring during the early phase of zygotic embryogenesis are not crucial for the establishment of the spatial organization of the developing embryo (Ilic-Grubor et al. 1998).

During zygotic embryogenesis, the suspensor appears to have different functions: it physically projects the embryo into the endosperm and provides hormones and nutrients for the developing embryo. In later stages of development the suspensor degenerates and is not present in the mature seed (Yeung and Meinke 1993). In the case of *Brassica* microspore-derived embryos suspensor formation is not essential for embryogenesis (Hause et al. 1994). An occasional formation of suspensors has been observed in microspore-derived embryos during the first 7–10 days in *Brassica* microspore cultures (Hause et al. 1994). During microspore embryogenesis, the suspensors were characterized as structural manifestation of the polarity established within the microspore embryo (Yeung et al. 1996).

In *Brassica* microspore cultures, after 4 days of cultivation, the culture contains a mixture of microspores, pollen grains and pro-embryos or multicellular structures with up to 12 nuclei. After 5 days, the pollen wall ruptures and the embryo undergoes a zygote-like development by the formation of an embryo proper. It is essential for the formation of pro-embryos that the former pollen wall stays intact during the first 5 days of cultivation. An earlier rupture of the pollen wall leads to vacuolization and elongation of the cells. Similar observations have been made in other species (A. Touraev and E. Heberle-Bors, unpubl. observ.). During pollen wall rupture, a polar distribution of starch grains was observed, suggesting that polarity of the microspore-derived embryos was induced by rupture of the pollen wall (Hause et al. 1994; Indrianto et al. 2001).

The first visible sign of polarity in microspore-derived embryos was apparent upon the rupture of the pollen wall and polarity seemed to be inherited from the gametophytic cell, which becomes structurally polarized as the nucleus moves laterally towards the pollen wall in the late microspore stage (Ilic-Grubor et al. 1998). The presence of the original cell wall on the surface of the future root pole of the embryo or on the tip of the suspensors, when they are formed, supports this view. Once released from the pollen wall two regions are observed in the embryogenic structures: a globular cluster of cells which gives rise to the embryo and a small region consisting presumably of one or a very few cells, covered with remnants of pollen wall. In the absence of the suspensors, remnants of the original pollen wall adhered to the surface on the future root pole of the embryos (Ilic-Grubor et al. 1998). When the suspensors were formed, they appeared as a result of transverse cell division in particular peripheral cell(s) still covered by remnants of the pollen wall. Similarly, the cell wall appears to direct the cell fate during thallus formation of the *Fucus* egg, and positional information seems to be provided by the cell wall surrounding the embryo (Berger et al. 1994).

Tracking wheat microspore embryos confirmed that the polarity precedes the rupture of the microspore wall and seems to be inherited from the gametophytic cell (Indrianto et al. 2001). As in *Brassica*, also in wheat starch grains accumulated in the cells in a restricted zone close to the germ pore in all multicellular structures, and the release of the embryo from the microspore wall was always opposite the site of starch granule accumulation where the microspore nucleus was situated. The presence of starch granules around the germ pore indicates that starch accumulation may be caused by facilitated nutrient uptake through the germ pore or they may represent amyloplasts required for gravity perception, a defining feature of roots (Indrianto et al. 2001). Taking into consideration that the single germ pore in wheat is in direct contact with the tapetum, and thus the polarity axis in the microspore is perpendicular to the interior surface of the anther wall, a transfer of positional information can be postulated (Indrianto et al. 2001). The transfer of positional information takes place from the sporophytic anther tissues to the gametophytic microspore and to the sporophytic embryo (Fig. 4).

Recently, Supena et al. (2003) have developed a new *Brassica napus* microspore culture system for studying suspensor function. Under optimal conditions (basically a different heat shock treatment regime; Jan Custers, pers. comm.), after 8–9 days of microspore culture, the microspore wall ruptured and a filamentous structure appeared, consisting of a single file of cells. After two more days, the distal tip cell of the filamentous structure began to divide longitudinally to form an embryo-like structure, as the result of a series of divisions similar to those in early zygotic embryos. Embryos forming via this route appeared much faster than by the conventional route. Similar results have recently been obtained by the authors of this chapter (T. Aionesei, E. Heberle-Bors and A. Touraev, unpubl. observ.). Supena et al. (2003) proposed that the filaments of cells observed in these *Brassica* microspore cultures are suspensor-like structures. Thus, in this system embryogenesis is highly similar to zygotic embryogenesis. The difference is that these suspensor-like structures develop before their distal cell undergoes the first longitudinal division, resulting in embryo formation, while in zygotic embryogenesis the

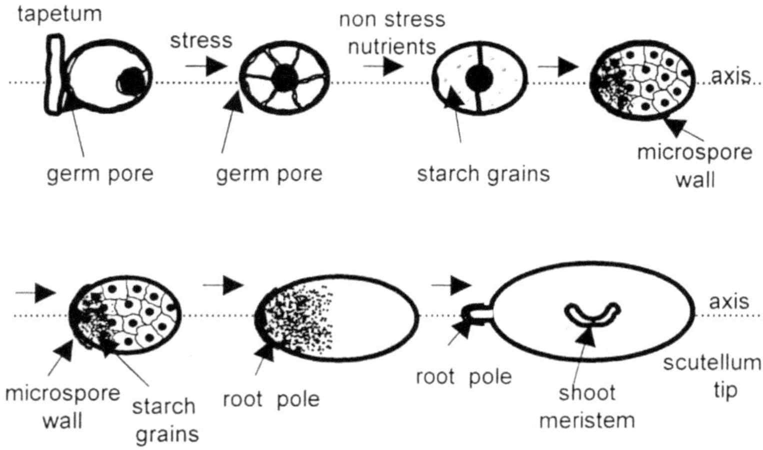

Fig. 4. Transfer of positional information from sporophytic anther tissues to the wheat gametophytic microspore and to the sporophytic embryo

embryo initial cell and the suspensor initial cell arise simultaneously upon asymmetrical division of the zygote.

The majority of these suspensor-bearing embryos had the suspensor attached to the root pole, indicating that it was the suspensor rather than the embryo itself that determined the apical–basal polarity of the embryo (Supena et al. 2003). This is reminiscent of the finding in embryos lacking suspensors, described earlier in this section, that the remains of the original pollen wall were attached to the root pole of the embryos. It is as if in these latter species the pollen wall takes over the role of the suspensor in determining apical–basal polarity.

Some of these suspensor-like structures initiated more than one embryo. Up to five embryos could be seen to appear along the suspensor structure, forming at bends in the elongated suspensor structures (Supena et al. 2003). Possibly the bends impeded the flow of nutrients or hormones, giving rise to physiologically isolated suspensor domains. Apparently as long as an embryo is absent it is the suspensor that determines where a new embryo emerges (Supena et al. 2003).

In these microspore cultures, in addition, embryos were found with two suspensors attached. These embryos developed very slowly. The two suspensors apparently had a mutually inhibitory effect on each other, causing failure of effective embryo axis establishment and arrest of embryo apical dome differentiation (Supena et al. 2003). Two suspensors formed mainly when the embryo was initiated from a cell in the middle of the filamentous structure, and further development of the embryo depended on the length of the two filament parts attached left and right (Supena et al. 2003).

8 Conclusion

Light and electron microscopy analysis of microspore cultures revealed many insights into the early events of microspore embryogenesis. The developmental switch of microspores from gametophytic to sporophytic pathways seems to involve a wholesale structural reorganization of the microspore or pollen grain during stress treatment in addition to reorganization at the biochemical and molecular levels which seems to be essential for the formation of embryogenic microspores.

The wheat cell tracking experiments revealed that theoretically all microspores can be reprogrammed to become embryogenic but are in different physiological states which do not allow them to respond synchronously to the stress trigger in a given time period. The three types of wheat microspores represent stages in a continuous process and not distinct classes of embryogenic or non-embryogenic microspores.

In *Brassica* and wheat, the first visible sign of polarity in microspore-derived embryos was apparent upon the rupture of the pollen wall and polarity seems to be inherited from the gametophytic cell, which becomes structurally polarized as the nucleus moves laterally towards the pollen wall in the late microspore. The transfer of positional information takes place from the sporophytic anther tissues to the gametophytic microspore and to the sporophytic embryo.

References

Assani A, Bakry F, Kerbellec F, Haicour R, Wenzel G, Foroughi-Wehr B (2003) Production of haploids from anther culture of banana [*Musa balbisiana* (BB)]. Plant Cell Rep 21:511–516

Barnabas B, Obert B, Kovacs G (1999) Colchicine, an efficient genome doubling agent for maize (*Zea mays* L.) microspores cultured in anthers. Plant Cell Rep 18:858–862

Benito-Moreno RM, Macke F, Hauser MT, Alwen A, Heberle-Bors E (1988) Sporophytes and gametophytes from in vitro cultured, immature tobacco pollen. In: Cresti M, Jori P, Paccini E (eds) Sexual reproduction in higher plants. Springer, Berlin Heidelberg New York, pp 137–142

Berger F, Taylor A, Brownlee C (1994) Cell fate determination by the cell wall in early *Fucus* development. Science 263:1421–1423

Binarova P, Straatman K, Hause B, Hause G, van Lammeren AMM (1993) Nuclear DNA synthesis during the induction of embryogenesis in cultured microspores and pollen of *Brassica napus* L. Theor Appl Genet 87:9–16

Binarova P, Hause G, Cenklova V, Cordevener JHG, van Lookeren Campagne MM (1997) A short severe heat shock is required to induce embryogenesis in late bicellular pollen of *Brassica napus* L. Sex Plant Reprod 10:200–208

Bolik M, Koop HU (1991) Identification of embryogenic microspores of barley (*Hordeum vulgare* L.) by individual selection and culture and their potential for transformation by microinjection. Protoplasma 162:61–68

Chaturvedi R, Razdan MK, Bhojwani SS (2003) Production of haploids of neem (*Azadirachta indica* A. Juss.) by anther culture. Plant Cell Rep 21:531–537

Chen C (1977) In vitro development of plants from microspores of rice. In Vitro 13:484–489
Coenen C, Lomax TL (1997) Auxin–cytokinin interactions in higher plants: old problems and new tools. Trends Plant Sci 2(9):351–356
Custers JBM, Cordewener JHG, Nöllen Y, Dons JJM, van Lookeren-Campagne MM (1994) Temperature controls both gametophytic and sporophytic development in microspore culture of *Brassica napus*. Plant Cell Rep13:267–271
Dunwell JM, Sunderland N (1974a) Pollen ultrastructure in anther cultures of *Nicotiana tabacum* I. Early stages of culture. J Exp Bot 25:352–361
Dunwell JM, Sunderland N (1974b) Pollen ultrastructure in anther cultures of *Nicotiana tabacum* II. Changes associated with embryogenesis. J Exp Bot 25:363–373
Dunwell JM, Sunderland N (1975) Pollen ultrastructure in anther cultures of *Nicotiana tabacum* III. The first sporophytic division. J Exp Bot 26:240–252
Eady C, Lindsey K, Twell D (1995) The significance of microspore division and division symmetry of vegetative cell-specific transcription and generative cell differentiation. Plant Cell 7:65–74
Fan Z, Armstrong KC, Keller WA (1988) Development of microspores in vivo and in vitro in *Brassica napus* L. Protoplasma 147:191–199
Ferrie AMR, Palmer CE, Keller WA (1995) Haploid embryogenesis. In: Thorpe TA (ed) In vitro embryogenesis in plants, vol 20. Kluwer, Dordrecht, pp 309–344
Gaillard A, Vergne P, Beckert M (1991) Optimization of maize microspore isolation and conditions for reliable plant regeneration. Plant Cell Rep 10:55–58
Garrido D, Vicente O, Heberle-Bors E, Rodriquez-Garcia MI (1995) Cellular changes during the acquisition of embryogenic potential in isolated pollen grains of *Nicotiana tabacum*. Protoplasma 186:220–230
Germana MA, Chiancone B (2003) Improvement of *Citrus clementina* Hort. ex Tan. microspore-derived embryoid induction and regeneration. Plant Cell Rep 22:181–187
Goldberg RB, de Paiva G, Yadegari R (1994) Plant embryogenesis: zygote to seed. Science 266:605–614
Gomez A, Pintos B, Aguiriano E, Manzanera JA, Bueno MA (2001) SSR markers for *Quercus suber* tree identification and embryo analysis. J Hered 92(3):292–295
Guha S, Maheshwari SC (1964) In vitro production of embryos from anthers of *Datura*. Nature 204:497
Guo Y-D, Pulli S (2000) An efficient androgenic embryogenesis and plant regeneration method through isolated microspore culture in timothy (*Phleum pratense* L.). Plant Cell Rep 19:761–767
Guo ZS, Sun AC, Wang YY, Gui YL, Gu SR, Miao SH (1978) Studies on induction of pollen plants and androgenesis in maize. Acta Bot Sin 20:204–209
Hansen M (1994) Gametic embryogenesis in *Brassica*: optimization of production and germination of embryos. In: Javornik B, Bohanec B, Kreft I (eds) Proc Int Colloquium on Impact of Plant Biotechnology on Agriculture, Rogla, Slovenia. Centre for Plant Biotechnology and Breeding, University of Ljubljana, Slovenia, pp 15–18
Hause B, van Veenendaal WLH, Hause G, van Lammeren AMM (1994) Expression of polarity during early development of microspore-derived and zygotic embryos of *Brassica napus* L. cv. Topas. Bot Acta 107:369–472
Heberle-Bors E (1983) Induction of embryogenic pollen grains in situ and subsequent in vitro pollen embryogenesis in *Nicotiana tabacum* L. by treatments of pollen donor plants with feminizing agents. Physiol Plant 59:67–72
Heberle-Bors E (1989) Isolated pollen culture in tobacco: plant reproductive development in a nutshell. Sex Plant Reprod 2:1–10
Heberle-Bors E (1999) Microspore culture, totipotency, and doubled haploids in plant breeding. In vitro Cell Dev Biol-Plant 35:165–167
Hoekstra S, van Zijderveld MH, Heidekamp F, van der Mark F (1992) Anther and microspore culture of *Hordeum vulgare* L. cv. Igri. Plant Sci 86:89–96

Höfer M, Touraev A, Heberle-Bors E (1999) Induction of embryogenesis from isolated apple microspores. Plant Cell Rep 18:1012–1017

Ilic-Grubor K, Attre SM, Fowke LC (1998) Comparative morphological study of zygotic and microspore-derived embryos of *Brassica napus* L. as revealed by scanning electron microscopy. Ann Bot 82:157–165

Indrianto A, Barinova J, Touraev A, Heberle-Bors E (2001) Tracking individual wheat microspores in vitro: identification of embryogenic microspores and body axis formation in the embryo. Planta 212:163–174

Iyer RD, Raina SK (1972) The early ontogeny of embryoids and callus from pollen and subsequent organogenesis in anther cultures of *Datura metel* and rice. Planta 104:146–156

J'Aiti F, Benlhabib O, Sharma HC, El Jaafari S, El Hadrami I (1999) Genotypic variation in anther culture and effect of ovary coculture in durum wheat. Plant Cell Tissue Organ Cult 59:71–76

Kasha KJ, Hu TC, Oro R, Simion E, Shim YS (2001) Nuclear fusion leads to chromosome doubling during mannitol pretreatment of barley (*Hordeum vulgare* L.) microspores. J Exp Bot 52(359):1227–1238

Koltunov AM (1993) Apomixis: embryo sac and embryos formed without meiosis or fertilization in ovules. Plant Cell 5:1425–1437

Kumlehn J, Lörz H (1999) Monitoring sporophytic development of individual microspores of barley (*Hordeum vulgare* L.). In: Clement C, Audran C-J (eds) Anther and pollen, from biology to biotechnology. Springer, Berlin Heidelberg New York, pp 183–189

Kyo M, Harada H (1986) Control of the developmental pathway of tobacco pollen in vitro. Planta 168:427–432

Lindsey K, Topping JF (1993) Embryogenesis: a question of pattern. J Exp Bot 44(259):359–374

Löschenberger F, Pfosser M, Heberle-Bors E (1995) Genetic variability for the frequency of double haploid green plants is correlated with the ratio of green to albino plants in wheat (*Triticum aestivum* L.) microspore-derived plants. J Genet Breed 49:37–44

Mayer U, Jürgens G (1998) Pattern formation in plant embryogenesis: a reassessment. Semin Cell Dev Biol 9:187–193

Miao S-H, Kuo C-S, Kwei Y-L, Sun A-T, Ku S-Y, Lu W-L, Wang Y-Y, Chen M-L, Wu M-K, Hang L (1978) Induction of pollen plants of maize and observations on their progeny. In: Proc Symp on Plant Tissue Culture. Science Press, Peking, pp 22–33

Misoo S, Yoshida K, Mastubayashi M (1979) Studies on the mechanisms of pollen embryogenesis. III. Mitotic responses of the pollen to varied sucrose concentrations and the process of embryoid formation in tobacco anther culture. Sci Rep Fac Agric Kobe Univ 13:193–202

Nitsch JP, Nitsch C (1969) Haploid plants from pollen grains. Science 163:85–87

Pal A (1983) Isolated microspore culture of the winged bean, *Psophocarpus tetragonolobus* (L) DC – growth, development and chromosomal status. Indian J Exp Biol 21:597–599

Pan JL, Gao GH, Ban H (1983) Initial patterns of androgenesis in wheat anther culture. Acta Bot Sin 25:34–39

Qu RD, Chen Y (1984) Pathways of androgenesis and observations on cultured pollen grains in rice (*Oryzia sativa* subsp. Keng). Acta Bot Sin 26:580–587

Raghavan V (1976) Role of the generative cell in androgenesis in henbane. Science 191:388–389

Raghavan V (1978) Origin and development of pollen embryoids and pollen calluses in cultured anther segments of *Hyoscyamus niger* (henbane). Am J Bot 65:984–1002

Raghavan V (1979) An autoradiographic study of RNA synthesis during pollen embryogenesis in *Hyoscyamus niger* (henbane). Am J Bot 66:784–795

Raghavan V (1986) Embryogenesis in angiosperms. A development and experimental study. Cambridge University Press, Cambridge

Raghavan V (1997) Molecular embryology of flowering plants. Cambridge University Press, Cambridge

Raina SK, Irfan ST (1998) High frequency embryogenesis and plantlet regeneration from isolated microspores of indica rice. Plant Cell Rep 17:957–962

Rao PS, Suprasanna P (1996) Methods to double haploid chromosome numbers. In: Mohan Jain M, Sopory SK, Veilleux RE (eds) In vitro haploid production in higher plants, vol 1. Kluwer, Dordrecht, pp 317–340
Rashid A, Siddiqui AW, Reinert J (1981) Ultrastructure of embryogenic pollen of *Nicotiana tabacum* var. Badischer Burley. Protoplasma 107:375–385
Rashid A, Siddiqui AW, Reinert J (1982) Subcellular aspects of origin and structure of pollen embryo of *Nicotiana*. Protoplasma 113:202–208
Reinert J (1959) Über die Kontrolle der Morphogenese und die Induktion von Adventiveembryonen in Gewebekulturen aus Karotten. Planta 53:318–333
Reynolds TL (1993) A cytological analysis of microspores of *Triticum aestivum* (Poaceae) during normal ontogeny and induced embryogenic development. Am J Bot 80:569–576
Reynolds TL (1997) Pollen embryogenesis. Plant Mol Biol 33:1–10
Sangwan RS, Camefort H (1984) Cold-treatment related structural modifications in the embryogenic anthers of *Datura*. Cytologia 49:473–487
Sangwan RS, Sangwan-Norreel BS (1987a) Biochemical cytology of pollen embryogenesis. Int Rev Cytol 107:221–272
Sangwan RS, Sangwan-Norreel BS (1987b) Ultrastructural cytology of plastids in pollen grains of certain androgenic and nonandrogenic plants. Protoplasma 138:11–22
Schumann G (1990) In vitro production of haploids in *Triticale*. In: Bajaj YPS (ed) Wheat. Biotechnology in agriculture and forestry, vol 13. Springer, Berlin Heidelberg New York, pp 348–402
Simmonds DH, Keller WA (1999) Significance of preprophase bands of microtubules in the induction of microspore embryogenesis of *Brassica napus*. Planta 208:383–391
Sinha S, Jha KK, Roy RP (1978) Segmentation pattern of pollen in anther culture of *Solanum surattrense*, *Luffa cylindrica* and *Luffa echinata*. Phytomorphology 28:43–49
Steward FC, Mapes MO, Smith J (1958) Growth and organized development of cultured cells. I. Growth and division of freely suspended cell. Am J Bot 45:693–703
Strabala TJ, Wu VH, Li Y (1996) Combined effects of auxin transport inhibitors and cytokinin: alterations of organ development in tobacco. Plant Cell Physiol 37(8):1177–1182
Sun CS (1978) Androgenesis of cereal crops. In: Proc Symp on Plant Tissue Culture. Science Press, Peking, pp 117–123
Sunderland N (1973) Pollen and anther culture. In: Street HE (ed) Plant tissue and cell culture. University of California Press, Berkeley, pp 205–239
Sunderland N, Evans LJ (1980) Multicellular pollen formation in cultured barley anthers. II. The A, B, C pathways. J Exp Bot 31:501–514
Sunderland N, Wicks FM (1971) Embryoid formation in pollen grains of *Nicotiana tabacum*. J Exp Bot 22:213–226
Sunderland N, Collins GB, Dunwell JM (1974) The role of nuclear fusion in pollen embryogenesis of *Datura innoxia* Mill. Planta 117:227–241
Sunderland N, Roberts M, Evans LJ, Wildon DC (1979) Multicellular pollen formation in cultured barley anthers. I. Independent division of the generative and vegetative cells. J Exp Bot 30:1133–1144
Supena EDJ, Liu CM, Custer J (2003) *Brassica napus* microspore culture as biological model for studying suspensor function. In: Proc Conf on Embryogenesis and Development Regulation in Plants, Book of Abstracts, Villa Gualino, Torino, Italy, 6–7 March, pp 14–17
Taylor RL (1967) The foliar embryos of *Malaxis padulosa*. Can J Bot 45:1553–1556
Telmer CA, Newcomb W, Simmonds DH (1993) Microspore development in *Brassica napus* and the effect of high temperature on division in vivo and in vitro. Protoplasma 172:154–165
Telmer CA, Newcomb W, Simmonds DH (1995) Cellular changes during heat shock induction and embryo development of cultured microspores of *Brassica napus* cv. Topas. Protoplasma 185:106–112
Toonen MAJ, Hendriks T, Schmidt EDL, Verhoeven HA, van Kammen A, de Vries SC (1994) Description of somatic-embryo-forming cells in carrot suspension cultures employing video cell tracking. Planta 194:565–572

Touraev A, Lezin F, Heberle-Bors E, Vicente O (1995) Maintenance of gametophytic development after symmetrical division in tobacco microspore culture. Sex Plant Reprod 8:70–76

Touraev A, Ilham A, Vicente O, Heberle-Bors (1996a) Stress induced microspore embryogenesis from tobacco microspores: an optimized system for molecular studies. Plant Cell Rep 15:561–565

Toureav A, Indrianto A, Wratschko I, Vicente O, Heberle-Bors E (1996b) Efficient microspore embryogenesis in wheat (*Triticum aestivum*. L) induced by starvation at high temperatures. Sex Plant Reprod 9:209–215

Touraev A, Vicente O, Heberle-Bors E (1997) Initiation of microspore embryogenesis by stress. Trends Plant Sci 2:285–303

Touraev A, Pfosser M, Heberle-Bors E (2001) The microspore: a haploid multipurpose cell. Adv Bot Res 35:53–109

Twell D, Howden R (1998) Mechanisms of asymmetric division and cell fate determination in developing pollen. In: Chupeau Y, Caboche M, Henry Y (eds) Androgenesis in haploid plants. Springer, Berlin Heidelberg New York, pp 69–97

Wang C-C, Chu C-C, Cun C-S, Wu S-H, Yin K-C, Hsü C (1973) The androgenesis in wheat (*Triticum aestivum*) anthers cultured in vitro. Sci Sin 16:218–222

Yao QA, Simion E, William M, Krochko J, Kasha KJ (1997) Biolistic transformation of haploid isolated microspores of barley *Hordeum vulgare* L. Genome 40:570–581

Yeung EC, Meinke DW (1993) Embryogenesis in angiosperms: development of the suspensor. Plant Cell 5:1371–1381

Yeung EC, Rahman MH, Thorpe TA (1996) Comparative development of zygotic and microspore-derived embryos in *Brassica napus* L. cv. Topas. I. Histodifferentiation. Int J Plant Sci 157:27–39

Zaki MAM, Dickinson HG (1990) Structural changes during the first divisions of embryos resulting from anther and microspore culture in *Brassica napus*. Protoplasma 156:149–162

Zaki MAM, Dickinson HG (1991) Microspore-derived embryos in *Brassica*: the significance of division symmetry in pollen mitosis I to embryogenic development. Sex Plant Reprod 4:48–55

Zarsky V, Garrido D, Rihova L, Tupy J, Vicente O, Heberle-Bors E (1992) Depression of the cell cycle by starvation is involved in the induction of tobacco pollen embryogenesis. Sex Plant Reprod 5:189–194

Zhao JP, Simmonds DH, Newcomb W (1996) Induction of embryogenesis with colchicines instead of heat in microspores of *Brassica napus* L. cv. Topas. Planta 198:433–439

Zhu ZQ, Sun JS, Wang JJ (1978) Cytological investigation on androgenesis of *Triticum aestivum*. Acta Bot Sin 20:6–12

Zonia LE, Tupy J (1995) Lithium treatment of *Nicotiana tabacum* microspores blocks polar nuclear migration, disrupts the partitioning of membrane-associated Ca^{2+}, and induces symmetrical mitosis. Sex Plant Reprod 8:152–160

I.3 The Role of Stress in the Induction of Haploid Microspore Embryogenesis

SVETLANA ZORINIANTS, ALISHER S. TASHPULATOV, ERWIN HEBERLE-BORS, and ALISHER TOURAEV[1]

1 Introduction

Plants display a remarkable potential for cellular totipotency. In fact, any differentiated plant cell has the ability to revert to the embryogenic condition and form an entire new plant. In evolutionary terms this ability to reproduce from a somatic cell has been an important survival adaptation to predation.

One of the most striking examples of cellular totipotency is in vitro microspore or pollen embryogenesis (reviewed by Touraev et al. 2001). Although pollen grains exhibit determinate differentiation during microsporogenesis to form gametes and pollen tubes during microgametogenesis, immature pollen or microspores can also enter an alternative developmental pathway in which they initiate indeterminate growth, giving rise to mature plants with a haploid (gametic) number of chromosomes. The first haploid-derived embryos were obtained by culturing isolated *Datura* anthers on solid medium, which contained mineral salts, sucrose and the phytohormones auxin and cytokinin (Guha and Maheshwari 1964). Further investigations showed, however, that stress, applied to male gametophytes, is the major factor that effectively switches microspore development from a gametophytic to a sporophytic mode of development, not only in model species but also in a wide variety of important crop species (Touraev et al. 1997).

In this chapter we summarize recent data concerning the role of various stresses in the induction of microspore embryogenesis and possible mechanisms of action of these stresses at cellular and molecular levels.

2 Stress and Microspore Embryogenesis

The switch of cultured microspores and immature pollen from a gametophytic to a sporophytic mode of development has been induced by various stresses applied either in vivo or in vitro (reviewed by Touraev et al. 1997). Sucrose and nitrogen starvation applied to isolated young bicellular pollen of

[1] Max F. Perutz Laboratories, University Departments at the Vienna Biocenter, Institute of Microbiology and Genetics, Dr. Bohrgasse 9/4, 1030 Vienna, Austria

Biotechnology in Agriculture and Forestry, Vol. 56
Haploids in Crop Improvement II (ed. by C.E. Palmer, W.A. Keller, and K.J. Kasha)

Nicotiana tabacum induced the formation of embryogenic pollen grains, which, after transfer to a simple sucrose- and nitrogen-containing medium, divide repeatedly and produce large numbers of embryos (Kyo and Harada 1986; Garrido et al. 1991). A heat-shock treatment is not effective at this stage of development but can induce embryogenesis at an earlier stage when unicellular microspores are used (Touraev et al. 1996a). A combination of starvation and heat stress can induce embryogenesis in nearly all viable microspores, while the microspores develop into mature fertile pollen under non-stress conditions in rich medium (Benito-Moreno et al. 1988).

Similarly, a combination of starvation and heat stress induces the formation of embryogenic microspores in wheat and allows large numbers of embryos to be produced (Touraev et al. 1996b), while culture under non-stress conditions leads to the formation of mature fertile pollen (Stauffer et al. 1992).

In *Brassica napus*, a heat-shock treatment for 8 h at 32 °C induces up to 40 % of the isolated microspores and young bicellular pollen grains to undergo sustained divisions, and thousands of plantlets can be obtained in a single experiment, while at 18 °C the microspores develop into apparently normal mature pollen grains (Custers et al. 1994).

To date, four stress pretreatments have been shown to induce microspore embryogenesis at reasonable frequencies: cold shock in maize (Gaillard et al. 1991), wheat (Gustafson et al. 1995), barley (Sunderland and Xu 1982), rice (Cho and Zapata 1988) and many other species; heat shock in rapeseed (Custers et al. 1994), wheat (Touraev et al. 1996b) and tobacco (Touraev et al. 1996a); carbohydrate and nitrogen starvation in tobacco (Kyo and Harada 1986; Touraev et al. 1996a), wheat (Touraev et al. 1996b), rice (Raina and Irfan 1998) and barley (Hoekstra et al. 1992); and colchicine treatment in rapeseed (Zhao et al. 1996). Some other stresses, such as ethanol and gamma irradiation (Pechan and Keller 1989), have not received a wide application.

These stresses were applied to isolated microspores (Custers et al. 1994; Touraev et al. 1996a), excised anthers (Hoekstra et al. 1992; Touraev et al. 1996b), buds or spikes severed from the plant body (Gaillard et al. 1991; Gustafson et al. 1995) or to whole donor plants (Heberle-Bors 1983).

Stress, in general, seems to be able to trigger dramatic developmental changes, i.e. whole-cell phase changes during the life cycle of organisms. Sporulation in bacteria (Msadek 1999), mating in yeast (Shinozaki and Russell 1996), cell aggregation and differentiation in *Dictyostelium* (Aubry and Firtel 1999) or the switch between the gametophytic and sporophytic generations as in homosporous ferns (Banks 1994), for example, are all induced by a lack of nutrients. Serum starvation seems to be required to release totipotency in differentiated adult mammalian cells (Wilmut et al. 1997).

3 Morphological Characteristics of Stress-Induced Embryogenic Microspores

Young tobacco microspores released from tetrads inside the anther are surrounded by a thin-walled primexine (reviewed by Twell 2002). The intine begins to form with a small delay, when the inner layer of the exine begins to develop. The nucleus occupies a central position. Then the exine is fully formed, a vacuole emerges that fills the whole lumen of the microspore, and the nucleus is oriented to a specific position at the periphery until the microspore nucleus divides. After first pollen mitosis, the small generative cell detaches from the pollen wall and is suspended in the cytoplasm of the vegetative cell. The vacuole is resorbed, and the vegetative cell in the maturing pollen grains accumulates materials required for pollination (RNA, protein, carbohydrates and lipids).

Upon stress treatment, as shown for tobacco, isolated microspores swell and their cytoplasm undergoes structural reorganization (Garrido et al. 1995). The nucleus moves to a central position, and cytoplasmic strands are formed that pass through the vacuole and connect the perinuclear with the subcortical cytoplasm. The same features can be seen in young stressed bicellular pollen grains and seem to occur irrespective of which developmental stage the microspores/pollen grains were isolated or the type of stress applied. For a more detailed description see Aionesei et al. (this Vol.).

Similar morphological changes have been seen in other plant cells, such as tobacco cells induced to reversible apoptosis by drugs (O'Brien et al. 1998), the cortex of legume roots that prepare for *Rhizobium* infection (Kijne 1992), hormone-treated protoplast and embryogenic *Cichorium* leaf cells (Dubois et al. 1991), and isolated egg cells in culture (Ponya et al. 1999) before fertilization. These features thus seem to be characteristic of mitotically active cells.

These so-called star-like microspores (Touraev et al. 1997) have also been observed in other species at the start or early on in microspore cultures, i.e. in rice (Raina and Irfan 1998), rapeseed (Zaki and Dickinson 1991), tobacco (Kyo and Harada 1986; Garrido et al. 1995; Touraev et al. 1996a) and wheat (Touraev et al. 1996b). However, these observations did not lead to a clear definition of embryogenic pollen because the formation of embryogenic microspores had not been followed throughout the whole culture period. Recent studies using tracking of the entire process of direct embryogenesis from single selected wheat microspores (Indrianto et al. 2001) clarified the situation. In vitro cultured microspores were shown to fall into three morphologically different types, no matter what kind of treatment was used. As described in another chapter of this book (Aionesei et al.), the three types of microspores turned out to be stages in a continuous process that takes place in response to stress and which leads to the formation of the 'star-like' microspores. These 'star-like' microspores fill with cytoplasm upon transfer to a rich medium under non-stress conditions and start to divide, leading to the formation of embryos and haploid plants.

4 Mechanism of the Stress-Induced Switch from the Gametophytic to the Sporophytic Mode of Development

Observations made in different systems have shown that regardless of the applied stress, the formation of embryogenic microspores is accompanied by the following common events: (1) swelling; (2) passage through either DNA-replication or mitosis, followed by a cell cycle arrest; (3) autophagy of the cytoplasm; (4) cytoskeletal and cytoplasmic rearrangement leading to the displacement of the microspore nucleus from a peripheral to a central position; (5) formation of a new cell wall below the gametophytic cell wall; (6) compaction of the chromatin; and (7) changes in gene expression (see Aionesei et al., this Vol.). The mechanisms that induce such changes in microspores and how they define the further (sporophytic) mode of development are still obscure. However, these morphological, physiological and molecular rearrangements also occur in other eukaryotic cells during various stress treatments. This provides an opportunity to propose some common events in plant stress response and, based on results obtained for different microspore culture systems, possible mechanisms for the reprogramming of microspores towards sporophytic development.

4.1 Heat Shock

Heat shock has been shown to be an effective trigger in switching the gametophytic development of microspores to the sporophytic pathway. It is effective when applied alone (e.g. *Brassica napus*) or in combination with starvation (*Nicotiana tabacum* and *Triticum aestivum*). In rapeseed, mild heat shock of 33 °C for 8 h is sufficient to induce sporophytic pathway in microspores, while strong heat shock at 42 °C has been applied to bicellular pollen. In general, the degree and duration of heat pretreatment are stage- and species-specific. For example, some species of Solanaceae are very sensitive to elevated temperatures: incubation of pepper microspores at 33 °C leads to lethality at all stages used when the treatment is longer than 12 h (A. Ribarits, E. Heberle-Bors, A. Touraev, unpubl. observ.). Maximal yield of multicellular structures in rice has been observed after applying a mild (33 °C) heat shock (Raina and Irfan 1998).

Heat-stressed microspores exhibit essentially the same features described above (see Sect. 4). Heat stress induces a wide spectrum of changes in eukaryotic cells such as large-scale but not total repression of protein synthesis, sensing of protein damage or membrane perturbation, induction of heat shock protein synthesis, stimulation of membrane ATPase and of Ca^{2+}-activated pathways, transient cell cycle arrest, endocytosis, etc. (reviewed by Piper 1993).

A heat shock treatment also dramatically changes the protein spectrum of the cell (reviewed by Mager and Hofmann 1997) accompanied by the synthesis of a highly conserved group of heat shock proteins (HSPs). The synthesis of these proteins is characteristic for both stressed prokaryotic and eukaryotic cells. HSPs are synthesized in plant cells not only during heat shock but also under osmotic (Almoguera et al. 1992), oxidative (Sun et al. 2001), cold (Sabehat et al. 1998) and heavy metal (Györgyey et al. 1991) treatments. Activation of HSP70 synthesis has been shown in maize (Gagliardi et al. 1995) and *Brassica napus* (Binarova et al. 1997) microspores subjected to high temperatures. A small HSP is activated in early bicellular pollen grains of tobacco, reprogrammed to sporophytic development by starvation (Zarsky et al. 1995).

The majority of HSPs are molecular chaperones. Some of the important functions attributed to molecular chaperones are the folding of proteins and the degradation of unstable molecules (Parcellier et al. 2003). Molecular chaperones also play a role in developmental processes. In plants, the synthesis of small HSPs is restricted to certain stages of the plant life cycle, such as embryogenesis (Löw et al. 2000), germination of seeds (Wehrmeyer et al. 1996), pollen development (Parcellier et al. 2003) and fruit maturation (Löw et al. 2000). These examples illustrate the role of HSPs in 'regular' developmental processes.

HSPs also participate in reprogramming cell fate of microspores. Telmer et al. (1993) suggested that HSP synthesis in microspores interferes with the synthesis of proteins necessary for their differentiation into pollen. Thus, once HSPs have blocked the pollen differentiation program, sporophytic development is possible. This model has been confirmed recently by immunogold-labeling of HSP70 and HSP90 during the induction of embryogenic microspores in *Brassica napus* (Segui-Simarro et al. 2003). HSP70 was detected in the nucleus in association with ribonucleoprotein structures in the interchromatin region and in the nucleolus, whereas HSP90 was mostly found in the interchromatin region. Translocation of HSP70 from the cytoplasm to the nucleus was observed in induced and non-induced microspores. However, the nuclear/cytoplasmic labeling density ratio was twofold higher in induced compared to non-induced (gametophytic) microspores exposed to the same inductive treatment but not committed to embryogenesis. This suggests a possible role for HSP70 in the switch to embryogenesis. The enrichment of HSP70 in the nucleus after the induction of microspore embryogenesis by elevated temperature was also observed in tobacco and pepper (Testillano et al. 2000).

Activation of HSP synthesis may also prevent apoptosis of microspores. Usually a significant number of microspores die when subjected to a stress (see Aionesei et al., Chap. I.3, this Vol.), as could be expected, whereas those that survive exhibit features of necrosis (swelling, autophagy) and apoptosis such as intranucleosomal cleavage of DNA (Wang et al. 1999). In contrast to animals, in plants a distinction between necrosis and apoptosis is sometimes not easy, and combinations of both types of cell death may occur (Kratsch and Wise 2000). Apoptosis or programmed cell death (PCD) is a genetically encoded suicide pathway, which begins with the expression of unique proteins involved in the systematic dismantling of the cell (Parcellier et al. 2003). This type of cellular death can occur both as a response to environmental stress (e.g. heat, cold, salinity etc.) and as a way to regulate the growth and development of the organism (Groover et al. 1997). An anti-apoptotic function of HSPs is well established in mammalian systems. HSP70 was shown to inhibit late caspase-dependent events such as activation of cytosolic phospholipase A2 in human cervical carcinoma cells (Jäättelä et al. 1998), while HSP90 inhibited apoptosome formation in human U-937 cells (Pandey et al. 2000).

Thus, taking into consideration the highly conserved nature of HSPs and their involvement in stress-induced cellular reprogramming of microspores, it seems reasonable to assume that they may also have an anti-apoptotic function in microspores. Most likely they are required for microspore survival under stress conditions while the stress reprograms the surviving microspores by a mechanism that may share features with necrosis and/or apoptosis. Blocking the apoptotic process by HSPs may result in the survival of cells with 'a reprogrammed' genome which switches the development from a gametophytic to a sporophytic mode. The changes in chromatin structure caused by apoptosis (among other events) may result in changes in gene expression.

HSPs not only function as molecular chaperones in the cell. The identified *E. coli* heat shock-induced protein FtsJ (Ogura et al. 1991) has different functions. FtsJ is a methyl-transferase and is well conserved from bacteria to humans. The important role of this protein for survival under elevated temperatures is confirmed by the fact that null mutations in the FtsJ gene produce a dramatically altered ribosome profile, a severe growth disadvantage, and a temperature-sensitive phenotype (Bügl et al. 2000). There are no data concerning the function of this protein in microspores. However, FtsJ methylates 23S rRNA and, therefore, may regulate translation (Bügl et al. 2000). The general translation pattern is indeed changed during the transition from the gametophytic to the sporophytic pathway (Garrido et al. 1993; Testillano et al. 2000). Therefore, FtsJ could play a role in microspore embryogenesis via the regulation of translation.

Another possible way that microspores may be reprogrammed by heat shock is via an influence on the cell cycle events in stressed microspores (see Aionesi et al., Chap. I.3, this Vol., for details). Microspore reprogramming involves important changes in the ultrastructural organization of nuclei, as was shown with *Brassica*, pepper and tobacco (Testillano et al. 2000). Embryogenic nuclei were characterized by the presence of small chromatin patches forming a reticulum, connected by chromatin fibers. Nucleoli showed a structural organization similar to that observed in plant interphase nucleoli of cycling cells (Testillano et al. 2000). A significant component visualized by immunolabeling in nuclei and cytoplasm of heat shock-induced embryogenic microspores of tobacco, *Brassica* and pepper was the Ntf6-MAP kinase (Testillano et al. 2000). This MAP kinase has been isolated from tobacco (Wilson et al. 1995) and has been related to the entry of quiescent cells into cell division and cytokinesis (Calderini et al. 1998; Prestamo et al. 1999). Moreover, the pattern of distribution of the Ntf6-MAPK in the embryogenic nuclei is similar to that reported in proliferating meristematic cells (Prestamo et al. 1999). The presence of PCNA (proliferating cell nuclear antigen) was also observed in embryogenic microspores of *Brassica*, tobacco and pepper (Testillano et al. 2000). The identification of components connected with both cell cycle signaling (i.e. MAP kinases) and cell cycle progression machinery (i.e. PCNA) in reprogrammed microspores may indicate that heat shock can affect microspores at one or both of these levels.

4.2 Cold Shock

Cold shock has long been known to increase doubled haploid formation in anther and microspore cultures (Nitsch and Norreel 1974; Duncan and Heberle 1976). For example, incubation of excised spikes of barley at 7 °C (Xu et al. 1981), *Triticale* at 4 °C (Marciniak et al. 2003) and buds of *Citrus clementina* at 4 °C (Germana and Chiancone 2003) significantly increased the yield of microspore embryos in anther cultures. Such pre-treatment was effective also for other important crops (Dunwell 1996). In shed microspore cultures of barley (Sunderland and Xu 1982), tobacco (Sunderland and Roberts 1979) and wheat (Wei 1982) the frequency of multicellular structures formed was dramatically increased by a pretreatment of excised spikes and buds at low temperatures (7 °C for barley and tobacco and 4 °C for wheat).

While the treatments of anthers, buds or inflorescences provided some valuable methods to produce doubled haploids and have also shed some light on the mechanism of stress action on embryo induction in microspores, the analyses of anthers, buds, inflorescences and whole plants to understand these mechanisms have always been hampered by the difficulty in distinguishing primary effects on the microspore from secondary ones mediated by the surrounding tissues of the microspore donor plant. Chilling may simply slow down degradation processes in the anther tissues – which may cause browning of in vitro cultured anthers – and thus prevent microspores from being exposed to toxic compounds produced in the decaying anther (Duncan and Heberle 1976). In cereals, microspores are physically attached to the tapetum and take up nutrients from the tapetum. Sunderland and Xu (1982) observed that microspores are detached in cold pretreated barley anthers and are located free inside the anther locule. The authors put forward the hypothesis that the 'true effect' of cold stress is this detachment of the microspores resulting in a lack of necessary nutrients and in starvation. However, microspores in many dicots, such as tobacco in which a cold treatment is also effective, are not attached to the tapetum and are freely distributed inside the anther locule in non-stressed anthers in vivo.

Cold pretreatment of spikes and buds has also increased efficiency of embryogenesis in isolated microspore cultures of wheat (Indrianto et al. 1999) and maize (Gaillard et al. 1991), but, again, the cold treatment was applied on anthers before microspore isolation. Despite numerous reports on cold shock-induced microspore embryogenesis, almost no research has been published on the direct effect of cold stress on isolated microspores. In only one such experiment, isolated and cold-treated tobacco microspores, even when cultured in a starvation medium, developed into mature pollen when cultured further in a maturation medium at normal temperature (Touraev et al. 1996c). Other experiments performed by the authors' group on the effect of cold stress alone and/or in combination with starvation confirmed that cold stress alone is not sufficient to switch wheat, tobacco and pepper microspores from the gametophytic pathway towards sporophytic development. After release from the cold, wheat and pepper microspores started to accumulate starch quickly during further incubation at normal temperature, and ultimately died (A. Ribarits, A. Indrianto, E. Heberle-Bors, A. Touraev, unpubl. observ.). A combination of cold stress and starvation could induce sporophytic division in tobacco microspores but only after a treatment of more than 14 days (Touraev et al. 1996c) at 40 °C. These experiments together with data on cold pretreated spikes and buds described above allow us to conclude that subjection to a cold treatment simply slows down physiological processes in isolated microspores, anther, buds and spikes, and the main factor that induces embryogenic microspore is starvation of the microspores inside spikes, buds and anthers, caused by the excision of these organs and organ systems from the plant. Nutrient deprivation appears to be the stress in these situations, which reprograms the microspores while the cold treatment, and the subcellular changes it causes, may alleviate the microspore from the stress and allow a higher frequency of microspores to survive and be reprogrammed. Thus, rather than acting as a stress, the cold-treatment seems to act as an anti-stress treatment.

Cytological analyses have indeed shown changes in microspores isolated from cold-stressed excised anthers that were quite similar to those observed in in vitro cultured anthers and embryogenic pollen grains formed in vivo (Rashid et al. 1981). Such microspores were decreased in size, the cytoplasm was poor in ribosomes, plastids were regressed and starch grains degraded. The electron microscopic study of microspores in cold-treated excised *Datura metel* flower buds (Sangwan and Camefort 1984) showed a non-homogeneous content of vacuoles, diminishing of starch grains in plastids, distinct and disorganized endoplasmic reticulum and more Golgi bodies compared to non-stressed microspores. Nuclei were not dramatically changed during the cold stress, but in some cases a decrease of their electron density and condensation of the nucleoli were detected (Sangwan and Camefort 1984). Similarly, barley microspores isolated from inflorescences pretreated for 28 days at 4 °C are slowed down in their passage through S phase and the arrest of microspores at G2 (Shim and Kasha 2003). Late microspores with a 2C DNA content divided asymmetrically but cytokinesis was delayed. An asymmetric division of microspores subjected to cold treatment has also been observed in wheat (Hu and Kasha 1999). In contrast, pretreatment of barley anthers at 7.5 °C over 14 days resulted in both symmetrical and asymmetrical divisions (Sunderland and Evans 1980).

Cold-induced structural rearrangements occur also in other cell types and are part of the general cold stress response. Karpilova et al. (1980) observed a general retardation in growth processes in *Cucumis sativus* leaves under non-freezing chilling conditions. Temperature-sensitive *Arabidopsis* mutants differ from chilling-resistant wild-type plants by their slower growth rate (Wu et al. 1997). It is also known that the cytoskeleton is affected by cold. For example, incubation of *Arabidopsis thaliana* plants at 4 °C induced changes in the β-tubulin isoform pattern in mesophyll cells (Chu et al. 1993).

The mechanisms that reprogram microspore fate through an apoptotic and/or molecular chaperone activity after heat shock may be claimed also for the cold stress. The changes in cell morphology and physiology observed in microspores during cold stress are also characteristic of programmed cell death in other tissues (e.g. DNA condensation, swelling of organelles, activation of Ca^{2+} pathways, etc.; Kratsch and Wise 2000). Cold stress also induces the expression of small HSPs in various tissues of tomato, similar to a heat shock (Sabehat et al. 1998).

4.3 Sugar Starvation

Sugar starvation is an efficient method to induce embryogenesis in isolated microspores of many important crops such as tobacco, wheat, barley and rice (Touraev et al. 2001). Starvation has been applied not only to isolated microspores but also to excised anthers, buds or the whole plant (Heberle-Bors 1989).

The carbohydrate status of the plant is one of the most important factors defining the normal development of microspores and the formation of fertile pollen. In *Arabidopsis*, increased leaf carbohydrate export and starch mobilization are required for flowering, suggesting that phloem carbohydrates have a critical function in floral transition (Corbesier et al. 1998). Interestingly, the addition of sucrose can rescue the late-flowering phenotype of several mutants (Roldan et al. 1999) and even promotes leaf morphogenesis and flowering in the dark (Roldan et al. 1999).

Physiological and structural changes that occur during sugar starvation have been extensively studied in somatic cells. Experiments performed in excised maize root tips, cultured sycamore cells and cultured rice suspension cells have shown that sugar starvation generally triggers sequential changes in the following cellular events: (1) an arrest of cell growth, (2) the rapid consumption of cellular carbohydrate content and a decrease in respiration rate, (3) the degradation of lipids and proteins, (4) an increase in accumulation of

Pi, phosphorylholine and free amino acids, and (5) a decline in glycolytic enzymatic activities (reviewed by Yu 1999).

The decline in cellular sugar and starch content is coupled with a decline in metabolic activity and an increase in vacuolar autophagy (Aubert et al. 1996). Triggering such autophagic processes presumably involves regression of the cytoplasm, including organelles, and recycling of respiratory substrates. Due to the presence of intracellular pools of carbohydrates and the ability to control the autophagic process, a plant cell can survive for some time after carbohydrate starvation (Yu 1999).

Some morphological and physiological changes typical of 'autophagic' cells were shown for sucrose-starved early- or mid-bicellular *Nicotiana tabacum* pollen under sugar starvation such as dedifferentiation of plastids, degradation of the lamellar structure, disappearance of starch, appearance of the large vacuole, etc. (Garrido et al. 1995).

One of the changes in starving microspores may be the regulation of the energy and respiratory status of microspores by the carbohydrates in the culture medium and the rate of their uptake by the microspores. The culture of isolated barley microspores in medium with a high concentration of sucrose resulted in a high rate of metabolism (Scott et al. 1995), leading to hypoxia, significant accumulation of ethanol and a decrease of the ATP level. Substitution of sucrose by maltose, which is utilized much slower, made sufficient oxygen available and allowed cells to undergo embryogenesis. The same effect of substitution of sucrose by maltose has been shown in wheat (Indrianto et al. 1999). In these experiments maltose was effective in combination with heat, cold, starvation and even without any pre-treatment. Therefore the hypothesis has been put forward that the slow uptake of maltose may exert a starvation stress on the microspores and that it was this starvation stress that induced microspore embryogenesis. Later, when maltose is taken up, it is used to feed the developing microspore embryo.

In support, we have recently found that carbohydrate utilization depends on medium pH (J. Barinova et al., submitted). Tobacco microspores, cultured in a sucrose-containing medium at low pH values (5.0–6.0), efficiently took up sucrose, had high invertase activity, and converted glucose into starch, resulting in the formation of gametophytic pollen. However, sucrose uptake and invertase activity were strongly reduced when microspores were cultured at pH values of 8.0 or higher, resulting in a complete inhibition of starch accumulation. These microspores were embryogenic and, after transfer to a conventional embryogenesis medium at an intermediate pH of 7, formed embryos. Thus it seems that microspores seem to starve at a high pH due to the lack of sucrose cleavage and uptake, despite the presence of a carbon source in the medium.

Sugar starvation also changes the structure of the plasmalemma, in particular phospholipid composition. About six phosphoproteins were isolated from embryogenic pollen grains of *Nicotiana tabacum* and *Nicotiana rustica* (Kyo and Harada 1990a,b). The fractionation of organelles showed that the most likely candidate for the localization of these phosphoproteins is the plasma membrane (Kyo and Ohkawa 1991).

In vitro phosphorylation assays in extracts of mid-bicellular and embryogenic tobacco pollen also showed qualitative and quantitative changes in protein kinase activities during the starvation treatment (Garrido et al. 1993). This suggests that protein kinases are likely to be involved in the transduction of the hunger signal, mediating the effect of starvation on gene expression and cell cycle regulation.

As has been mentioned above, the expression of small HSP gene in early bicellular pollen grains of tobacco is induced by starvation (Zarsky et al. 1995), which further supports the idea that molecular chaperons may play a role in cellular reprogramming via preventing apoptosis. This point of view has indirectly been argued for by data obtained from cultured barley anthers (Wang et al. 1999). The microspores from freshly isolated, non-starved anthers demonstrated features of apoptosis such as intranucleosomal cleavage of DNA. Pretreatment of anthers by mannitol (starvation+osmotic stress) prevented this DNA degradation.

4.4 Colchicine Treatment

Colchicine, a microtubule-depolymerizing agent, is the only chemical treatment effectively inducing embryogenesis from microspores of *Brassica napus* (Zaki and Dickinson 1991; Zhao et al. 1996). Approximately 15% of rapeseed microspores divided and formed embryos when incubated in the presence of 25 μM colchicine for 42 h (Zhao et al. 1996). The treatment seems to be specifically effective on microspores since the yield of embryos after treatment of unicellular *Brassica* microspores is ten times higher than after treatment of mitotic and bicellular pollen (Zhao et al. 1996). The requirement for disruption of the cytoskeleton at a particular developmental stage is not surprising. Colchicine treatment resulted in complete depolymerization of microtubules only in unicellular microspores whereas microtubules in bicellular pollen were not affected (Simmonds 1994; Zhao et al. 1996).

The primary action of colchicine seems to be microtubule depolymerization, which releases the anchored nucleus and is thus disrupting microspore asymmetry (Simmonds 1994). The nucleus is thus free to migrate away from the edge of the cell. In addition, the preprophase band also appears after colchicine treatment. Changes in cell polarity may participate in the reprogramming of microspore fate.

Another action of colchicine is on cytoskeleton dynamics. Colchicine binds to α- and β-tubulin heterodimers, which inhibits further dimer addition to microtubules and results eventually in microtubule depolymerization (Sternlicht et al. 1983). The elevated concentration of free tubulins acts to depress the synthesis of new α- and β-tubulins, which includes the synthesis of pollen-specific tubulins (Carpenter et al. 1992) and thus prevents the progression of pollen development. This effect may convert microspores to the sporophytic pathway.

It is also possible that the tubulin cytoskeleton participates indirectly in the reprogramming of microspore by interacting with the cell cycle machinery and cell cycle-dependent kinases. Phosphorylation and dephosphorylation are key regulatory processes involved in changing the distribution of microtubules through the G2 phase to the M phase in higher plant cells (Katsuta and Shibaoka 1992). In tobacco BY-2 cells a treatment by staurosporine, a kinase inhibitor, causes a decrease in the rate of development of PPBs, and inhibits the disappearance of PPBs once they have developed, with the resultant accumulation of cells with a PPB (Katsuta and Shibaoka 1992).

Microtubules have been shown to associate with A-type cyclin-dependent kinase (cdc2), which is central to the orderly progression of the cell cycle (Weingartner et al. 2001). During interphase and prophase cdc2 is associated mainly with the chromatin but later migrates into the spindle. Cdc2 is involved in many important processes such as nuclear membrane breakdown and chromatin condensation. The level of cdc2 accumulation correlates with proliferating activity of the cells. For example, in wheat leaves the maximal level was reached in the meristematic zone (John et al. 1990). Association of cdc2 protein with microtubules may induce feedback links between these two components. One may speculate that the inhibition of spindle formation by colchicine may affect cdc2 biosynthesis or change the balance between chromatin-associated and non-associated cdc2-protein in the cell and lead to changes in cell cycle transition.

The knowledge about the mechanism of colchicine-induced microspore embryogenesis is confined to the cytoskeleton and related processes as described above. Unfortunately, nothing is known about the effect of colchicine on processes characteristic to embryogenic microspores induced by heat, cold and starvation. Therefore, it is difficult to characterize colchicine as a 'classical' stress to induce microspore embryogenesis. Moreover, colchicine-induced microspore embryogenesis is restricted to *Brassica napus*. No other report on microspore embryogenesis induction by colchicine exists. In the hand of the authors, tobacco microspores, cultured in the presence of colchicines, divided symmetrically and even formed multicellular structures with cells similar to callus cells but never developed further to embryos or plants (A. Touraev, E. Heberle-Bors, unpubl. observ.).

5 Conclusions and Perspectives

The data presented in this chapter show that stress applied to microspores affects different processes (e.g. cell cycle regulation, cytoskeleton organization, carbohydrate metabolism, protein biosynthesis and phosphorylation) and structures (membrane, plastid and mitochondrion structure) simultaneously, resulting in dedifferentiation of the microspore with features typical of actively proliferating cells. Moreover, cross-talk between signal transduction pathways for different stresses may take place. In some cases, the signal transduction pathways triggered by different stresses are common to more than one stress type. Most stresses have shown to elicit an increase in cytosolic free calcium levels and to involve protein phosphatase and kinase [including mitogen-activated protein kinase (MAPK) cascades (Pearce and Humphrey 2001)]. Components of these cascades, such as MAP kinases, were identified in stress-induced embryogenic microspores (Testillano et al. 2000).

Stress in general changes the morphology of microspores: a vacuolated and polarized microspore is transformed into a depolarized and dedifferentiated cell, which has a centrally positioned and decondensed nucleus, typical of cells preparing for a change in development. After transfer to a suitable rich medium, the 'microspore' becomes enriched in cytoplasm and the cell cycle is reactivated. The mechanisms integrating stress responses and cell-cycle checkpoint pathways in microspores are still unknown. However, some data indicating a tight link between these two chains of events were obtained in other eukaryotic systems (Pearce and Humphrey 2001).

Another important conclusion is that we are still far from knowing exactly what stress means in the context of microspore reprogramming. While heat and starvation treatment shared certain features in their effect on microspore reprogramming in a number of different species, cold cannot really be considered a stress, but rather an anti-stress, while for colchicine we lack evidence for its general relevance in microspore embryogenesis and for its stress effect. However, high pH can now be added to the list of stresses effectively reprogramming microspores (Touraev et al. 2001). A better understanding of stress in general, and of the events underlying cellular reprogramming, will surely help to identify new stresses which alone, or in combination, may lead to more efficient induction of microspore embryogenesis for doubled haploid production or allow induction in still recalcitrant species.

Stress-induced dedifferentiation and cell divisions with subsequent formation of embryos are not limited to isolated microspores. High osmolality, salinity and heavy metal ion treatment have been shown to induce somatic embryogenesis in shoot-apical–tip explants of carrot (Kamada et al. 1993) and *Arabidopsis* (Ikeda-Iwai et al. 2003). Thus, despite the different induction mechanisms, the pathways and targets of all stresses induce in microspores the same processes, leading to dedifferentiation, which is an obligatory condition for reprogramming their development.

References

Almoguera C, Jordano J (1992) Developmental and environmental concurrent expression of sunflower dry-seed-stored low-molecular-weight heat-shock protein and Lea mRNAs. Plant Mol Biol 19:781–792

Aubert S, Gout E, Bligny R, Marty-Mazars D, Barrieu F, Alabouvette J, Marty F, Douce R (1996) Ultrastructural and biochemical characterization of autophagy in higher plant cells subjected to carbon deprivation: control by the supply of mitochondria with respiratory substrates. J Cell Biol 133:1251–1263

Aubry L, Firtel R (1999) Integration of signaling networks that regulate *Dictiostelium* differentiation. Annu Rev Cell Dev Biol 15:231–237

Banks JA (1994) Sex-determining genes in the homosporous fern *Ceratopteris*. Development 120:1949–1958

Benito-Moreno RM, Macke F, Hauser MT, Alwen A, Heberle-Bors E (1988) Sporophytes and gametophytes from in vitro cultured, immature tobacco pollen. In: Cresti M, Jori P, Paccini E (eds) Sexual reproduction in higher plants. Springer, Berlin Heidelberg New York, pp 137–142

Binarova P, Hause G, Cenklova V, Cordevener JHG, van Lookeren Campagne MM (1997) A short severe heat shock is required to induce embryogenesis in late bicellular pollen of *Brassica napus* L. Sex Plant Reprod 10:200–208

Bügl H, Fauman EB, Staker BL, Zheng F, Sidney RK, Saper MA, Bardwell JCA, Jacob U (2000) RNA methylation under heat shock control. Mol Cell 6:349–360

Calderini O, Börge L, Vicente O, Binarova P, Heberle-Bors E, Wilson C (1998) A cell cycle regulated MAP kinase with a possible role in cytokinesis in tobacco cells. J Cell Sci 111:3091–3100

Carpenter JL, Ploense SE, Snustad DP, Silflow CD (1992) Preferential expression of α-tubulin gene of *Arabidopsis* in pollen. Plant Cell 4:557–571

Chu B, Snustad DP, Carter JV (1993) Alteration of β-tubulin gene expression during low-temperature exposure in leaves of *Arabidopsis thaliana*. Plant Physiol 103:371–377

Cho MS, Zapata FJ (1988) Callus formation and plant regeneration in isolated pollen culture of rice (*Oryza sativa* L., cv. Taipei 309). Plant Sci 58:239–244

Corbesier L, Lejeune P, Bernier G (1998) The role of carbohydrate in the induction of flowering in *Arabidopsis thaliana*: comparison between wild type and starch-less mutant. Planta 206:131–137

Custers JBM, Gordewener JHG, Nöllen Y, Dons JJM, van Lookeren-Campagne MM (1994) Temperature controls both gametophytic and sporophytic development in microspore culture of *Brassica napus*. Plant Cell Rep13:267–271

Dubois T, Guedrira M, Dubois J, Vasseur J (1991) Direct somatic embryogenesis in leaves of *Chicorium*: a histological and SEM study of early stages. Protoplasma 162:120–127

Duncan EJ, Heberle E (1976) Effect of temperature shock on nuclear phenomena in microspores of *Nicotiana tabacum* and consequently on plantlet production. Protoplasma 90:173–177

Dunwell JM (1996) Microspore cultures. In: Mohan Jain S, Sopory SK, Veilleux RE (eds) In vitro haploid production in higher plants, vol 1. Kluwer, Dordrecht, pp 205–216

Gagliardi D, Breton C, Chaboud A, Vergne P, Dumas C (1995) Expression of heat shock factor and heat shock protein 70 genes during maize pollen development. Plant Mol Biol 29:841–856

Gaillard A, Vergne P, Beckert M (1991) Optimization of maize isolation and conditions for reliable plant regeneration. Plant Cell Rep 10:55–58

Garrido D, Charvat B, Benito Moreno RM, Alwen A, Vicente O, Heberle-Bors E (1991) Pollen culture for haploid plant production in tobacco. In: Negrutiu I, Gharti-Chhetri G (eds) A laboratory guide for cellular and molecular plant biology. Birkhäuser, Basel, pp 59–69

Garrido D, Eller N, Heberle-Bors E, Vicente O (1993) De novo transcription of specific messenger RNAs during the induction of tobacco pollen embryogenesis. Sex Plant Reprod 6:40–45

Garrido D, Vicente O, Heberle-Bors E, Rodriquez-Garcia MI (1995) Cellular changes during the acquisition of embryogenic potential in isolated pollen grains of *Nicotiana tabacum*. Protoplasma 186:220–230

Germana MA, Chiancone B (2003) Improvement in *Citrus clementina* ort. Ex Tan. microspore derived embryoids induction and regeneration. Plant Cell Rep 22:181–187

Groover A, DeWitt N, Heidel A, Jones A (1997) Programmed cell death of plant tracheary elements differentiating in vitro. Protoplasma 196:197–211

Guha S, Maheshwari SC (1964) In vitro production of embryos from anthers of *Datura*. Nature 204:497

Gustafson V, Baenziger PS, Wright MS, Stroup WW, Yen Y (1995) Isolated wheat microspore culture. Plant Cell Tissue Organ Cult 42:207–213

Györgyey J, Gartner A, Nemeth K, Magyar Z, Hirt H, Heberle-Bors E, Dudits D (1991) Alfalfa heat shock genes are differentially expressed during somatic embryogenesis. Plant Mol Biol 16:999–1007

Heberle-Bors E (1983) Induction of embryogenic pollen grains in situ and subsequent in vitro pollen embryogenesis in *Nicotiana tabacum* L. by treatments of pollen donor plants with feminizing agents. Physiol Plant 59:67–72

Heberle-Bors E (1989) Isolated pollen in tobacco: plant reproductive development in a nutshell. Sex Plant Rep 2:1–10

Hoekstra S, van Zijderveld MH, Heidekamp F, van der Mark F (1992) Anther and microspore culture of *Hordeum vulgare* L. cv. Igri. Plant Sci 86:89–96

Hu T, Kasha KJ (1999) A cytological study of pretreatments used to improve isolated microspore cultures of wheat (*Triticum aestivum* L.) cv. Chris. Genome 42:432–441

Ikeda-Iwai M, Umehara M, Satoh S, Kamada H (2003) Stress-induced somatic embryogenesis in vegetative tissues of *Arabidopsis thaliana*. Plant J 34:107–114

Indrianto A, Heberle-Bors E, Touraev A (1999) Assessment of various stresses and carbohydrates for their effect on the induction of embryogenesis in isolated wheat microspores. Plant Sci 143:71–79

Indrianto A, Barinova J, Touraev A, Heberle-Bors E (2001) Tracking individual wheat microspores in vitro: identification of embryogenic microspores and body axis formation in the embryo. Planta 212:163–174

Jäättelä M, Wisseng D, Kokholm K, Kallunki T, Egeblad M (1998) Hsp 70 exerts its anti-apoptotic function downstream of caspase3-like proteases. EMBO J 17:6124–6134

John PC, Sek FJ, Carmichael JP, McCurdy DW (1990) p34cdc2 homologue level, cell division, phytohormone responsiveness and cell differentiation in wheat leaves. J Cell Sci. 97:627–630

Kamada H, Ishikawa K, Saga H, Haada H (1993) Induction of somatic embryogenesis in carrot by osmotic stress. Plant Tissue Culture Lett 10:38–44

Karpilova I, Chugunova N, Bil' K, Chermnykh L (1980) Ontogenetic changes of chloroplast ultrastructure, photosynthates and photosynthate outflow from the leaves in cucumber plants under conditions of reduced night temperature. Soviet Plant Physiol 29:113–120

Katsuta J, Shibaoka H (1992) Inhibition by kinase inhibitors of the development and the disappearance of the preprophase band of microtubules in tobacco BY-2 cells. J Cell Sci 103:397–405

Kijne JW (1992) The *Rhisobium* infection process. In: Stacey G, Burris RH, Evans HJ (eds) Biological nitrogen fixation. Chapman and Hall, New York, pp 349–398

Kratsch HA, Wise RR (2000) The ultrastructure of chilling stress. Plant Cell Environ 23:337–350

Kyo M, Harada H (1986) Control of the developmental pathway of tobacco pollen in vitro. Planta 168:427–432

Kyo M, Harada H (1990a) Phosphorylation of proteins associated with embryogenic dedifferentiation of immature pollen grains of *Nicotiana rustica*. J Plant Physiol 136:716–722

Kyo M, Harada H (1990b) Specific phosphoproteins in the initial period of tobacco pollen embryogenesis. Planta 182:58–63

Kyo M, Ohkawa T (1991) Investigation of subcellular localization of several phosphor-proteins in embryogenic pollen grains in tobacco. J Plant Physiol 137:525–529

Löw D, Brändle L, Nover C, Forreiter C (2000) Cytosolic heat-stress proteins Hsp 17.7 class I and Hsp 17.3 class II of tomato act as molecular chaperones in vivo. Planta 211:575–582

Mager WH, Hohmann S (1997) Stress response mechanisms in the yeast *Saccharomyces cerevisiae*. In: Hohmann S, Mager WH (eds) Yeast stress responses. Landes, New York, pp 75–99

Marciniak K, Kaczmarek Z, Adamski T, Surma M (2003) The anther-culture response of triticale line × tester progenies. Cell Mol Biol Lett 8:343–351

Msadek T (1999) When the going gets tough: survival strategies and environmental signaling networks in *Bacillus subtilis*. Trends Microbiol 7:201–207

Nitsch C, Norreel B (1974) La culture de pollen isolé sur milieu synthétique. CR Acad Sci (Paris) Ser D 278:1031–1034

O'Brien IEW, Baguley BC, Murray BG, Morris BAM, Ferguson IB (1998) Early stages of the apoptotic pathway in plant cells are reversible. Plant J 13:803–814

Ogura T, Tomoyasu T, Yuki T, Morimura S, Begg KJ, Donachie WD, Mori H, Niki H, Higara S (1991) Structure and function of the ftsH gene in *Escherichia coli*. Res Microbiol 142:279–282

Pandey P, Saleh A, Nakazawa A, Shalendra K, Srinavasa MS, Vijay K, Weicheselbaum R, Nalin C, Alnemri ES, Kufe D, Kharbanda S (2000) Negative regulation of cytochrome c-mediated oligomerization of Apaf-1 and activation of preocaspase-9 by heat shock protein 90. EMBO J 19:4310–4322

Parcellier A, Gurbuxani S, Schmitt E, Solary E, Garrido C (2003) Heat shock proteins, cellular chaperones that modulate mitochondrial cell death pathway. Biochem Biophys Res Commun 304:505–512

Pearce AK, Humphrey TC (2001) Integrating stress-response and cell-cycle checkpoint pathways. Trends Plant Sci 11:426–433

Pechan PM, Keller WA (1989) Induction of microspore embryogenesis in *Brassica napus* L. by gamma irradiation and ethanol stress. In Vitro Cell Dev Biol 25:1073–1074

Piper PW (1993) Molecular events associated with acquisition of heat tolerance by the yeast *Saccharomyces cerevisiae*. FEMS Microbiol Rev 11:1–11

Ponya Z, Timar I, Szabo L, Kristof Z, Barnabas B (1999) Morphological characterization of wheat (*T. aestivum* L.) egg cell protoplasts isolated from immature and over-aged caryopsis. Sex Plant Rep 11:357–359

Prestamo G, Testillano PS, Vicente O, Gonzalez-Melendi P, Coronado MJ, Wilson C, Heberle-Bors E, Risueno MC (1999) Ultrastructural distribution of MAP kinase and transcripts in quiescent and cycling plant cells and pollen grains. J Cell Sci 112:1065–1076

Raina SK, Irfan ST (1998) High frequency embryogenesis and plantlet regeneration from isolated microspores of indica rice. Plant Cell Rep 17:957–962

Rashid A, Siddiqui AW, Reinert J (1981) Ultrastructure of embryogenic pollen of *Nicotiana tabacum* var. Badischer Burley. Protoplasma 107:375–385

Roldan M, Gomez-Mena C, Ruiz-Garcia L, Salinas J, Martinez-Zapater JM (1999) Sucrose availability on the aerial part of the plant promotes morphogenesis and flowering of *Arabidopsis* in the dark. Plant J 20:581–590

Sabehat A, Lurie S, Weiss D (1998) Expression of small heat-shock proteins at low temperature: possible role in protecting against chilling injuries. Plant Physiol 117:651–658

Sangwan RS, Camefort H (1984) Cold-treatment related structural modifications in the embryogenic anthers of *Datura*. Cytologia 49:473–487

Scott P, Lyne RL, ap Rees T (1995) Metabolism of maltose and sucrose by microspores isolated from barley (*Hordeum vulgare* L.) Planta 197:435–441

Segui-Simarro JM, Testillano PS, Risueno MC (2003) Hsp70 and Hsp90 change their expression and sub-cellular localization after microspore embryogenesis induction in *Brassica napus* L. J Struct Biol 142:379–391

Shim YS, Kasha KJ (2003) The influence of pretreatment on cell stage progression and the time of DNA synthesis in barley (*Hordeum vulgare* L.) uninucleate microspores. Plant Cell Rep 21:1065–1071

Shinozaki K, Russel P (1996) Conjugation, meiosis and osmotic stress response are regulated by Spc1 kinase through Atf1 transcription factor in fission yeast. Genes Dev 10:2276–2288

Simmonds DH (1994) Mechanism of induction of microspore embryogenesis in *Brassica napus*: significance of preprophase band of microtubules in the first sporophytic divisions. In: Akkas

N (ed) Biomechanics of active movement and division of cells. Springer, Berlin Heidelberg New York, pp 569–574

Stauffer C, Benito Moreno RM, Ylstra B, Vicente O, Heberle-Bors E (1992) Seed set after pollination with in-vitro-maturated, isolated pollen of *Triticum aestivum*. Theor Appl Genet 81:576–580

Sternlicht H, Ringel I, Szasz J (1983) Theory for modelling the copolymerization of tubulin and tubulin-colchicine complex. Biophys J 42:255–267

Sun W, Bernard C, van de Cotte B, Van Montagy M, Verbruggen N (2001) At-HSP17.6A encoding a small heat-shock protein in *Arabidopsis*, can enhance osmotolerance upon overexpression. Plant J 27:407–415

Sunderland N, Evans LJ (1980) Multicellular pollen formation in cultured barley anthers. II. The A, B and C pathways. J Exp Bot 31(121):501–540

Sunderland N, Roberts M (1979) Cold-pretreatment of excised flower buds in float culture of tobacco anthers. Ann Bot 43:405–414

Sunderland N, Xu ZH (1982) Shed pollen culture in *Hordeum vulgare*. J Exp Bot 136:1086–1095

Telmer CA, Newcomb W, Simmonds DH (1993) Microspore development in *Brassica napus* and the effect of high temperature on division in vivo and in vitro. Protoplasma 172:154–165

Testillano PS, Coronado MJ, Segui JM, Domenech J, Gonzalez-Melendi P, Raska I, Risueno MC (2000) Defined nuclear changes accompany the reprogramming of microspore to embryogenesis. J Struct Biol 129:223–232

Touraev A, Ilham A, Vicente O, Heberle-Bors (1996a) Stress induced microspore embryogenesis from tobacco microspores: an optimized system for molecular studies. Plant Cell Rep 15:561–565

Touraev A, Indrianto A, Wratschko I, Vicente O, Heberle-Bors E (1996b) Efficient microspore embryogenesis in wheat (*Triticum aestivum*. L) induced by starvation at high temperatures. Sex Plant Rep 9:209–215

Touraev A, Pfosser M, Vicente O, Heberle-Bors E (1996c) Stress as the major signal controlling the developmental fate of tobacco microspores: towards a unified model of induction of pollen embryogenesis. Planta 2000:144–152

Touraev A, Vicente O, Heberle-Bors E (1997) Initiation of microspore embryogenesis by stress. Trends Plant Sci 2:285–303

Touraev A, Pfosser M, Heberle-Bors E (2001) The microspore: a haploid multipurpose cell. Adv Bot Res 35:54–109

Twell D (2002) The developmental biology of pollen. In: O'Neill SD, Roberts JA (eds) Plant reproduction. Academic Press, Sheffield, pp 86–153

Wang M, Hoekstra S, van Bergen S, Lamers GEM, Oppedijk BJ, van der Heijden MW, de Preister W, Schilperoort RA (1999) Apoptosis in developing anthers and the role of ABA in this process during androgenesis in *Hordeum vulgare* L. Plant Mol Biol 39:489–501

Wehrmeyer N, Hernandez LD, Finkelstein RR, Vierling E (1996) Synthesis of small heat-shock proteins in part of the developmental program of late seed maturation. Plant Physiol 112:747–757

Wei ZM (1982) Pollen callus culture in *Triticum aestivum*. Theor Appl Genet 63:71–73

Weingartner M, Binarova P, Drykova D, Schweighofer A, David JP, Heberle-Bors E, Doonan J, Börge L (2001) Dynamic recruitment of cdc2 to specific microtubule structures during mitosis. Plant Cell 13:1929–1943

Wilmut I, Schnieke AE, McWhir J, Kind AJ, Campbell KH (1997) Viable offspring derived from fetal and adult mammalian cells. Nature 385:810–813

Wilson C, Anglmayer R, Vicente O, Heberle-Bors E (1995) Molecular cloning, functional expression in *Escherichia coli*, and characterization of multiple mitogen-activated-protein kinases from tobacco. Eur J Biochem 233:249–257

Wu J, Lightner J, Warwick N, Browse J (1997) Low-temperature damage and subsequent recovery of fab 1 mutant *Arabidopsis* exposed to 2 °C. Plant Physiol 113:347–356

Xu ZH, Huang B, Sunderland N (1981) Culture of barley anthers in conditioning media. J Exp Bot 32:767–778

Yu SM (1999) Cellular and genetic responses of plants to sugar starvation. Plant Physiol 121:687–693

Zaki MAM, Dickinson HG (1991) Microspore-derived embryos in *Brassica*: the significance of division symmetry in pollen mitosis I to embryogenic development. Sex Plant Rep 4:48–55

Zarsky V, Garrido D, Eller N, Tupy J, Vicente O, Schöffl F, Heberle-Bors E (1995) The expression small heat shock gene is activated during induction of tobacco pollen embryogenesis by starvation. Plant Cell Environ 18:139–147

Zhao JP, Simmonds DH, Newcomb W (1996) Induction of embryogenesis with colchicines instead of heat in microspores of *Brassica napus* L. cv. Topas. Planta 198:433–439

I.4 Microspore Embryo Induction and Development in Higher Plants: Cytological and Ultrastructural Aspects

CHRISTOPHE CLÉMENT[1], RAJBIR S. SANGWAN[2], and BRIGITTE SANGWAN-NORREEL[2]

1 Introduction

The development of doubled haploids (DH) through anther or microspore culture represents a modern tool for the improvement of cultivated species, enabling plant breeders to produce homozygous lines in a few months. Several strategies may be used to produce DH lines from haploid cells of cultivated angiosperms: (1) gynogenesis consists of obtaining in vitro haploid plants from female haploid cells in the embryo sac. This technique has been optimized in several species such as *Beta vulgaris* L., though many technical obstacles limit its industrial development; (2) the ' *bulbosum* method' by interspecific crosses (Devaux 1992), which consists of artificially crossing *Hordeum vulgare* L. as the female parent and *Hordeum bulbosum* L. as the male parent. After fertilization, the genome provided by the *Hordeum bulbosum* L. male gamete is progressively eliminated during the subsequent embryogenesis (Devaux et al. 1996), resulting in a haploid plantlet containing exclusively the haploid genome of the *Hordeum vulgare* L. female gamete. Similar phenomena occur when crossing *Triticum aestivum* L. (female parent) with *Zea mays* L. (male parent) to obtain haploids of wheat; and (3) microspore embryogenesis, which consists of enabling microspores to switch their initial gametophytic program to an alternative sporophytic development (Touraev et al. 1997), giving rise to haploid embryos which further develop into haploid plants.

Since the initial description of microspore embryo production from *Datura* anther culture was reported about 40 years ago (Guha and Maheshwari 1964, 1966), this unique developmental potential has been recognized both as an important pathway for haploid plant production and as a potential model for studying early regulatory and morphogenetic events in plant embryogenesis. The morphology, histology and ultrastructure of angiosperm microspore embryogenesis have been characterized in a large number of species (for reviews see Sangwan and Sangwan-Norreel 1987a; Touraev et al.

[1] Université de Reims Champagne Ardenne, UFR Sciences, Biologie et Physiologie Végétales, BP 1039, 51687 Reims Cedex 2, France
[2] Université de Picardie Jules Verne, Androgenèse et Biotechnologies, 33, rue Saint-Leu, 80039 Amiens, France

Biotechnology in Agriculture and Forestry, Vol. 56
Haploids in Crop Improvement II (ed. by C.E. Palmer, W.A. Keller, and K.J. Kasha)

1997). Moreover, microspore embryogenesis is one of the most commonly used methods for developing doubled haploids in many crops including *Nicotiana tabacum* (Touraev et al. 1997), *Brassica napus* (Binarova et al. 1997), *Triticum aestivum* (Chu et al. 1990), *Oryza sativa* (Xie et al. 1995) and *Hordeum vulgare* (Jacquard et al. 2003). To date, microspore embryogenesis has been successfully combined with genetic transformation in *Datura* and *Nicotiana* (Sangwan et al. 1993) along with success also in a few other species including *Zea mays* (Jardinaud et al. 1995a,b), *Triticum* (Loeb and Reynolds 1994) and *Hordeum vulgare* (Kasha et al. 1995; Yao et al. 1997).

Microspore embryos are formed from cultured microspores by a relatively simple manipulation of the culturing conditions. Microspore embryogenesis can be performed using either anther or isolated microspore culture. In both cases, the process results in the regeneration of haploid plantlets from microspores, which are initially destined to develop into pollen grains (Touraev et al. 1997). As summarized in Fig. 1, the process of microspore embryogenesis can be divided into four steps: (1) the pretreatment, or induction phase, destined to switch the pollen fate from the initial gametophytic program to the alternative sporophytic embryogenic program; (2) the culture phase, which results in the embryogenic development of the reoriented microspore; (3) the regeneration phase allowing the development of androgenetic embryos into haploid plantlets; and (4) chromosome doubling, which may occur spontaneously or after colchicine application.

A number of technical factors influencing the response of microspores in pollen embryogenesis has been identified. These include the growth conditions of the donor plant (Lyne et al. 1986; Luckett and Smithard 1992; Jähne and Lörz 1995), the developmental stage of microspores (Hoekstra et al. 1992; Pickering and Devaux 1992; Salmenkallio-Martilla and Kauppinen 1995; Devaux et al. 1996), the type of pretreatment (Hoekstra et al. 1992; Mordhorst and Lörz 1993; Cistué et al. 1994) and the nature of carbohydrates in the culture medium (Cai et al. 1992; Pickering and Devaux 1992; Kao 1993; Mordhorst and Lörz 1993).

The microspore enters the embryogenic process following two possible pathways. In the first case, the microspore develops into a haploid callus from which haploid plantlets can be regenerated through indirect embryogenesis or organogenesis. In the second case the microspore develops directly into a haploid embryo (direct microspore embryogenesis) which then regenerates into a haploid plantlet. The indirect microspore embryogenesis method was mostly used in the 1970s, but the successive optimizations of the protocols have progressively led to use of direct embryogenesis in most cases now.

Regarding direct microspore embryogenesis, the reorientation of the microspore towards the sporophytic program requires dramatic changes in microspore physiology which remains mostly uncharacterized and not well understood. Cytological observations have provided basic information over a long time period (Norreel 1970; for a review see Sangwan and Sangwan-Norreel 1987a) which help us to understand the underlying cell physiology.

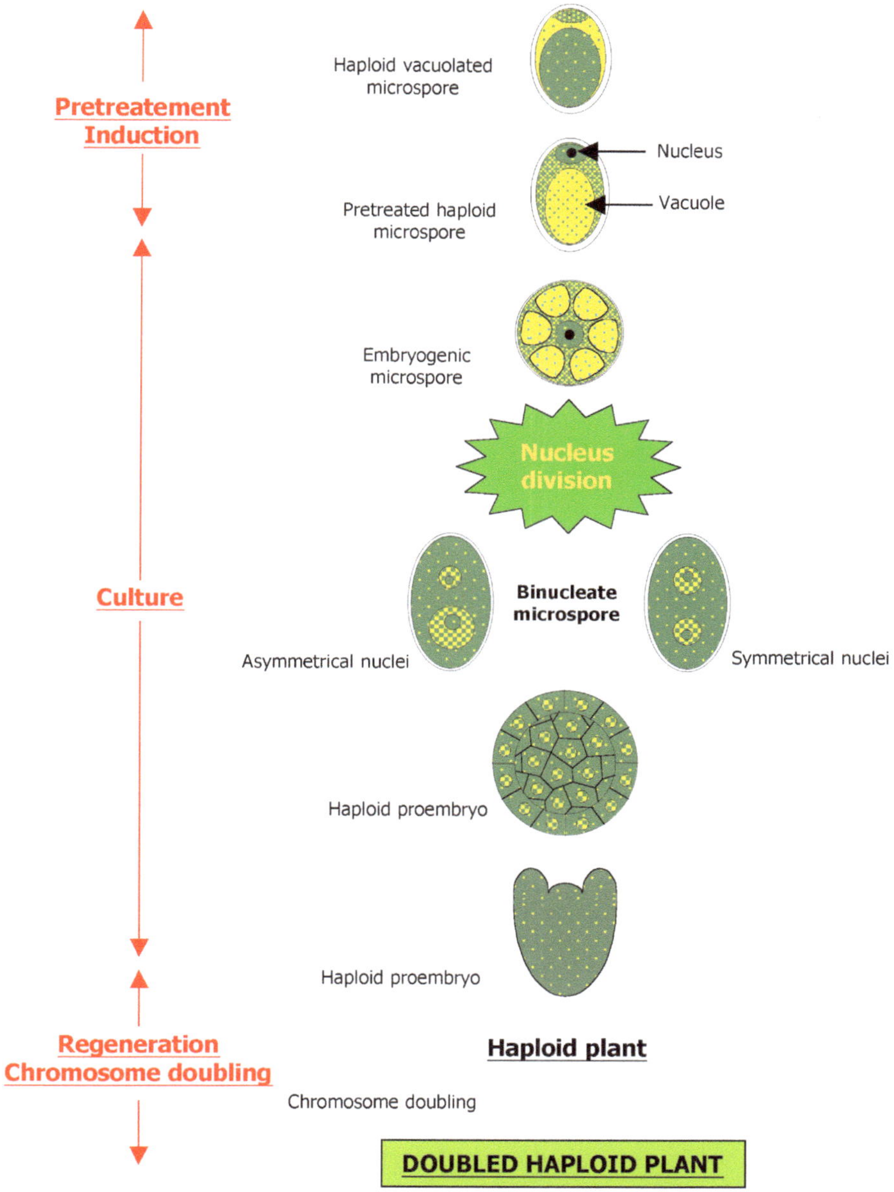

Fig. 1. Pathway of microscope embryogenesis and haploid plant production

Therefore, in this chapter, we have tried to follow the microspore during the entire embryogenesis process, from the donor plant through to haploid embryo formation. The various steps of the protocol will be described and special attention will be paid to the contributions of microscopy to the understanding of current fundamental problems related to microspore embryogenesis, such as the origin of albinism or the interference of apoptosis with the process of microspore reorientation.

2 The Microspore at the Sampling Stage

In most species, sampling of anthers for in vitro microspore embryogenesis is performed during microspore vacuolation (Fig. 2A) and before microspore mitosis; otherwise no haploid embryos can be obtained in most species. Microspores develop in the anther after meiosis (Fig. 2B,C). The gametophytic program leads the microspore to divide and form a pollen grain with two cells: the generative cell that will give rise to the male gametes and the vegetative cell whose fate is restricted to the elaboration of the pollen tube. Some specific genes are expressed during the microspore vacuolation step, indicating a specific physiological process that can be approached using microscopy.

The dynamism of the microspore at sampling is the determinant for its ability to follow the embryogenic alternative pathway. Indeed, detailed cytological observations of microspores in *D. innoxia* lead to the conclusion that embryogenic microspores are characterized by strong organelle activity when pretreatment occurs (Sangwan-Norreel 1978).

During in vivo vacuolation, the microspores undergo extensive rearrangement of the microspore cytoplasm due to the transition to the new haploid state of the nucleus (Sangwan 1986; Sangwan et al. 1989). This phenomenon is initiated during meiosis and is characterized by the progressive hydrolysis of the cytosol and organelles in lysosomes that form and fuse together to form the vacuole. The competence of the microspore to embryogenesis can be identified through transmission electron microscopy using the tonoplast feature as a marker. Indeed, sensitive microspores are characterized by the deposition of a tannin coating in the internal face of the tonoplast (Sangwan and Camefort 1983).

Moreover, using a combination of electron microscopic and cytochemical techniques, Sangwan et al. (1989) studied the activity of the lytic enzymes during male meiosis and microsporogenesis in *D. innoxia*. They presented evidence for digestion of cytoplasmic structures such as ribosomes, organelles and membranes during vacuolated microspores and lytic activity within the vacuoles. In the bicellular pollen or mature pollen the lytic activity was mainly observed in the small vacuoles. The authors discussed the significance of this cellular autophagy at the vacuolated stage of the microspore in

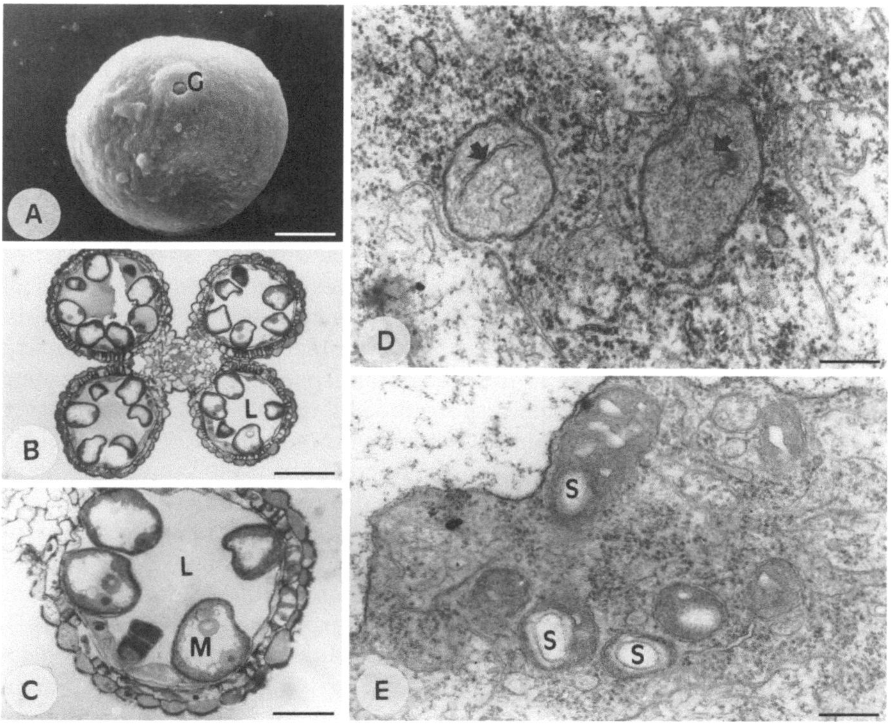

Fig. 2. **A** Scanning electron microscope (*SEM*) of barley microspore at sampling. *G* Germinating pore. *Bar* 10 μm. **B** Light microscopy (*LM*) anther section of barley at the sampling stage showing the target microspores shared in the four loculi (*L*). *Bar* 9 μm. **C** Higher magnification showing the vacuolated microspores (*M*). *L* Loculus. *Bar* 40 μm. **D** Plastids in the microspore of the Igri cultivar characterized by lack of starch and presence of thylakoids (*arrows*). *Bar* 0.25 μm. **E** Plastids in the microspore of the Cork cultivar characterized by lack of thylakoids and presence of starch (*S*) grains in the stroma. *Bar* 0.33 μm

relation to the sporophyte–gametophyte transition. It appears that an autophagic flow may discharge old cell components into the central vacuole by a protrusion process, and so contribute to the 'cellular turnover', and are probably related to the sporophytic to gametophytic transition; this is consistent with the arguments of Mackenzie et al. (1967). They showed that, during prolonged meiotic prophase, elimination of ribosomes or cytoplasmic RNA occurs, an observation probably related to the transition from the sporophytic to the gametophytic phase. The sporophytic gene expression is thus partly suppressed and the new haploid genome controls gametophytic rather than sporophytic development. However, little is known of the enzymes involved in the turnover of cytoplasmic RNA, although some evidence does suggest that the hydrolytic enzymes such as the phosphatases may be involved (Sangwan et al. 1989). Our studies have also shown the formation of 'autophagic-type' nuclear vacuoles during the meiotic prophase (Sangwan

1986). Although mostly ribosomes have been found to be hydrolysed in autophagic vacuoles, all types of organelles are affected by this cytoplasm cleaning, including the semi-autonomous plastids and mitochondria. This situation generates instability of the whole cell physiology which may confer to the microspore the competence to switch from the gametophytic towards the sporophytic pathway following external stimuli applied during pretreatment.

Plastids deserve special attention with respect to microspore embryogenesis, since they may represent a specific marker to identify androgenic species (Sangwan and Sangwan-Norreel 1987b). In androgenic species, plastids remain undifferentiated during microspore vacuolation and are devoid of starch reserves at the time of sampling for further pollen embryogenesis. After microspore mitosis, these plastids accumulate starch reserves during pollen maturation. However, in the recalcitrant species, plastids differentiate early into amyloplasts in the microspore and still exhibit amylaceous reserves when the anther is collected for microspore embryogenesis.

In Poaceae, the state of plastids in the microspore at the sampling stage is correlated with the regeneration of albino androgenic plantlets (Caredda and Clément 1999). Albinism affects all the Poaceae genotypes in a different manner. In barley (*Hordeum vulgare* L.), the fine structure of plastids was studied in the microspores of an albino- and a non-albino-producing cultivar (Caredda et al. 2000, 2004). As early as the microspore stage (sampling stage), noticeable differences were observed in the structure and behavior of plastids. In the non-albino-producing cultivar Igri (Fig. 2D), plastids remain quite undifferentiated and mere dense, exhibited thylakoids in the stroma and divided extensively. In contrast, the plastids in microspores of albino-producing cultivars such as Cork (Fig. 2E), Madras, Scarlett, Prisma and Douchka were scarce in the cytoplasm, seldom divided and had no thylakoids but huge starch reserves. Moreover, plastid DNA estimation in the microspores of various cultivars showed that microspore plastids in the albino-producing cultivars had much less DNA than those of a non-albino-producing cultivar (Caredda et al. 2000). This information indicates that plastids may be affected as early as the microspore stage in albino-producing cultivars, affecting the quality of the regenerated haploid plantlets.

At the time of microspore sampling the whole anther is affected by programmed cell death. In *Hordeum vulgare*, apoptotic DNA fragmentation naturally exists in the whole anther during microsporogenesis, without distinction between cytoplasmic and nuclear genomes (Wang et al. 1999a,b). The phenomenon could be localized in the anther using the apoptosis-specific TUNEL reaction. It was shown that the microspore nucleus is also affected by these events and exhibits positive reaction to the TUNEL test. In most cases, the fate of the vegetative cell in vivo is limited since it is destined to disappear soon after fertilization, suggesting that DNA degradation may occur during in vivo pollen development (Mogensen 1996). A crucial point for the induction of microspore embryogenesis consists in stopping this phenomenon and restoring DNA integrity (Wang et al. 1999a,b). This means that the microspore is engaged in a programmed cell death that must be stopped in microspores following embryogenesis. Interestingly, Sangwan et al. (1989) showed lytic enzyme activity during male meiosis and sporogenesis in *D. innoxia*. They also presented evidence for digestion of cytoplasmic structures such as ribosomes, organelles and membranes during microspore vacuolation and lytic activity within the vacuoles.

With respect to microspore embryogenesis, the pollen wall is a key structure, since the first steps of embryogenesis occur within the microspore wall and also because the microspore-derived embryo has to break through the wall in order to develop further. At sampling, the microspore wall consists of a double layer represented by the intine and the exine. When the microspore is collected for embryogenesis, the intine appears as a thin layer of pectin and cellulose, which is not sufficient to interfere with the development of the microspore-derived embryo. When the microspore follows the gametophytic pathway in vivo, the intine further thickens after pollen mitosis. By contrast, the exine is already formed at sampling. It is composed of sporopollenin, a complex aggregate of phenolic compounds and lipids, which are synthesized in the tapetum and secreted into the loculus during early microspore vacuolation.

3 The Pretreated Microspore

Although the pretreatment is the key point of the microspore embryogenesis protocol, very little information is available concerning the effects of the inducing stress on microspore cytology and physiology (see Chap. 3, this Vol.). Unfortunately, the reported data are scarce and insufficient to provide an overall idea of microspore behavior following pretreatment. The lack of information may be due to the diversity of pretreatments and species studied, and also because of the lack of investigations. Indeed, the reported parameters mainly focus on the yield of microspore embryogenesis following the optimization of the protocol.

The microspore pretreatment has strong physiological consequences for microspores using either microspore or anther culture. First, the pretreatment provokes the death of a significant portion of the microspores. Cold treatment is one of the most frequently used types of stress applied to microspores in order to modify the fate of the microspore. Though efficient, a long cold treatment (3–4 weeks) generates cytological alterations, affecting all the cell organelles, e.g. teratological plastid formation (Sangwan and Camefort 1984; Caredda et al. 1999). However, this longer cold treatment also causes a high level of cell death. Currently, the length of cold treatment is reduced to a few days and is usually associated with another type of stress such as osmotic stress or starvation (Touraev et al. 1997). In barley, the optimization of a protocol using copper sulfate during pretreatment enabled up to 88.2%

survival of microspores following the inductive stress (Wojnarowiez et al. 2002). In addition, this pretreatment favors the synchronization of the developmental stage of the microspores. Indeed, during in vivo development, microspores within the anther are not perfectly synchronous in their cell cycle. It seems that the stress undergone by microspores during pretreatment generates a global harmonization of cell cycles in the microspores (Hu and Kasha 1999).

When the microspore survives the pretreatment, its metabolism is strongly perturbed. In tobacco, combined cytological and biochemical studies have shown that during the starvation stress, starch is hydrolysed and the concentration of both soluble sugars and proteins decreases. In the meantime, the chromatin structure is modified and the nucleolus and the level of RNA in the cytoplasm are reduced (Zarsky et al. 1990; Garrido et al. 1995). This indicates that there is an arrest of RNA synthesis and a subsequent reduction of exchanges between the nucleus and the cytoplasm in the microspore (Garrido et al. 1995).

Plastids are sensitive to pretreatment conditions. In tobacco, plastids dedifferentiate and starch is mobilized after starvation pretreatment (Zarsky et al. 1990; Garrido et al. 1995). In studies with barley, focusing on the putative origin of albinism, plastids have been followed in the winter cultivar Igri and spring cultivars that produce green and albino plantlets, respectively, following microspore embryogenesis (Caredda et al. 2000, 2004). After the combined osmotic, cold and starvation stress, the density of plastids is dramatically reduced in the microspores of all cultivars, but plastid behavior was different according to the cultivar type. In Igri, the frequency of plastid divisions was increased whereas starch was mobilized. In the albino-producing cultivars, plastids rarely divided while starch accumulated in the stroma. These data confirm that the plastid physiology is different in each type of cultivar and may be correlated with the occurrence of albino haploid plantlets.

The stress pretreatment modifies the course of the apoptotic process in the anther (Wang et al. 1999a,b). These authors performed a 4-day pretreatment of anthers in mannitol at 4°C and observed that apoptosis was accelerated (within the first 24 h). After pretreatment, the remaining microspores that were not affected by programmed cell death may develop into embryos.

The effects of pretreatment are poorly investigated from a cytological point of view, although much information can be obtained from complement and biochemical/molecular studies (Shim and Kasha 2003). In particular, most of the pretreatments include cold treatment associated with or without chemical or osmotic shock. Since the pollen gene expression has been well defined following cold stress (Lee and Lee 2003), in situ hybridization using mRNA probes can be used for further cytological studies regarding the effects of these pretreatments.

4 Embryogenic Development of the Microspore

After pretreatment, the microspore undergoes several steps leading to the formation of a haploid embryo. The first step consists of the switch of microspore physiology generated by the stress of pretreatment giving rise to a typical embryogenic microspore. The second step is characterized by the first haploid division of the embryogenic microspore, and the third one by the development of a haploid embryo that may further develop into a haploid plant. It has been suggested that the induction stimulating factor(s) is not provided by the culture medium as the induction of embryos can take place in a simple medium (for review see Sangwan and Sangwan-Norrell 1987a) and the culture medium is only necessary for embryo development and not for induction.

4.1 Behavior of the Pretreated Microspore

The pretreated microspore remains highly vacuolated with a flat nucleus. The behavior of stressed microspores was followed in detail using cell tracking in tobacco, barley and wheat (Touraev et al. 1997; Kumlehn and Lörz 1999; Indrianto et al. 2001). Progressively, the nucleus moves to the centre of the microspore, likely oriented by the cytoskeleton. Simultaneously, the vacuole is fragmented and subvacuoles are regularly distributed around the central nucleus, giving rise to a typical 'star-like' microspore. These events are the first signs of microspore reorientation and represent an obliged way in the process of microspore embryogenesis (Indrianto et al. 2001). It is likely that the 'star-like microspore' develops an embryogenic program.

The switch towards the sporophytic pathway following pretreatment is revealed by several changes in microspore physiology as evident from the behavior of the organelles. Reactive microspores are characterized by an increase in nuclear DNA contents after a few days of culture (Sangwan-Norreel 1983), indicating that the cell cycle is reactivated, leading to the first embryogenic mitosis. Regarding gene transcription, the embryogenic phase begins with a reduced stainability of cytoplasmic RNAs, followed by an increase in the RNA levels (Sangwan and Sangwan-Norreel 1987b). In *Zea mays*, the nuclear pore increases in size, indicating a reactivation of exchanges between the nucleus and the cytoplasm in the microspore (Alché et al. 2000). In *Capsicum annuum*, there is a strong reorganization of the nucleus during pollen embryogenesis, as revealed by modifications of the chromatin density and the nucleolar activity (Testillano et al. 1995). This strongly suggests a drastic change in gene transcription, corresponding to a change in the microspore developmental program.

During the days after pretreatment, there is a continuation of the organelle degradation and lytic activities correlated with the change of ploidy in the nucleus (Sangwan and Sangwan-Norreel 1987b; Garrido et al. 1995). This indicates that the program of cytoplasm reorganization that was initiated in the microspore in vivo following meiosis is maintained under in vitro conditions despite the drastic treatment of microspores.

The period between the pretreatment and the first haploid division remains poorly investigated, from a cytological point of view. However, this step corresponds with the occurrence of the sporophytic program in competent microspores and should be further studied in order to understand the initial steps of microspore embryogenesis. Indeed, numerous molecular tools are now available to study the different steps of developmental (especially the embryos), cell cycle and cell differentiation. The genes specifically expressed in pollen grains have been recently identified (Honys and Twell 2003) and this could represent a strong basis to investigate the differential behavior of in vivo and in vitro developing microspores. Thus, probes could be used to perform in situ hybridization and obtain complementary information concerning the cytological effects of gene expression.

4.2 The First Division

According to the species and the type of culture (anther or isolated microspores), in vitro microspore mitosis occurs, for example, after 1 day of culture in *Brassica* (Custers et al. 1994), 5 days in *Hordeum* (Wojnarowiez et al. 2002) and 10 days in *Nicotiana* (Cordewener et al. 1996).

The first division may provide different bicellular structures, depending upon the type of nucleus. The ideal situation is when the microspores behave like a zygote. In this case, the different steps of embryo development are mostly identical to those followed by a zygotic embryo. This occurs when the microspore has changed its structure to 'star like' (Kumlehn and Lörz 1999; Indrianto et al. 2001). Otherwise, the first division generates two different nuclei: vegetative and the generative nuclei/cells that appear when the microspore develops in vivo. In this case, several scenarios are possible and have been extensively reviewed earlier (Sangwan Norreel 1981; Goralski et al. 1999). Cytological studies allowed the definition of all the possible pathways to form a haploid embryo from the unicellular microspore (Sangwan and Sangwan-Norreel 1987b). Either the microspore nucleus divides and provides all the cells of the microspore, or the first mitosis is close to pollen mitosis in vivo, leading to different cell types in the microspore-derived structure with various nuclear origins. In most cases, only the cells originating from the vegetative nucleus survive and provide the cells of the embryos, whereas those originating from the generative nucleus are destined to disappear during embryo development. Initially, it was thought that the generative cell did not form the embryo alone, but Raghavan (1976a,b) showed that the generative cell does produce embryos in *Hyoscyamus niger*.

4.3 The Microspore-Derived Embryos

Several types of structures develop from microspores that are called embryoids. In barley the proportion of embryo development into haploid plants varies according to the cultivar and the culture conditions (Paire et al. 2003). The analysis of DNA content in pollen embryos showed that the initiation of pollen embryos is spread out over time and that the quality of embryos differs according to the precocity of their appearance (Sangwan-Norreel 1981), explaining that all the embryoids do not give rise to haploid plantlets. In this chapter, we will mainly consider embryos whose development conforms to zygotic embryos, i.e. those embryos that will develop into haploid plantlets.

Morphologically, it has been shown that microspore-derived embryos have a similar anatomical pathway to zygotic embryos. This was perfectly well demonstrated in *Brassica napus* (Ilic-Grubor et al. 1998): scanning electron microscopy enabled the authors to show that each step of zygotic embryo development was followed by the pollen embryo, from the globular embryo up to the mature embryo. In some cases, it was shown that the microspore-derived embryo exhibits suspensor-like structures, though it is usually admitted that pollen embryos are without suspensors (Sangwan and Sangwan-Norreel 1987b).

The initiation of the pollen embryo is characterized by the accumulation of RNA-rich cytoplasmic bodies and their dispersal after the globular stage (Sangwan and Camefort 1982), although pollen embryos have less RNA than zygotic embryos (Sangwan and Sangwan-Norreel 1987b). These RNA-rich cytoplasmic bodies are supposed to be the sites of the synthesis of new enzymes. The fate of the pollen wall during in vitro pollen embryogenesis is well described (Sangwan and Sangwan-Norreel 1980, 1987b). It has been shown that many gametophytic enzymes/proteins are present in the pollen wall (Knox 1971). What would be the fate of these gametophytic enzymes during the sporophytic development (embryogenesis) of the pollen? It has been convincingly shown that the pollen wall does not participate in the embryonic development (Sangwan and Sangwan-Norreel 1980). Depending upon the species, the microspore-derived structure progressively grows and is released from the pollen wall after a week. For example, in *Datura*, pollen wall elimination occurred within 6–7 days of culture and simultaneously a new thin wall is formed around the embryos (Sangwan and Sangwan-Norreel 1980). However, in *Nicotiana* the pollen wall remains longer (usually after 16 cells/about 12 days) around the pollen embryo (Vazart 1973). Pollen embryos formed a new cell wall within the old one before the disintegration of the pollen wall started. The structure of the new embryonic wall resembles that of a classical plant cell wall.

Prior to in vitro culture, the microspore has a large central vacuole and parietal cytoplasm. During embryogenesis the central vacuole does not disappear, but divides simultaneously with the whole cell, providing one vacuole per cell at each stage of early embryogenesis (Sangwan and Sangwan-Norreel

1987b). A similar situation occurs in the somatic cells of carrot and *Cichorium* undergoing dedifferentiation or embryogenesis (Sangwan and Sangwan-Norreel 1987b and references within). This suggests that during embryogenesis or dedifferentiation there is fragmentation of the central vacuole, while during differentiation or maturation there is regression of the central vacuole. The tonoplast seems to play an active but unknown role during in vitro embryogenesis.

In vivo, double fertilization leads to the formation of endosperm and zygotic embryo. It has recently been shown in maize that developing pollen embryos can present either 'embryo-like' or 'endosperm-like' features of development (Testillano et al. 2002). In the former case, pollen embryos show small polygonate cells with typical meristematic features, whereas in the latter case, cells are characterized by coenocytic organization, synchronous divisions, vacuolated cytoplasm, starch granules, incomplete walls containing callose and differential tubulin organization. These results were supported by the discovery of the expression in pollen embryos of genes specifically expressed in maize endosperm (Magnard et al. 2000). This confirms that the reoriented microspore really acts as a zygote that can provide both types of embryos. In addition, these data also suggest that following in vivo fertilization the determinism of zygote fate towards the 'embryo-like' or the 'endosperm-like' pathway is due at least partly to the male parent.

Cytological investigations should help to elucidate the occurrence of albinism in androgenic plantlets. Therefore, plastid development and differentiation have been followed closely during microspore embryo development in several barley cultivars, such as the albino (Cork, Douchka, Madras, Prisma, Scarlett) and green/non-albino (Igri) cultivars (Caredda et al. 2000, 2004). In Igri at early stages of haploid embryo development (Fig. 3A–C), several types of plastids are observed, revealing the plasticity and the drastic reorganization of the plastidome. One population of plastids are strictly amyloplasts characterized by exclusive accumulation of starch within the stroma that were degenerating (Fig. 3D) or not (Fig. 3E). Another population is composed of plastids that appear to be able to modify their fate during the embryonic development and provide chloroplasts in the haploid plantlet. These chloroplasts keep their ability to divide (Fig. 3F) and are characterized by the presence of prolamellar bodies (Fig. 3F), thylakoid differentiation (Fig. 3F,G) and plastoglobules (Fig. 3G). In these plastids, starch is usually less important than in amyloplasts sensu stricto (Fig. 3F,G,H). In some cases, plastids have an elongated shape and developing thylakoids are organized in parallel, similar to what happens when a proplastid differentiates into a chloroplast (Caredda et al. 1999).

At the end of the culture phase, embryos are present throughout the anther (Fig. 4A). Their cells exhibit meristematic features (Fig. 4B) and the aspect of plastids is well defined, thus suggesting clearly the developmental pathway they followed: some of them are teratological and will soon disappear (Fig. 4C), whereas most of the plastids show evidence of differentiation

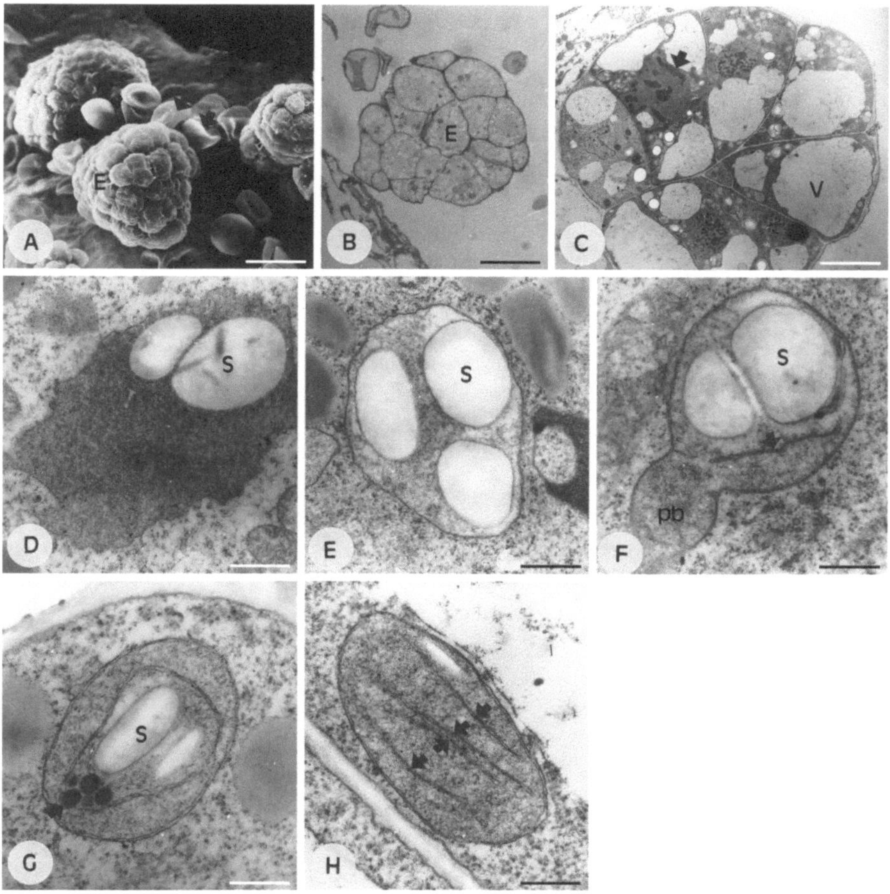

Fig. 3. **A** SEM view of barley young globular microspore-derived embryos (*E*) beside unreactive microspores (*arrows*) after 12 days of anther culture. *Bar* 62.5 μm. **B** LM view of the globular microspore-derived embryo (*E*) showing the multicellular structure but without protoderm. *Bar* 40 μm. **C** Transmission electron microscopy (*TEM*) view of the globular embryo showing nucleus division (*arrow*) and fragmentation of vacuole (*V*) originating from the initial microspore. *Bar* 7.7 μm. **D–H** Plastid features in the globular microspore-derived embryo. **D** Plastid close to degradation with irregular shape and solely including starch grains (*S*). *Bar* 0.6 μm. **E** Amyloplast with regular shape but including solely starch grains (*S*). *Bar* 0.45 μm. **F** Dividing plastid including starch (*S*) but also thylakoid (*arrow*) and a prolamellar body (*pb*). *Bar* 0.35 μm. **G** Plastid exhibiting plastoglobules, reserves of lipids for further formation of thylakoid membranes (*arrow*). *S* Starch. *Bar* 0.25 μm. **H** Elongated plastid with a small starch grain and parallel thylakoids (*arrows*). *Bar* 0.25 μm

towards the chloroplast type. This is revealed by the reduction of starch in the stroma (Fig. 4D,E), the intensification of divisions (Fig. 4D,E), and the enhancement of thylakoid development from the internal membrane (Fig. 4D) and from the prolamellar body (Fig. 4E). Some plastids remain full

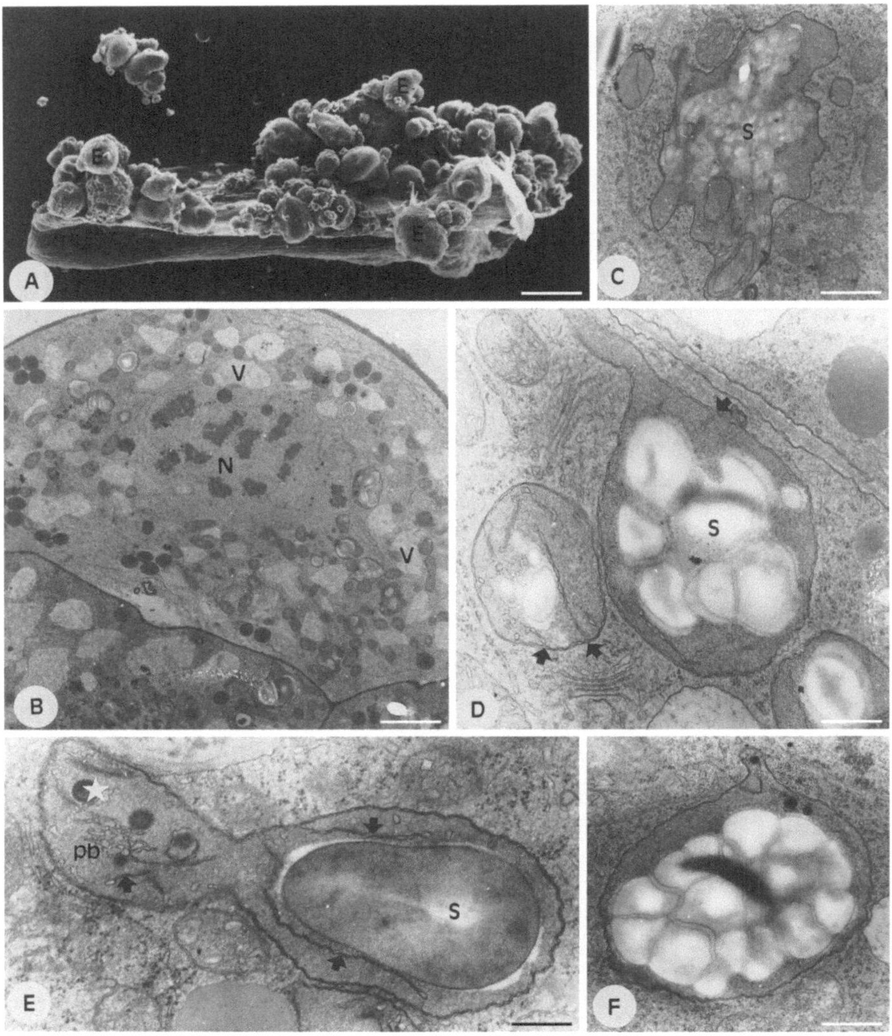

Fig. 4. **A** SEM view of a responding barley anther showing numerous microspore-derived embryos (*E*) after 28 days of anther culture. *Bar* 90 µm. **B** General TEM view of an embryo cell showing typical meristematic features, with a dense cytoplasm including small vacuoles (*V*) and a dividing nucleus (*N*). *Bar* 2.5 µm. **C–F** Plastids in 28-day-old microspore-derived embryos. **C** Abnormal plastid destined to disappear. *S* Starch. *Bar* 0.67 µm. **D** Plastids showing division feature, remaining starch grains (*S*) and developing thylakoids from the internal membrane (*arrows*). *Bar* 0.25 µm. **E** Typical dividing prechloroplast including a remaining starch grain (*S*), developing thylakoids (*arrows*), a prolamellar body (*pb*) and many plastoglobules (*white star*). *Bar* 0.25 µm. **F** Amyloplast whose fate is uncertain. *S* Starch. *Bar* 0.5 µm

Table 1. Plastid shapes in 28-day-old microspore-derived embryos of various barley cultivars. *ND* Not determined

Cultivars	Percentage of green regenerated plants	Density (per 100 μm^2)	Thylakoid numbers	Starch (% volume)	Divisions (per 100 μm^2)	Plastid sections showing DNA (%)
Igri	87.8	3.8 ±0.5	2.66±0.5	26 ±3.3	6.4 ±0.25	21.3±2.1
Cork	2.9	0.37±0.1	0.5 ±0.2	90.3±4.3	0.09±0.4	2.6±1.3
Douchka	4.7	0.56±0.2	0.7 ±0.3	84 ±3	0.12±0.06	7.6±1.5
Madras	3.8	0.65±0.2	0.5 ±0.3	81.5±1.9	0.11±0.07	3.4±1.6
Prisma	9.1	1.14±0.4	1.5 ±0.4	87.5±3.3	0.18±0.05	4.6±1.1
Scarlet	2.8	0.34±0.2	0.8 ±0.4	85.6±5.9	0.14±0.03	ND

of starch (Fig. 4F) with no clear sign of differentiation. Embryos transferred to the regeneration medium have plastids with numerous thylakoids and starch representing 26% of the plastid volume (Table 1). Moreover, DNA was immunodetected in 21% of plastid sections. Plastid behavior was drastically different in the albino-producing cultivars. During microspore embryo development, plastids were scarce, few divisions occurred and plastids mainly differentiated into amyloplasts with starch representing up to 90% of the plastid volume. In each plastid of the mature microspore embryo, a range of 0.5 thylakoids could be found and only 3% contained DNA. Finally, 87.8% of green haploid plants can be obtained in Igri as against 2.9% in Cork. These results show a strong correlation between the state of the plastids during the culture phase of microspore embryogenesis and the occurrence of albinism. In addition, the origin of albinism takes place at very early stages of microspore development and it is likely that the opportunity for plastids to switch their developmental program into green plastids is determined at the uninucleate microspore stage. However, this phenomenon is partly reversible under certain conditions. For example, using copper sulfate during the pretreatment or in the culture medium the global yield of microspore embryogenesis increases in all cultivars tested and the proportion of green plantlets is enhanced up to 90% in Igri and 27% in Cork (Wojnarowiez et al. 2002). The cytological characterization of the copper effects during the whole process is currently under study and will be of great help in the understanding of the phenomenon.

5 Conclusions

The experimental studies reviewed here on microspore embryogenesis, particularly the cytophysiological aspects, have been directed to determine the causes and modalities of microspore embryogenesis. In other words, what makes the microspore or pollen (male gametophyte) follow an in vitro sporophytic (microspore embryogenesis) pathway instead of its normal gametophytic pathway. This corresponds to a change in the genetic program of the pollen. Clearly, the deviation from the gametophytic to the sporophytic pathway is due to changes in the normal functioning of the pollen and to the acquisition of new morphogenetic abilities. The cytological modifications (reviewed here) occurring in pollen at the time of embryogenesis have thrown some light on the mechanism of embryogenic induction. From these studies, it was also evident that the induction of microspore embryogenesis is provoked by a shock or stress treatment given to the microspore. In addition, ultrastructural, cytochemical and biochemical studies have shown important modifications in the nuclei and cytoplasmic organelles in the embryogenic competent microspores. The ultrastructural studies of the microspore embryogenesis have shown the formation of two equal nuclei, their fusion and the formation of diploid embryos. In certain cases degeneration of the generative cell and the formation of embryo from the vegetative cell were observed. In the developing embryos, pollen wall elimination occurred rapidly (in about a week) and simultaneously a new thin wall was formed around the embryos.

Moreover, the fine cytology and molecular biological approach to the induction mechanism of microspore embryogenesis still have to be developed in the future in order to increase our knowledge of microspore embryo development. The structural aspects were extensively described by many authors, but several functional aspects remain to be studied in detail using cytological methods associated with complementary techniques of visualization, biochemistry and molecular biology (Eady et al. 1995; Magnard et al. 2000; Testillano et al. 2002). Moreover, the molecular biological approach is one of the major mileposts to obtain a large number of genes that show expression specific to pollen differentiation, i.e. the inductive phase of embryogenic competence.

In addition, several aspects should be followed up during microspore embryogenesis from the pretreatment period which would help to increase our understanding of this phenomenon. These include spatial distribution of gene expression in the microspore, the derived structures and the intimate plastid behavior for albinism.

References

Alché JD, Castro AJ, Solymoss M, Timar I, Barnabas B, Rodríguez-García MI (2000) Cellular approach to the study of androgenesis in maize anthers: immunocytochemical evidence of the involvement of the ubiquitin degradative pathway in androgenesis induction. J Plant Physiol 156:146–155

Binarova P, Hause G, Cenklova V, Cordewener JHG, van Lookeren Campagne MM (1997) A short severe heat shock is required to induce embryogenesis in late bicellular pollen of *Brassica napus* L. Sexual Plant Reprod 10:200–208

Cai Q, Szarejko I, Polok K, Maluszynski M (1992) The effect of sugars and growth regulators on embryoid formation and plant regeneration from barley anther culture. Plant Breed 109:218–226

Caredda S, Clément C (1999) Androgenesis and albinism in Poaceae: influence of genotype and carbohydrates. In: Clément C, Pacini E, Audran JC (eds) Anther and pollen: from biology to biotechnology. Springer, Berlin Heidelberg New York, pp 211–226

Caredda S, Devaux P, Sangwan RS, Clément C (1999) Differential development of plastids during microspore embryogenesis in barley. Protoplasma 208:248–256

Caredda S, Doncoeur C, Devaux P, Sangwan RS, Clément C (2000) Plastid differentiation during androgenesis in albino and non albino-producing cultivars of barley (*Hordeum vulgare* L.). Sexual Plant Reprod 13:95–104

Caredda S, Devaux P, Sangwan RS, Proult I, Clément C (2004) Plastid ultrastructure and DNA related to albinism in androgenetic embryos of various barley (*Hordeum vulgare* L.) cultivars. Plant Cell Tissue Organ Cult 76(1):35–43

Chu CC, Hill RD, Brule-Babel A (1990) High frequency of pollen embryoid formation and plant regeneration in *Triticum aestivum* L. on monosaccharide containing media. Plant Sci 66:255–262

Cistué L, Ramos A, Castillo AM, Romagosa I (1994) Production of large number of doubled haploid plants from barley anthers pretreated with high concentrations of mannitol. Plant Cell Rep 13:709–712

Cordewener JHG, Custers JBM, Dons HJM, van Lookeren Campagne M (1996) Molecular and biochemical events during the induction of microspore embryogenesis. In: Jain MS, Sopory SK, Veilleux RE (eds) In vitro production of higher plants, vol 1. Kluwer, Dordrecht, pp 111–124

Custers JBM, Cordewener JHG, Nöllen Y, Dons JJM, van Lookeren Campagne M (1994) Temperature controls both gametophytic and sporophytic development in microspore cultures of *Brassica napus*. Plant Cell Rep 13:267–271

Devaux P (1992) Haploidy in barley and wheat improvement. In: Dattée A (ed) Reproductive biology and plant breeding. Springer, Berlin Heidelberg New York, pp 139–151

Devaux P, Zivy M, Kilian A, Kleinhofs A (1996) Doubled haploids in barley. In: Scoles G, Rossnagel B (eds) Proc 5th Int Oat Conf and 7th Int Barley Genet Symp. University Extenson Press, Saskatoon, Saskatchewan, Canada, pp 213–222

Eady C, Lindsey K, Twell D (1995) The significance of microspore divisions and division symmetry for vegetative cell specific transcription and generative cell differentiation. Plant Cell 7:65–74

Garrido D, Vicente O, Heberle-Bors E, Rodriguez-Garcia I (1995) Cellular changes during the acquisition of embryogenic potential in isolated pollen grains of *Nicotiana tabacum*. Protoplasma 186:220–230

Goralski G, Matthys-Rochon E, Vergne P, Przywara L (1999) Androgenic development: a fascinating embryo formation process. Acta Biol Cracov 41:51–65

Guha S, Maheshwari SC (1964) In vitro production of embryos anther of *Datura*. Nature 204:497–498

Guha S, Maheshwari SC (1966) Cell division and differentiation of embryos in the pollen of *Datura* in vitro. Nature 212:97–98

Hoekstra S, van Zijdervzld MH, Louwerse JD, Heidelkamp F, van der Mark F (1992) Anther and microspore culture of *Hordeum vulgare* L. Plant Sci 8:89–96

Honys D, Twell D (2003) Comparative analysis of the *Arabidopsis* pollen transcriptome. Plant Physiol 132:640–652

Hu T, Kasha KJ (1999) A cytological study of pretreatments used to improve isolated microspores cultures of wheat (*Triticum aestivum* L.) cv. Chris. Genome 42:432–441

Ilic-Grubor K, Attree M, Fowke LC (1998) Comparative morphological study of zygotic and microspore-derived embryos of *Brassica napus* L. as revealed by scanning electron microscopy. Ann Bot 82:157–165

Indrianto A, Barinova I, Touraev A, Heberle-Bors E (2001) Tracking individual wheat microspores in vitro: identification of embryogenic microspores and body axis formation in the embryo. Planta 212:163–174

Jacquard C, Wojnarowiez G, Clément C (2003) Anther culture in barley. In: Kasha K, Maluszinski M (eds) Doubled haploid manual. Springer, Berlin Heidelberg New York, pp 21–28

Jähne A, Lörz H(1995) Cereal microspore culture. Plant Sci 109:1–12

Jardinaud MF, Souvré A, Alibert G, Beckert M (1995a) *uidA* gene transfer and expression in maize microspores using the biolistic method. Protoplasma 187:138–143

Jardinaud MF, Souvré A, Alibert G, Beckert M (1995b) Optimisation of DNA transfer and transient ß-glucuronidase expression in electroporated maize (*Zea mays* L.) microspores. Plant Cell Rep 15:55–58

Kao KN (1993) Viability, cell division and microcallus formation of barley microspores in culture. Plant Cell Rep 12:366–369

Kasha KJ, Yao Q, Simion E, Hu T, Oro R (1995) Production and applications of haploids in crops. Use of induced mutation and molecular techniques of crop improvement. International Atomic Energy Agency, Vienna, Austria, pp 23–37

Knox RB (1971) Pollen wall proteins: localization, enzymic and antigenic activity during development in *Gladiolus*. J Cell Sci 9:209–237

Kumlehn J, Lörz H (1999) Monitoring sporophytic development of individual microspores of barley (*Hordeum vulgare* L.). In: Clément C, Pacini E, Audran JC (eds) Anther and pollen: from biology to biotechnology. Springer, Berlin Heidelberg New York, pp 183–190

Lee JY, Lee DH (2003) Use of serial analysis of gene expression technology to reveal changes in gene expression in *Arabidopsis* pollen undergoing cold stress. Plant Physiol 132:517–529

Loeb TA, Reynolds TL (1994) Transient expression of the uidA gene in pollen embryoids of wheat following microprojectile bombardment. Plant Sci 104:81–91

Luckett DJ, Darvey NL (1992) Utilisation of microspore culture in wheat and barley improvement. Aust J Bot 40:807–828

Luckett DJ, Smithard RA (1992) Doubled haploid production by anther culture for Australian barley breeding. Aust J Agric Res 43:67–78

Lyne RL, Bennett RI, Hunter CP (1986) Embryoid and plant production from cultured barley anthers. In: Withers LA, Alderson PG (eds) Plant tissue culture and its agricultural application. Butterworths, London, pp 405–410

Mackenzie A, Heslop-Harrison J, Dickinson HG (1967) Elimination of ribosomes during meiotic prophase. Nature 215:997–999

Magnard JL, Le Deunff E, Domenech J, Rogowsky PM, Testillano PS, Rougier M, Risueno MC, Vergne P, Dumas C (2000) Genes normally expressed in the endosperm are expressed at early stages of microspore embryogenesis in maize. Plant Mol Biol 44:559–574

Mogensen HL (1996) The hows and whys of cytoplasmic inheritance in seed plants. Am J Bot 83:383–404

Mordhorst AP, Lörz H (1993) Embryogenesis and development of isolated barley (*Hordeum vulgare* L.) microspores are influenced by the amount and composition of nitrogen sources in culture media. J Plant Physiol 142:485–492

Norreel B (1970) Etude cytologique de l'androgenèse expérimentale chez *Nicotiania tabacum* et *Datura innoxia*. Bull Soc Bot Fr 117:461–478

Paire A, Devaux P, Lafitte C, Dumas C, Matthys-Rochon E (2003) Proteins produced by barley microspores and their derived androgenic structures promote in vitro zygotic maize embryo formation. Plant Cell Tissue Organ Cult 73:167–176

Pickering RA, Devaux P (1992) Haploid production: approaches and use in plant breeding. In: Shewry PR (ed) Barley: genetics, biochemistry, molecular biology and biotechnology. CAB International, Oxford, pp 519–547

Raghavan V (ed) (1976a) Experimental embryogenesis in vascular plants. Academic Press, London

Raghavan V (1976b) Role of generative cell in androgenesis in henbane. Science 191:388–389

Salmenkallio-Marttila M, Kauppinen V (1995) Efficient regeneration of fertile plants from protoplasts isolated from microspore cultures of barley (*Hordeum vulgare* L.). Plant Cell Rep 14:253–256

Sangwan RS (1986) Formation and cytochemistry of nuclear vacuoles during meiosis in *Datura*. Eur J Cell Biol 40:210–218

Sangwan RS, Camefort H (1982) Ribosomal bodies specific to both pollen and zygotic embryogenesis in *Datura*. Experientia 38:395–397

Sangwan RS, Camefort H (1983) The tonoplast, a specific marker of embryogenic microspores of *Datura* cultured in vitro. Histochemistry 78:473–480

Sangwan RS, Camefort H (1984) Cold-treatment related structural modifications in the embryogenic anthers of *Datura*. Cytologia 49:473–487

Sangwan RS, Sangwan-Norreel BS (1980) Biochemical and structural aspects of induction and development of pollen-embryos in *Datura*. Bull Soc Bot Fr 127:109–122

Sangwan RS, Sangwan-Norreel BS (1987a) Biochemical cytology of pollen embryogenesis. Int Rev Cytol 107:221–272

Sangwan RS, Sangwan-Norreel BS (1987b) Ultrastructural cytology of plastids in pollen grains of certain androgenic and nonandrogenic plants. Protoplasma 138:11–22

Sangwan RS, Mathivet V, Vasseur G (1989) Ultrastructural localization of acid phosphatase during male meiosis and sporogenesis in *Datura*: evidence for digestion of cytoplasmic structures in the vacuoles. Protoplasma 149:38–46

Sangwan RS, Ducrocq C, Sangwan-Norreel BS (1993) *Agrobacterium*-mediated transformation of pollen embryos in *Datura innoxia* and *Nicotiana tabacum*: production of transgenic haploid and fertile homozygous dihaploid plants. Plant Sci 95:99–115

Sangwan-Norreel BS (1978) Cytochemical and ultrastructural peculiarities of embryogenic pollen grains and of young androgenic embryos in *Datura innoxia*. Can J Bot 56(7):805–817

Sangwan-Norreel BS (1981) Evolution in vitro du contenu en ADN nucléaire et de la ploïdie des embryons polliniques du *Datura innoxia*. Can J Bot 59(4):508–517

Sangwan-Norreel BS (1983) Male gametophyte nuclear DNA content evolution during androgenic induction in *Datura innoxia* Mill. Z Pflanzenphysiol 111:47–54

Shim YS, Kasha KJ (2003) The influence of pretreatment on cell stage progression and the time of DNA synthesis in barley (*Hordeum vulgare* L.) uninucleate microspores. Plant Cell Rep 21:1065–1071

Testillano PS, Gonzalez-Melendi P, Ahmadian P, Fadon B, Risueno MC (1995) The immunolocalization of nuclear antigens during the pollen developmental programme and the induction of pollen embryogenesis. Exp Cell Res 221:41–54

Testillano PS, Ramirez C, Domenech J, Coronado MJ, Vergne P, Matthys-Rochon E, Risueno MC (2002) Young microspore-derived embryos show two domains with defined features also present in zygotic embryogenesis. Int J Dev Biol 46:1035–1047

Touraev A, Vicente O, Heberle-Bors E (1997) Initiation of microspore embryogenesis by stress. Trends Plant Sci 2:297–302

Vazart B (1973) Formation d'embryoídes à partir de microspores de tabac: évolution de l'infrastructure des cellules au cours de la première semaine de culture. Mémoire de la Société Botanique Française, Colloque de Morphologie, Paris, pp 243–260

Wang M, van Bergen S, Lamers GEM, Oppedjik BJ, Schilperroort RA (1999a) Programmed cell death during androgenesis in *Hordeum vulgare* L. In: Clément C, Pacini E, Audran JC (eds) Anther and pollen: from biology to biotechnology. Springer, Berlin Heidelberg New York, pp 201–209

Wang M, Hoekstra S, van Bergen S, Lamers GEM, Oppedijk BJ, de Priester W, Schilperroort RA (1999b) Apoptosis in developing anthers and the role of ABA in this process during androgenesis in *Hordeum vulgare* L. Plant Mol Biol 39:489–501

Wojnarowiez G, Jacquard C, Devaux P, Sangwan RS, Clément C (2002) Influence of copper sulfate on anther culture in barley (*Hordeum vulgare* L.). Plant Sci 162:843–847

Xie JH, Gao MW, Cai QH, Cheng X, Shen Y, Liang ZQ (1995) Improved isolated microspore culture efficiency in medium with maltose and optimized growth regulator combination in japonica rice (*Oryza sativa*). Plant Cell Tissue Organ Cult 42:245–250

Yao QA, Simion E, William M, Krochko J, Kasha KJ (1997) Biolistic transformation of haploid microspores of barley (*Hordeum vulgare* L.). Genome 40:570–581

Zarsky V, Rihova L, Tupy J (1990) Biochemical and cytological changes in young tobacco pollen during in vitro starvation in relation to pollen embryogenesis. In: Van der Plas LHW, van Aartri JK (eds) Progress in plant cellular and molecular biology. Kluwer, Dordrecht, pp 228–233

Zarsky V, Garrido D, Rihova L, Vincente O, Heberle-Bors E (1992) Derepression of the cell cycle by starvation is involved in the induction of tobacco pollen embryogenesis. Sexual Plant Reprod 4:189–194

I.5 Biochemical and Molecular Aspects of Haploid Embryogenesis

KIM BOUTILIER, MARTIJN FIERS, CHUN-MING LIU, and APOLONIA (LONNEKE) H.M. VAN DER GEEST[1]

1 Introduction

Microspore-derived embryo (MDE) cultures provide a unique system to study fundamental aspects of plant tissue culture and embryo development. In this culture system microspores and immature pollen grains (referred to here for simplicity as microspores) are induced to form haploid embryos by exposing them to a stress treatment, such as increased temperature, starvation or exogenous growth regulators. Under the appropriate culture conditions these embryos undergo numerous rounds of cell division and differentiate to form the sequential embryonic stages that are typical for zygotic embryos of that species. The developmental pathway from microspore to MDE often differs between plant species; however, it is likely that common processes underlie the switch in developmental pathways. As summarized by Yeung (1995), developmental concepts such as competence, induction and determination can be used to describe and understand in vitro embryo formation from haploid microspores. Likewise, the developmental processes that characterize zygotic embryogenesis, i.e. morphogenesis and differentiation, where the embryo divides and forms the basic tissue and organ systems of the adult plant (Laux and Jürgens 1997), and maturation, where storage products accumulate and the seed becomes desiccation-tolerant (Wobus and Weber 1999), can also take place during MDE development.

MDE culture has a number of advantages over other in vitro embryo systems or in planta models as an experimental system for studying concepts related to embryo formation and differentiation. Zygotic embryos develop within the confines of the maternal tissues and are therefore difficult or tedious to isolate, especially at the zygote and pro-embryo stages. The difficulty in isolating young zygotic embryos is reflected by the small number of publications in which differential screening methods have been used to identify early zygotic embryo-expressed genes (Heck et al. 1995; Li and Thomas 1998). In contrast to zygotic embryos, MDEs are usually produced in large numbers in liquid cultures and can be isolated at different stages of development spanning few-celled to mature embryos. MDEs growing in liquid cultures are more amenable than excised zygotic embryos as an experimental

[1] Plant Research International. P.O. Box 16, 6700 AA Wageningen, The Netherlands

Biotechnology in Agriculture and Forestry, Vol. 56
Haploids in Crop Improvement II (ed. by C.E. Palmer, W.A. Keller, and K.J. Kasha)

system to study the effect of cell perturbation or exogenously applied substances, such as growth regulators, on embryo development (Liu et al. 1993; Fischer and Neuhaus 1996; Hays et al. 2000, 2002). Somatic embryo cultures also provide a valuable in vitro experimental system for studying aspects of totipotency and embryo development; however, most somatic embryo cultures require long induction periods, develop through an intermediate callus phase and contain both embryogenic and non-embryogenic cell clusters (Zimmerman 1993; Mordhorst et al. 1997). These characteristics can make it difficult to study the stages of competence, induction and determination that occur during the initiation of somatic embryo formation. It is not surprising then that MDE cultures, especially those in which embryos develop at a high frequency from single cells, have been used as an experimental system to study the molecular, biochemical and cellular aspects of plant cell totipotency and embryo development (Touraev et al. 1996; Goralski et al. 1999; Custers et al. 2001; Yeung 2002).

This chapter describes the application of MDE cultures as an experimental system to understand two fundamental aspects of plant development: the formation of embryogenic cells and the early differentiation of the embryo. We will focus on two aspects of the molecular and biochemical research being carried out in this area, namely the identification of genes that are preferentially expressed during early MDE development, and the role of extracellular signalling molecules in MDE development. Recent developments in this area will be highlighted using published data, as well as unpublished results from experiments in our group using the *Brassica napus* MDE culture system. Where possible, we have tried to place these developments in the broader context of somatic and zygotic embryo development.

2 Gene Identification Strategies for Early MDE Development

A number of gene expression studies have been performed over the years using MDE cultures of different monocot and dicot species. Some of these studies have examined expression patterns of specific genes and proteins known to be involved in zygotic embryogenesis or thought to play a specific role in MDE development (Crouch 1982; Boutilier et al. 1994; Cordewener et al. 1995; Perry et al. 1999; Smykal and Pechan 2000; de Faria Maraschin et al. 2003a,b). Other studies have been directed toward analysis of differential gene expression or protein profiles at a single time point, or a developmental time course spanning multiple stages. Some of the considerations involved in setting up a differential screen, as well as the outcomes of a number of specific screens that have been performed in our laboratory as well as in other groups, are described below.

Differential gene analysis strategies using MDE cultures have been largely targeted toward understanding the molecular processes underlying two key stages of MDE development: (1) the developmental switch, where the developmental fate of the immature pollen grains is altered by a stress treatment so that embryogenic rather than gametophytic cells are formed; and (2) the differentiation stage, where embryogenic cells continue to divide and sequentially establish the tissue and organ patterns that characterise the developing embryo. There are a number of difficulties involved in setting up good molecular screens to identify genes that are differentially expressed during the switch from microspore to MDE development. Ideally, a good molecular screen would allow one to isolate and compare cells from the same donor plants that have been exposed to the same induction treatment, but which differ only in their response to this treatment, e.g. embryogenic or non-embryogenic. In addition, the screen should take place at the earliest time point at which embryogenic cell formation has been established. However, practical problems relating to the asynchronous response of individual microspores to the induction treatment and in separating responsive from non-responsive cells at a very early stage make it difficult to obtain homogenous and specific cell populations for screening. The ability to enrich at a very early stage for embryogenic versus non-embryogenic cell types from the same culture requires good cellular or molecular markers. A number of candidate marker genes that could be used to predict embryogenic cell formation are now available. These marker genes are expressed during early zygotic or in vitro embryo development, and in some cases have been shown to predict embryogenic cell formation in somatic embryo cultures (Boutilier et al. 1994, 2002; Lotan et al. 1998; Stone et al. 2001; Chesnokov et al. 2002; Zhang et al. 2002). Transformation of donor plants with non-destructive promoter:reporter constructs could be used to identify and isolate viable cells in MDE cultures. Interestingly, the promoter of a gene that was isolated in a differential screen in *B. napus* MDE cultures, p22A1, proved to be a very useful marker for the first embryogenic cell division when fused to the GFP reporter gene (Fukuoka et al. 2003). Flow sorting has also been used in *B. napus* to sort embryogenic and non-embryogenic cells at a very early stage of development (Schulze and Pauls 1998, 2002).

Gene expression analysis of differentiating MDEs is relatively straightforward as compared to screens aimed at identifying genes involved in the switch from microspore to haploid embryo development. Morphologically, embryo cell clusters are first recognisable as bona fide embryos either by the presence of a protoderm layer at the globular stage or by the presence of a suspensor attached to the embryo proper at the proglobular stage. Post-globular stages of embryo development are recognisable by their overall morphology, such as, for example, the characteristic heart, torpedo and cotyledon stages of dicot embryos. Embryos corresponding to specific developmental stages can be purified by sieving through nylon mesh of different pore sizes or, at relatively young stages, by density gradient centrifugation. Proglobular stages of MDE embryo development are more difficult to identify solely on morphology, as unlike zygote embryos of certain species, which divide in a regular and recognisable pattern, in vitro cultured embryos tend to divide irregularly (Yeung et al. 1996). Recently, Custers et al. (2001) have succeeded in developing *B. napus* MDE cultures with a high frequency of suspensor-bearing embryos. These embryos undergo the same ordered divisions as their zygotic counterparts, making it possible to morphologically identify differentiating embryos as early as the two-celled embryo stage (J.B.M. Custers, pers. comm.). The presence of a long suspensor on these MDEs also makes it possible to separate them by sieving from non-embryogenic cells in the same culture.

2.1 Early Embryo-Expressed Genes in Monocot MDE Cultures

Differential gene expression analysis of monocot MDE cultures has been performed in barley, wheat and maize. Vrinten et al. (1999) identified three genes whose expression was strongly upregulated in 3-day-old barley MDE cultures as compared to untreated microspores. The identified cDNAs encode a non-specific lipid transfer protein (ECLTP), a glutathione S transferase (ECGST) and an unknown protein (ECA1). The *ECA1* gene likely encodes an arabinogalactan-like protein (AGP). AGPs are a class of secreted proteins that are thought to play a role in plant development (Kreuger and van Holst 1996; see below). Detailed expression analysis suggests that of the three genes that were identified, only *ECA1* is a specific marker for embryogenic cell formation in barley, as only *ECA1* expression levels were correlated with the embryo forming capacity of the culture.

Reynolds and colleagues took a similar approach to identify MDE-expressed genes in wheat anther cultures (Reynolds and Crawford 1996; Reynolds 2000), except that in this screen cDNA libraries were constructed from advanced-stage 'pollen embryoids'. Embryo-expressed clones were identified by screening the library with mature pollen and pollen embryoids. Subsequent expression analysis of individual genes was used to identify markers for the different stages of MDE development; however, the identity of only one of these clones, *EcMt*, was determined. *EcMt* encodes a cysteine-labelled metallothionein, and its expression is specifically induced in anther cultures starting as early as 6 h after embryo induction (Reynolds and Crawford 1996; Reynolds 2000). *EcMt* expression was reported to be a marker for MDE development; however, *EcMt* is induced by the plant hormone abscisic acid (ABA), both in MDE cultures and other tissues (Reynolds and Crawford 1996; Reynolds 2000), making it unclear whether *EcMt* expression actually marks the formation of embryogenic cells or simply the accumulation of ABA that accompanies the establishment of embryogenic cell formation in wheat. Analysis of ABA levels and *EcMt* expression in embryogenic and non-embryogenic cells from the same culture should answer this question.

2.2 Early Embryo-Expressed Genes in *B. napus* MDE Cultures

The majority of screens for early embryo-expressed genes have been performed using the *B. napus* MDE embryo culture system. The popularity of this system is largely due to the ease and efficiency of MDE embryo culture and the availability of good non-embryogenic controls (Custers et al. 2001; Yeung 2002). Another benefit of using *B. napus* as a model plant is that its genes are very similar at the nucleotide level to those of *Arabidopsis thaliana* (*Arabidopsis*; Brunel et al. 1999). This close similarity is also advantageous in that the molecular and functional genomics tools available for *Arabidopsis* research, including an efficient and effective transformation system, a fully sequenced genome, full genome microarrays, and a wide range of characterised mutants and mutant populations, can provide an initial platform for determining the identity and function of *B. napus* MDE-expressed genes (Hall et al. 2002).

2.2.1 mRNA Differential Display PCR Identifies a Small Secreted Signalling Peptide

Research on MDE development in our laboratory is mainly focused on the identification and functional characterisation of *B. napus* genes and proteins that are differentially regulated during the switch from microspore to haploid embryo development and during the early differentiation of the embryo. One of the first molecular screens we performed used mRNA differential display PCR to identify genes expressed during the switch from microspore to MDE development, and the transition from globular to heart-shaped embryos (Custers et al. 2001). MDE cultures were analysed at consecutive developmental stages beginning 8 h after the initiation of the embryo culture (developmental switch) through to the torpedo stage of embryo development (establishment of basic organ patterns). A large number of early embryo-expressed genes were identified that belong to a wide range of protein function categories and that show diverse temporal expression patterns (Custers et al. 2001). One of the genes that was isolated in the screen, *DD3–12*, encodes CLE19, a small 74 amino acid secreted protein that belongs to the CLAVATA3/ESR (CLE)-like family of proteins (Cock and McCormick 2001). CLE proteins are characterised by their small size (average size 8 kDa), an N-terminal secretion peptide, and a conserved C-terminal KRXVPXGPNPLHN motif, which does not show similarity to other known functional motifs. Only a few *CLE* genes have been characterised in detail. One well-characterised *CLE* gene is *CLV3*. *CLV3* is expressed in the stem cells of the shoot meristem and acts in a negative signalling loop with *WUSCHEL* (*WUS*) to maintain the stem cell population of the meristem (Laux 2003).

Our initial differential display-PCR analysis indicated that CLE19 is expressed at the globular to heart stage of development. This expression pattern was later confirmed by RNA gel blot and CLE19 promoter:reporter gene analyses in *Arabidopsis* and *B. napus*. In zygotic embryos, CLE19 expression marks the transition from the globular to heart stage of development: CLE19 is first expressed in the cotyledon primordia of triangular stage embryos and becomes gradually restricted to the L1 layer at the tip of the cotyledons and then to the base of the cotyledons surrounding the shoot apical meristem (Fiers et al. 2004). Functional analysis of CLE19 is underway; however, our preliminary results suggest that CLE19 may also function in a CLV3-WUS-like signalling loop.

2.2.2 Microarray Analysis of Early MDE Development

We have recently initiated microarray-based expression profiling experiments in *B. napus* MDE cultures to obtain a broader perspective on the molecular pathways that accompany the switch from pollen development to embryo formation in vitro, and to classify temporal patterns of gene expression during early embryo development. DNA microarray analysis is a high throughput technology that allows for the simultaneous and comparative analysis of expression profiles for thousands of genes (Aharoni and Vorst 2002). To date, two cDNA microarray analysis studies have examined changes in gene expression during early embryo development, both of which made use of somatic embryo cultures (Thibaud-Nissen et al. 2003; van Zyl et al. 2003). The cDNAs on these microarrays were derived from random EST collections. The use of such non-targeted arrays can be limiting in that genes preferentially involved in embryo development processes are generally not well represented in EST sequencing programs, and are therefore underrepresented on the array. This indeed appears to be the case, as in both studies a large number of genes were spotted on the arrays, but very few differentially expressed genes were detected. The lack of similar studies using *Arabidopsis* full genome arrays and very young zygotic embryos highlights the difficulty associated with isolating few-celled embryos from seeds.

To identify expression profiles associated with early MDE development, we developed a dedicated microarray carrying cDNAs for genes expressed at three major points of MDE development: (1) 1-day heat-stressed microspore cultures in which a fraction of the microspores have undergone their first sporophytic cell division; (2) 4-day heat-stressed microspore cultures containing a mixture of pre-globular embryos that have started to emerge from the microspore exine wall, developing pollen, and non-embryogenic cells; and (3) purified 10-day globular and heart stage MDEs. A PCR-based technique, suppression subtraction hybridisation (Diatchenko et al. 1996), was also applied to the 4-day and 10-day mRNA samples to normalise the mRNA abundance in these samples and to enrich for genes that are either up- or downregulated relative to non-embryogenic samples. We used the expression profiles of the approximately 1,600 unique arrayed cDNAs to characterise the

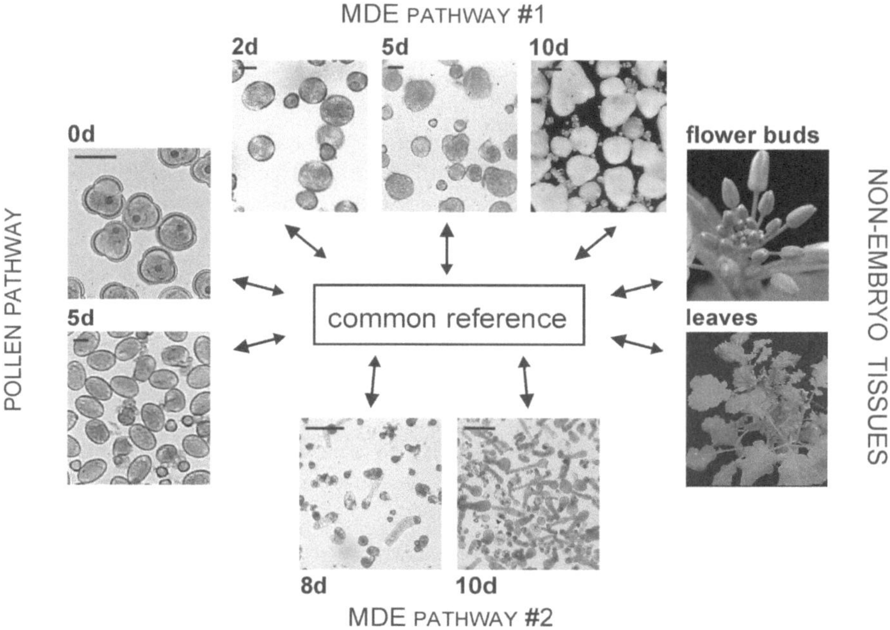

Fig. 1. Developmental stages and experimental design used to identify MDE-expressed genes on a *B. napus* cDNA microarray. The microarray contains cDNAs from *B. napus* MDE libraries spanning the first embryogenic cell division to the heart stage of development. The array was hybridised with nine samples corresponding to four major developmental groups: (1) MDEs at 2, 5 and 10 days after start of culture (MDE pathway #1); (2) suspensor-bearing MDEs at 8 and 10 days after start of culture (MDE pathway #2). Note that the developmental stage of the embryo proper in 8- and 10-day suspensor embryo cultures is approximately the same as that in embryos without suspensors from 2- and 5-day cultures, respectively; (3) developing microspores at start of culture (*0d*) and 5 days later (pollen pathway); and (4) non-embryo tissues. Each of the nine samples was hybridised to the microarray together with a common reference. The common reference sample was created by combining equal amounts of RNA from the embryo- and microspore culture samples. *Scale bar* is 20 μm in the pollen pathway *0d* and *5d* samples and MDE pathway #*2d* and *5d* samples, and 100 μm in the MDE pathway #1 *10d* sample and MDE pathway #2 *8d* and *10d* samples

development of three types of cultures: embryogenic cultures (three time points), embryogenic cultures containing suspensor-bearing MDEs (two time points) and control pollen cultures (two time points). An overview of the experimental setup and the developmental stages studied is shown in Fig. 1.

Our preliminary analysis of the microarray data focused on the identification of genes that are upregulated in embryogenic cultures relative to pollen cultures. Approximately 200 embryo upregulated genes were identified and further subdivided into categories based on their temporal expression patterns and/or preferential expression in suspensor-bearing or non suspensor-bearing MDEs. Five gene expression categories were identified including: (1) genes expressed in 2-day embryogenic cultures (11 genes); (2) genes expressed in 5-day embryogenic cultures (30 genes); (3) genes expressed in 10-day globular–heart stage embryos (38 genes); (4) genes with upregulated expression in suspensor-bearing embryos (14 genes); and (5) genes expressed during all stages of MDE development (115 genes). Most of the genes we identified have not been assigned a function, nor have they been annotated as being embryo-expressed ESTs. Of the annotated genes we identified, the majority fell into functional classes related to transcription, chromatin remodelling, protein degradation and signal transduction. Very few metabolic genes were identified.

This data set of early MDE-expressed cDNAs provides a rich source of potential targets for further expression and functional analysis. Detailed analysis of a number of these early embryo-expressed genes is in progress and should generate new insight into the factors controlling embryo development in MDE cultures and seeds. Conceivably, a subset of these genes could be further developed to serve as molecular markers for different stages of MDE culture, and may even provide new molecular tools for the improvement of different aspects of MDE culture.

2.2.3 Subtractive Hybridisation Uncovers a Key Regulator of Embryo Development

In a parallel approach to the differential display-PCR screen, we used *B. napus* MDE cultures to identify genes expressed during the 8- to 32-cell stage of embryo development (Boutilier et al. 2002). In our hands, these proglobular-staged embryos are typically found in microspore cultures subjected to a 4-day heat-stress treatment. However, cultures at this time point also contain a large proportion of non-embryogenic cells that are either following the pollen development pathway, are arrested in development or are non-viable. We therefore made use of a heat-stressed non-embryogenic sample to subtract these non-embryogenic mRNAs from our embryogenic mRNA population (Pechan et al. 1991).

Using this subtractive screen, we identified five cDNA sequences corresponding to four unique genes that were originally named BNM for *Brassica napus* microspore derived-embryo (Table 1). Two of the cDNAs, *BNM2A* and *BNM2B*, encode BURP domain proteins (Hattori et al. 1998); one of the cDNAs, *BNM4*, encodes the orthologue of the *Arabidopsis* inward rectifying K^+ channel protein, AtKAT1 (Sentenac et al. 1992); one of the cDNAs encodes BNM5, an unknown protein; and the last cDNA, *BNM3* or *BABY BOOM* (*BBM*), encodes a protein with similarity to the AP2/EREBP domain transcription factor family. AP2/EREBP transcription factors are characterised by the AP2/ERF domain, a DNA binding domain that was first identified in the APETALA2 (AP2) and ETHYLENE RESPONSE ELEMENT (ERE) BINDING PROTEINS (EREBP; Jofuku et al. 1994; Ohme Takagi and Shinshi 1995). AP2/ERF proteins are plant specific and mediate diverse developmental and stress-related responses (Riechmann and Meyerowitz 1998). The BBM transcription factor plays a central role in plant embryo development and is described in more detail below.

BBM expression pattern and function have been examined in *B. napus* and *Arabidopsis* (Boutilier et al. 2002). BBM is expressed throughout the major stages of embryo development in both seeds and MDEs, but is upregulated during early embryo development. Messenger RNA in situ hybridisation to developing MDEs and seeds showed that BBM is expressed throughout the developing embryo and transiently in the young endosperm. BBM expression does not appear to be seed specific as mRNAs are detected at low level in roots and in other organs. Analysis of gain-of-function mutants shows that BBM plays a key role in embryo development. Ectopic overexpression of BBM under control of the UBIQUITIN (UBI) and 35S promoters induces somatic embryogenesis and the formation of cotyledon-like structures from the vegetative tissues of young seedlings and occasionally from the leaves of mature plants. Both the UBI:BBM and the 35S:BBM transgenic plants also show a low penetrance of other pleiotropic phenotypes that, together with the induction of somatic embryos, point to a general role for BBM in promoting cell division and morphogenesis. This capacity for promoting cell growth and differentiation was confirmed by an experiment in which UBI:BBM and wild-type *Arabidopsis* explants were placed on minimal medium or on medium containing growth regulators that are normally used to induce regeneration. In both cases the UBI:BBM explants outperformed the wild-type explants, showing enhanced and accelerated regeneration under the standard hormone-induced regeneration protocol, and the ability to completely regenerate into plantlets in the absence of added growth regulators (Boutilier et al. 2002).

2.2.4 Transcriptional Regulation of Embryogenesis

Recently, a number of genes have been identified that show gain- or loss-of-function phenotypes similar to those seen in BBM gain-of-function plants. These genes include the *LEAFY COTYLEDON* (*LEC*) genes, *LEC1* and *LEC2* (Lotan et al. 1998; Stone et al. 2001), *WUS* (Zuo et al. 2002) and *PICKLE* (Ogas et al. 1997). The *LEC1* and *LEC2* genes both encode seed-expressed transcription factors: the LEC1 protein shows similarity to the HAP3 subunit of eukaryotic NF-Y/CCAAT box-binding factors (Lotan et al. 1998), while LEC2 contains a B3 domain, a plant-specific DNA binding domain (Stone et al. 2001). Both *LEC1* and *LEC2* were originally identified as loss-of-function mutants showing defects during the differentiation and maturation stages of embryo development (Meinke et al. 1994; West et al. 1994). The *lec* loss-of-function phenotypes (desiccation intolerance, failure to accumulate maturation-specific mRNAs, and cotyledons with post-germination leaf characteristics) suggest that wild-type *LEC* genes are required to specify embryo cell fate and maintain embryo-specific processes. Further support for this idea was obtained by gain-of-function studies in which *LEC1* and *LEC2* were ectopically expressed in *Arabidopsis* under control of the 35S promoter (Lotan et al. 1998; Stone et al. 2001). Both *35S:LEC1* and *35S:LEC2* transgenics spontaneously produce somatic embryos on seedlings, although somatic embryo formation in the *35S:LEC1* plants is considerably weaker than in *35S:LEC2* transgenics.

A second gene shown to induce spontaneous embryo development is *WUS*. *WUS* encodes a homeobox transcription factor that together with the *CLV1*, *CLV2* and *CLV3* genes is involved in the specification of stem cell identity in the shoot apical meristem (Laux et al. 1996; Schoof et al. 2000). Ectopic expression of *WUS* together with *STM* promotes ectopic organogenesis (Gallois et al. 2002), a phenotype that was confirmed by the discovery of *WUS* in a gain-of-function mutant screen for cytokinin-independent regeneration (Zuo et al. 2002). Zuo et al. (2002) showed that ectopic WUS expression induces cell proliferation and somatic embryo formation in the absence of added growth regulators. Using steroid-inducible *WUS* expression it was also shown that *Arabidopsis* seedlings retain the capacity to respond to the WUS signal and form somatic embryos up to 7 days after germination (Zuo et al. 2002). We obtained similar results using steroid-inducible BBM activity. The results obtained with WUS and BBM are in contrast to those obtained with steroid-inducible LEC transgenics, where somatic embryos are only formed when seeds are plated directly on medium containing the steroid inducer (Zuo et al. 2002). Thus specific factors present in a narrow developmental window are required for LEC-mediated somatic embryo induction, whereas WUS- and BBM-mediated somatic embryo formation relies on factors that are more broadly expressed in juvenile plants.

The *PICKLE* gene also plays a central role in the induction of embryo development; however, unlike *LEC*, *WUS* and *BBM*, somatic embryos are formed in a *pkl* loss-of-function background. *PKL* was identified in a genetic screen for mutants showing altered root development (Ogas et al. 1997), and later as a genetic enhancer of the *crabs claw* mutant phenotype (Eshed et al. 1999). *Pkl* mutants fail to repress embryonic programs after germination with the result that *pkl* plants develop greenish pickle-shaped roots that express embryonic characteristics and form somatic embryos when excised and placed on minimal tissue culture medium. *PKL*, like *WUS*, is expressed throughout the life cycle of the plant. These genes seem to play a broader role than *BBM* and *LEC* in promoting meristematic growth during plant development.

What can these observations tell us about embryo formation in MDE cultures? Successful MDE culture depends on many factors including the species or genotype under study, pretreatments used to optimise the competence of the microspores to respond to the induction treatment, the developmental stage used as starting material, the stress treatment used to induce embryogenesis, and the media composition and culture conditions (Ferrie et al. 1995;

Maluszynski et al. 2003). Many species and genotypes are recalcitrant to MDE formation, and even in genotypes that respond well to microspore culture only a fraction of the microspores will develop into embryos. Recalcitrance for MDE formation can occur at many steps in the process and may have many origins. For example, recalcitrance for MDE formation may arise from a lack of competence to respond to the induction signal, or the inability to sustain embryogenic cell formation or initiate embryo differentiation. Ectopic expression of wild-type LEC1/2, BBM or WUS proteins or expression of dominant negative forms of PKL under control of a microspore-specific promoter (Custers et al. 1997) could be used to induce meristematic or embryogenic cell formation in recalcitrant genotypes, or to increase the number of MDEs in responsive genotypes. However, the question remains as to whether microspores initially need to be competent to respond to these embryo-inducing signals, or whether expression of these signals is sufficient to both induce embryogenic competence and sustain embryo formation. Clearly a better understanding of how competence for embryogenesis is established in microspores and how embryogenesis-promoting genes function at a mechanistic level is an essential part of any strategy aimed at improving MDE formation in both recalcitrant and responsive genotypes.

3 Extracellular Signalling Molecules in MDE Development

Medium conditioned by rapidly growing embryo cultures or non-embryogenic nurse cells has been shown to stimulate the formation of embryogenic cells and embryos in both gametophytic and somatic-based culture systems (von Arnold et al. 2002). The effect of conditioned growth medium on embryogenesis has been best studied in carrot and conifer somatic embryo cultures, where it has been shown that certain secreted proteins and oligosaccharides can act as signalling molecules to promote cell proliferation and the growth of embryogenic cells. These signalling molecules include specific endochitinases (de Jong et al. 1992), arabinogalactan proteins (AGPs; Egertsdotter et al. 1993; Kreuger and van Holst 1993, 1995; Egertsdotter and von Arnold 1995; Toonen et al. 1997; van Hengel et al. 2001) and endogenous or bacterially produced lipophilic chitin oligosaccharides (de Jong et al. 1993; Egertsdotter and von Arnold 1998; Dyachok et al. 2000, 2002). The biological function and mode of action of chitinases, AGPs and lipophilic chitin oligosaccharides during somatic embryogenesis is not known; however, the ability of all three types of molecules to substitute for each other in promoting embryogenic cell formation is likely due to the ability of endochitinases to produce signalling molecules through AGP cleavage or lipophilic chitin oligosaccharide formation (Dyachok et al. 2002; van Hengel et al. 2002). How AGPs and lipophilic chitin oligosaccharides themselves promote cell division is also not known, but several studies suggest that they may function by inhibiting programmed cell death (McCabe et al. 1997; Dyachok et al. 2002).

Secreted proteins and oligosaccharides are also likely to stimulate the growth and development of MDE cultures, although direct evidence for this is just beginning to emerge (Zheng et al. 2002). Ovary co-culture and ovary conditioned medium (OVCM) have been used for many years to improve the efficiency of MDE formation in wheat and barley, while conditioned medium from wheat and barley MDE cultures has likewise been used to promote the growth of isolated zygotes in monocot species (Köhler and Wenzel 1985; Ziauddin et al. 1990; Bruins et al. 1996; Hu and Kasha 1997; Li and Devaux 2001). Recent attempts to identify the embryo-promoting molecules in MDE conditioned medium and to understand the role of endosperm-like nurse cells in MDE cultures have shed some light on putative roles for secreted molecules in promoting embryo development.

3.1 Ovary Conditioned Media Promotes MDE Development

A number of studies have demonstrated that co-cultivation of isolated barley and monocot microspores with live ovaries increases embryo yield and quality (Köhler and Wenzel 1985; Ziauddin et al. 1990; Bruins et al. 1996; Hu and Kasha 1997; Li and Devaux 2001). Recently, Zheng et al. (2002) showed that medium preconditioned by live ovaries was able to replace direct ovary co-culture in promoting MDE in responsive wheat genotypes. Microspores from responsive genotypes cultured with a single dose of ovary conditioned medium (OVCM) showed accelerated MDE formation as compared to ovary co-culture, most likely due to the early stimulation of cell division. A positive effect of ovary co-culture on cell division was observed earlier in barley-isolated microspore cultures (Li and Devaux 2001). In wheat, the greatest effect on cell division and embryo formation was obtained by OVCM that had been conditioned by ovaries for 7 days (Zheng et al. 2002). However, the final embryo yield was even greater when microspores were cultured continuously with live ovaries or with a combination of live ovaries and OVCM, suggesting that a sustained source of embryo-promoting factors increased embryo yields. In contrast to responsive genotypes, co-culture of microspores from recalcitrant genotypes with live ovaries had no effect on MDE formation. However, the addition of OVCM alone or together with live ovaries at an early stage of microspore culture dramatically enhanced MDE formation to levels observed in responsive genotypes. OVCM appears to stimulate microspore division within a narrow developmental window, as OVCM added at a later stage of development could not induce microspore division in recalcitrant genotypes. Together, these observations led Zheng et al. (2002) to suggest that microspores from responsive genotypes are initially able to secrete sufficient amounts of cell division promoting nursing factors, whereas recalcitrant genotypes lack this ability. These factors can be supplied to recalcitrant microspores by OVCM, but are released too slowly by live ovaries to be effective within the narrow developmental window in which they are needed.

Based on this hypothesis it should be possible to induce MDE in recalcitrant microspores by co-cultivating them with responsive microspores or MDEs.

While it is clear that ovaries have a nursing effect on embryogenesis, the nature of the conditioning factor is not known. Köhler and Wenzel (1985) examined the stimulatory effect of conditioned ovary medium on callus induction in shed pollen cultures of barley microspores and concluded that an auxin-like growth regulator could be the conditioning agent. Phenylacetic acid (PAA), a naturally occurring growth regulator with auxin-like activity, has been suggested to be the growth stimulating factor in OVCM (Ziauddin et al. 1990). However, in barley ovary co-culture appears to be more effective in promoting MDE development than PAA, suggesting that additional components contribute to the growth stimulating effect of OVCM (Li and Devaux 2001).

3.2 Signalling Molecules Secreted by Barley MDE Cultures

Conditioned media from barley MDE cultures has been shown to have a stimulating effect on the in vitro growth of zygotes excised from developing maize and barley seeds (Holm et al. 1994; Paire et al. 2003). As a first step toward identifying the specific components in MDE conditioned medium that contribute to the beneficial effect on embryo development, Paire et al. (2003) assessed conditioned medium from MDE cultures collected at different time points (4, 7, 11, 14, 18, 21 and 28 days) for its ability to stimulate embryo development in maize zygotes. When zygotes were cultured continuously on conditioned medium from a single time point, only conditioned medium from 18- and 21-day cultures supported embryo development. Conditioned medium from earlier stages of MDE development stimulated cell divisions, leading to small clusters of cells rather than embryos. MDE cultures at 18 and 21 days contain numerous, clearly formed MDEs that had been released from the exine wall of the microspore. The specific conditioned medium component responsible for stimulating zygotic embryogenesis was not identified in this study; however, preliminary evidence suggests that it is protein based. Firstly, the active component in conditioned medium is heat labile and protease sensitive. Secondly, the timing of appearance of significant amounts of secreted proteins in the conditioned medium correlates with the embryo-promoting capacity of the conditioned medium, and, further, non-embryogenic cultures that do not support growth of zygotic embryos do not secrete detectable amounts of protein. Finally, the addition of a greater than 50-kDa purified protein fraction to cultured zygotes was sufficient to stimulate embryo formation. The authors suggest that the active protein could be AGPs, since AGPs are detected in the conditioned medium at 18 days, and AGPs are of a size range that is compatible with the isolated protein fraction. Although there is not enough evidence at this moment to state which components in the barley MDE conditioned medium promote embryo development, further biochemical analysis should lead to the rapid identification of these factors.

3.3 Secreted Peptides in *B. napus* MDE Cultures

B. napus is considered a good model system to study MDE development, in part due to the high efficiency of embryo production. However, embryo yield in *B. napus* MDE cultures can vary from 0 up to 10%, possibly due to variability in the condition of the donor plants, the differences in developmental stages of the microspores used as starting material and toxic factors released by dead or dying microspores (Pechan and Smykal 2001).

We are using a proteomics-based approach to study the biochemical differences in the conditioned medium of high-yielding embryogenic cultures and non-responsive cultures. We are focusing our analysis on small proteins, as a number of small proteins, and even peptides, have been shown to play a role in plant cell proliferation and meristem growth (Takayama and Sakagami 2002). One example of a growth stimulating peptide is phytosulfokine. In *Asparagus* cell culture, mesophyll cell division only occurs when the cells are cultured at sufficiently high density. A sulfated pentapeptide (phytosulfokine-α) and a sulfated tetra-peptide (phytosulfokine-β) released by *Asparagus* cells cultured at high density were shown to stimulate mesophyll cell division in low-density cell cultures (Matsubayashi et al. 1997). Phytosulfokines were also identified as the compounds that contribute to the growth stimulating effect of conditioned medium from carrot somatic embryo cultures (Hanai et al. 2000).

We examined the differences in small protein profiles present in the conditioned medium from high-yield MDE cultures (HEC) in which 2% of microspores developed into embryos, and non-responsive MDE cultures (NRC) that did not produce any embryos. Both cultures were examined at about 10 days after culture initiation. HEC cultures contained a mix of developing embryos at the globular to torpedo stages, together with some arrested but viable microspores, as well as dead microspores. NRC cultures contained only viable-arrested and dead microspores. An isolation protocol was established in which the culture medium was acidified and protein fractions were isolated based on the differences in hydrophobicity and charge. This separation yielded 60 fractions, of which around 40 fractions showed detectable signals at 214 nm (mainly for proteins and peptides) and 254 nm (mainly for metabolites). Clear differences between the HEC and NRC medium could be detected using matrix-assisted laser desorption/ionisation time-of-flight mass spectrometry (MALDI-TOF-MS; Fig. 2). Two large non-protein peaks (metabolites) were specific for the HEC medium.

From the 60 fractions isolated we were able to identify 12 proteins in the NRC conditioned medium and 16 proteins in the HEC conditioned medium. Only 2 of these 28 proteins were found in the conditioned medium from both cultures. Fractions representing the most dramatic differences between HEC and NRC conditioned medium were subjected to trypsin digestion and de novo sequencing, from which six small proteins from different fractions of the HEC and NRC media were identified (Table 1). All of the small proteins listed in Table 1 are encoded by genes with predicted secretion signal peptides, which is consistent with their presence in the culture medium. Two of these small proteins are homologous to Bp4 and BAN54, which are known to be pollen-specific (Albani et al. 1990; Kim 1997). The *Bp4* gene was present on our *B. napus* MDE microarray (see Sect. 2.2.2) and our expression analysis also supports a late pollen-specific rather than embryo expression pattern for this gene. The presence of these proteins in the medium from 10-day MDE cultures, which no longer contain any viable pollen, suggests that these proteins are derived from developing pollen grains from earlier stages of culture, and thus are very stable. Fraction 12 contained a protein that was abundantly present in HEC but absent in the NRC medium. This protein corresponds to an EST sequence identified in *B. napus* flower buds (GenBank acc. CD838723). Mass spectrometry data suggested that there is a potential internal disulfide bond connecting two cysteine groups within the peptide. The protein encoded by the cDNA carries a putative signal sequence, suggesting that it is an extracellular protein. No clear homologue of this protein could be found in the *Arabidopsis* genome.

Table 1. Tryptic peptides identified by mass spectrometry in the media of high embryogenic cultures (*HEC*) and non-responsive cultures (*NRC*). Localisation of all proteins found was extracellular, based on sequence analysis

Mass (Da)	Identified sequences	Percentage of identity[a]	Protein description
(Da)			
1,105	VECDAICKPK	80	Unknown floral gene, *B. rapa*
1,377	GCKVECDAICKPK	77	Same as above
1,505	LPNSNWCCNTTPR	100	Same as above
1,503	LPNSNWC:CNTTPR[b]	100	Same as above
1,535	LPNSNWC^{X}C^{X}NTTPR[c]	100	Same as above
1,237	TYPYKLPLDK	Unknown	
1,253	PYKLPLDK	Unknown	
1,336	YCXDEQQLPVNK	73	BAN54, *B. rapa*
1,392	LMDEQQLPVNK	73	Same as above
1,426	YXDEQQLPVNK	64	Same as above
1,520	LFSCDEQQLPVNK	62	Same as above
1,292	IPITGSYCLPTK	100	Bp4-like, *B. oleracea*
1,382	IPITGSYXLPTK	92	Same as above
2,606	CIGYLTQNGPLPR	100	LTP, *B. napus*
1,451	TRTNLNNMAR	80	Same as above

[a] Sequence similarity with known proteins in GenBank
[b] C:C, Internal SS bridge
[c] C^X, Extra group(s) on cysteine; X, unknown

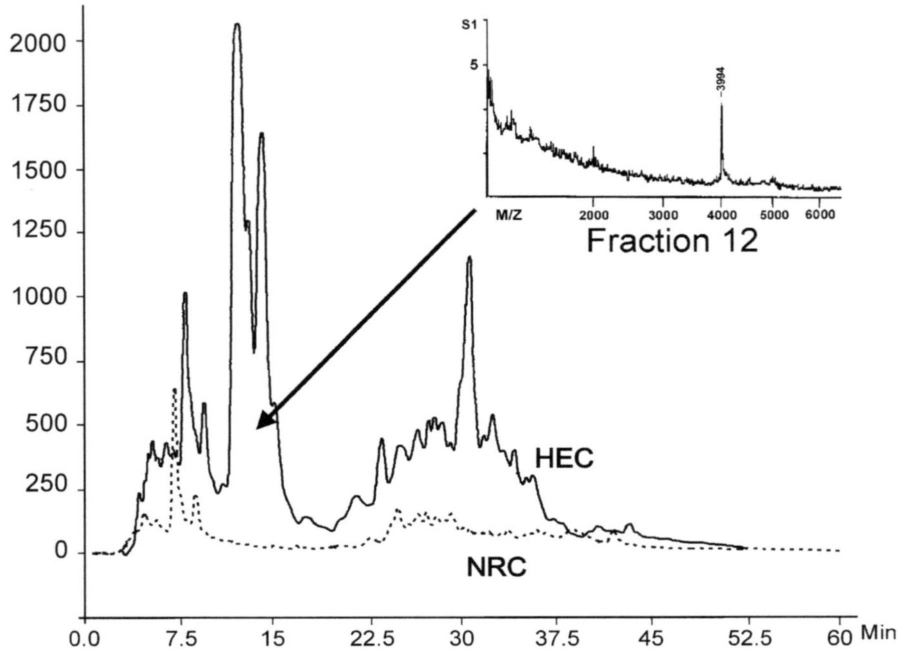

Fig. 2. Analysis of proteins present in the conditioned media of 10-day-old non-responsive (*NRC*) and high-yield (*HEC*) *Brassica napus* MDE cultures by high performance liquid chromatography. *Inset* shows the MALDI-TOF-MS profile of fraction 12 from HEC conditioned medium. The 4-kDa protein in this fraction shows similarity to an unknown protein from *B. napus*

In summary, a proteomics approach has been used as an efficient and sensitive way to analyse secreted proteins in *B. napus* culture media. Our future work will focus on the identification of small proteins present at earlier stages of high-yielding embryo and non-responsive cultures. Additional assays will be needed to elucidate the functions of these small proteins, and to determine if any of these proteins are responsible for the increased or decreased embryogenic capacity observed in different microspore cultures.

3.4 A Role for Endosperm-Like Nurse Cells in Monocot MDE Cultures

One interesting observation regarding growth stimulating molecules that are secreted into conditioned medium of embryo cultures is that non-embryogenic cells rather than embryogenic cells may actually be the source of these signalling molecules. This appears to be the case for the carrot EP3 protein. EP3 is a member of the class IV family of endochitinases, and was initially identified based on its ability to rescue embryo formation in the temperature-sensitive carrot cell line ts11 (de Jong et al. 1992). Localisation of

EP3 mRNAs in carrot somatic embryo cultures and seeds showed that the *EP3* genes are not expressed in developing embryos, but rather in non-embryogenic cell types in culture and in integument and endosperm tissues that surround the developing embryo in planta (van Hengel et al. 1998). This observation led to the suggestion that EP3-like chitinases are part of a nursing system that functions in embryo-surrounding cells during zygotic embryo development, and that this nursing system is mimicked by non-embryogenic suspension cells during somatic embryogenesis (van Hengel et al. 1998).

Indirect support for the role of non-embryogenic nurse cells in promoting MDE development arose during a screen for genes that are specifically expressed during early maize MDE development (Magnard et al. 2000). Two genes encoding small secreted proteins, *ZmAE1* and *ZmAE3*, were identified as being expressed in 5-day-old MDEs. However, subsequent expression analysis in developing seeds showed that both *ZmAE1* and *ZmAE3* mRNAs are localised to the endosperm rather than to the zygotic embryo (Opsahl-Ferstad et al. 1997). In seeds, *ZmAE1* is initially expressed in the endosperm in the so-called embryo-surrounding region (Opsahl-Ferstad et al. 1997) just as endosperm cellularisation begins. *ZmAE1* expression persists in this region as the endosperm develops, and later extends to a region of the endosperm adjacent to pedicel, the basal endosperm transfer layer. This layer is thought to be involved in nutrient transfer from the seed coat to the endosperm (Hueros et al. 1999). *ZmAE3* is also expressed in the embryo-surrounding region of the endosperm, but is restricted to a few cells adjacent to the adaxial side of embryo. These expression results were surprising, and suggested that maize MDEs may actually possess endosperm-like characteristics, and, by extension, that these non-embryogenic cells may play a nursing role by promoting the growth of MDEs. Ultrastructural examination of developing MDEs from the same stage as those used to isolate *ZmAE1* and *ZmAE3* showed that MDEs contain two distinct domains of different sizes and cellular characteristics (Magnard et al. 2000; Testillano et al. 2002). The larger domain shows cellular features that are characteristic of early endosperm cells, such as partially coenocytic organisation, synchronised cell division, a central vacuole and incomplete or 'free-growing' cell walls, whereas the smaller domain is embryo-like in that it is cellularised and consists of small proliferating or meristematic-like cells. If maize MDEs do indeed contain separate embryo- and endosperm-like regions then *ZmAE1* and *ZmAE3* mRNAs should be specifically localised to the endosperm-like region of the MDE. However, this is not the case as both *ZmAE* genes are expressed in both compartments of the MDE (Magnard et al. 2000). As suggested by Magnard et al. (2000), expression of the *ZmAE* genes in both compartments could indicate that embryo–endosperm gene expression is deregulated in vitro. Alternatively, the smaller domain may not be embryo-like, but rather homologous to the densely cytoplasmic, cellularised region of the endosperm that normally surrounds the early embryo. If both domains of the androgenic structures are indeed endosperm-like, then this would imply that MDEs first go through an endosperm-like phase and then later develop into embryos. This discrepancy could be resolved by examining spatial and temporal distribution of a large number of embryo- and endosperm-expressed genes in the androgenic structures.

Although there is no direct evidence for a role of endosperm-like nurse cells in promoting MDE development in maize, the similarities between these cells and the putative non-embryogenic nurse cells in carrot seeds and somatic embryo cultures are striking and deserve further attention. In this respect, laser or genetic ablation experiments in maize MDE cultures and other embryo culture systems could be used to demonstrate a direct role for these non-embryogenic nurse cells in promoting of embryo development in vitro.

The identification and further characterisation of the secreted molecules in MDE cultures will also answer some important questions about the role of secreted growth promoting factors in embryo development. For example, are the compounds found in MDE conditioned medium also found in developing seeds, and, if so, are these molecules secreted by the embryo or by the non-embryogenic cells and tissues that surround the embryo in the seed and in MDE cultures?

4 Conclusions and Perspectives

The use of MDE culture as a model system to study the molecular control of embryogenesis is resulting in an ever-expanding collection of early embryo-expressed genes. Such expression-based studies could be further refined and enhanced by the use of marker genes for the early detection and enrichment of specific cell types, and through the use of full genome microarrays. Isolation of the promoters for these early embryo-expressed genes and characterisation of transgenic lines carrying promoter:reporter constructs can be used to develop a collection of early embryo-expressed promoters. Such promoter sets are currently underrepresented in the databases, but if developed would provide valuable tools for tissue and stage-specific embryo expression both in vitro and in developing seeds. These genes and reporter constructs would also be valuable as markers to identify embryogenic cells in mutant screens and for mutant characterisation.

Functional analysis of this collection of MDE-expressed genes is clearly needed to understand the role of these genes during microspore-derived and zygotic embryo development. Functional analysis of MDE-expressed genes can be carried out in the homologous species using standard gain- and loss-of-function transgenic approaches; however, this is an enormous task, as most species used for MDE culture are less amenable to high throughput analysis. In this respect, it would be more efficient to identify early embryo-expressed genes using model species for MDE culture, and then to analyse the function of the orthologous genes in a heterologous species such as *Arabidopsis*, where tools are available for high-throughput functional genomics. More targeted approaches such as mutant screens and genetic genomics (Jansen and Nap 2001), although laborious, offer a more direct route to identifying genotype-based differences in embryogenic cell formation in MDE cultures of certain model species.

Another area where MDE culture is proving to be an extremely useful tool is in the area of cell–cell communication and the identification of factors in conditioned medium that promote cell division and differentiation. New developments in the high-throughput and ultra-sensitive detection of proteins and metabolites will facilitate further elucidation of the extracellular signals controlling growth and differentiation during embryogenesis.

Acknowledgements The research in our laboratory on MDE in canola was partly supported by the Dutch Ministry of Agriculture, Nature Management and Fisheries (DWK; programs 281 and 392). We thank Jan Custers and Ronny Joosen for comments on the manuscript.

References

Aharoni A, Vorst O (2002) DNA microarrays for functional plant genomics. Plant Mol Biol 48:99–118

Albani D, Robert LS, Donaldson PA, Altosaar I, Arnison PG, Fabijanski SF (1990) Characterization of a pollen-specific gene family from *Brassica napus* which is activated during early microspore development. Plant Mol Biol 15:605–622

Boutilier KA, Gines MJ, DeMoor JM, Huang B, Baszczynski CL, Iyer VN, Miki BL (1994) Expression of the BnmNAP subfamily of napin genes coincides with the induction of *Brassica* microspore embryogenesis. Plant Mol Biol 26:1711–1723

Boutilier K, Offringa R, Sharma VK, Kieft H, Ouellet T, Zhang L, Hattori J, Liu CM, van Lammeren AA, Miki BL, Custers JB, van Lookeren Campagne MM (2002) Ectopic expression of BABY BOOM triggers a conversion from vegetative to embryonic growth. Plant Cell 14:1737–1749

Bruins MBM, Rakoczy Trojanowska M, Snijders CHA (1996) Isolated microspore culture in wheat (*Triticum aestivum* L): the effect of co-culture of wheat or barley ovaries on embryogenesis. Cereal Res Comm 24:401–408

Brunel D, Froger N, Pelletier G (1999) Development of amplified consensus genetic markers (ACGM) in *Brassica napus* from *Arabidopsis thaliana* sequences of known biological function. Genome 42:387–402

Chesnokov YV, Meister A, Manteuffel R (2002) A chimeric green fluorescent protein gene as an embryogenic marker in transgenic cell culture of *Nicotiana plumbaginifolia* Viv. Plant Sci 162:59–77

Cock JM, McCormick S (2001) A large family of genes that share homology with CLAVATA3. Plant Physiol 126:939–942

Cordewener JHG, Hause G, Görgen E, Busink R, Hause B, Dons HJM, van Lammeren AAM, van Lookeren Campagne MM, Pechan P (1995) Changes in synthesis and localization of the 70 kDa class of heat shock proteins accompany the induction of embryogenesis in *Brassica napus* L. microspores. Planta 196:747–755

Crouch ML (1982) Non-zygotic embryos of *Brassica napus* L. contain embryo-specific storage proteins. Planta 156:520–524

Custers JBM, Oldenhof MT, Schrauwen JAM, Cordewener JHG, Wullems GJ, van Lookeren Campagne MM (1997) Analysis of microspore-specific promoters in tobacco. Plant Mol Biol 35:689–699
Custers JBM, Cordewener JHG, Fiers MA, Maasen BTH, van Lookeren Campagne MM, Liu CM (2001) Androgenesis in *Brassica*; a model system to study the induction of plant embryogenesis. In: Bhojwani SS, Soh WY (eds) Current trends in the embryology of angiosperms. Kluwer, Dordrecht, pp 451–470
De Faria Maraschin S, Lamers GEM, Wang M (2003a) Cell death and 14-3-3 proteins during the induction of barley microspore androgenesis. Biologia 58:59–68
De Faria Maraschin S, Lamers GE, de Pater BS, Spaink HP, Wang M (2003b) 14-3-3 isoforms and pattern formation during barley microspore embryogenesis. J Exp Bot 54:1033–1043
De Jong AJ, Cordewener J, Lo Schiavo FL, Terzi M, Vandekerckhove J, van Kammen A, de Vries S (1992) A carrot somatic embryo mutant is rescued by chitinase. Plant Cell 4:425–433
De Jong AJ, Heidstra R, Spaink HP, Hartog MV, Meijer EA, Hendriks T, Lo Schiavo F, Terzi M, Bisseling T, van Kammen A, de Vries SC (1993) *Rhizobium* lipooligosaccharides rescue a carrot somatic embryo mutant. Plant Cell 5:615–620
Diatchenko L, Lau YF, Campbell AP, Chenchik A, Moqadam F, Huang B, Lukyanov S, Lukyanov K, Gurskaya N, Sverdlov ED, Siebert PD (1996) Suppression subtractive hybridization: a method for generating differentially regulated or tissue-specific cDNA probes and libraries. Proc Natl Acad Sci USA 93:6025–6030
Dyachok JV, Tobin AE, Price NPJ, von Arnold S (2000) Rhizobial nod factors stimulate somatic embryo development in *Picea abies*. Plant Cell Rep 19:290–297
Dyachok JV, Wiweger M, Kenne L, von Arnold S (2002) Endogenous nod-factor-like signal molecules promote early somatic embryo development in Norway spruce. Plant Physiol 128:523–533
Egertsdotter U, von Arnold S (1995) Importance of arabinogalactan proteins for the development of somatic embryos of Norway spruce (*Picea abies*). Physiol Plant 93:334–345
Egertsdotter U, von Arnold S (1998) Development of somatic embryos in Norway spruce. J Exp Bot 49:155–162
Egertsdotter U, Mo Lars H, von Arnold S (1993) Extracellular proteins in embryogenic suspension cultures of Norway spruce (*Picea abies*). Physiol Plant 88:315–321
Eshed Y, Baum SF, Bowman JL (1999) Distinct mechanisms promote polarity establishment in carpels of *Arabidopsis*. Cell 99:199–209
Ferrie AMR, Palmer CE, Keller WA (1995) Haploid embryogenesis. In: Thorpe TA (ed) In vitro embryogenesis in plants, vol 20. Kluwer, Dordrecht, pp 309–344
Fiers M, Hause G, Boutilier K, Casamitjana-Martinez E, Weijers D, Offringa R, van der Geest L, van Lookeren Campagne M, Liu C-M (2004) Mis-expression of the CLV3/ESR-like gene CLE19 in *Arabidopsis* leads to a consumption of root meristem. Gene 327:37–49
Fischer C, Neuhaus G (1996) Influence of auxin on the establishment of bilateral symmetry in monocots. Plant J 9:659–669
Fukuoka H, Tsuwamoto R, Nunome T, Ohyama A, Takahata Y (2003) Isolation and characterization of an embryogenesis-specific promoter p22a1 in Brassicaceae. In: Proc 7th Int Congr on Plant Molecular Biology, AOPC/ISPMB 2003 Congress Secretariat, Barcelona, Book of Abstracts, p 41
Gallois JL, Woodward C, Reddy GV, Sablowski R (2002) Combined SHOOT MERISTEMLESS and WUSCHEL trigger ectopic organogenesis in *Arabidopsis*. Development 129:3207–3217
Goralski G, Matthys Rochon E, Vergne P, Przywara L (1999) Androgenic development: a fascinating embryo formation process. Acta Biol Cracov Ser Bot 41:51–65
Hall AE, Fiebig A, Preuss D (2002) Beyond the *Arabidopsis* genome: opportunities for comparative genomics. Plant Physiol 129:1439–1447
Hanai H, Matsuno T, Yamamoto M, Matsubayashi Y, Kobayashi T, Kamada H, Sakagami Y (2000) A secreted peptide growth factor, phytosulfokine, acting as a stimulatory factor of carrot somatic embryo formation. Plant Cell Physiol 41:27–32

Hattori J, Boutilier K, van Lookeren Campagne MM, Miki BL (1998) A conserved BURP domain defines a novel group of plant proteins with unusual primary structures. Mol Gen Genet 259:424–428

Hays DB, Reid DM, Yeung EC, Pharis RP (2000) Role of ethylene in cotyledon development of microspore-derived embryos of *Brassica napus*. J Exp Bot 51:1851–1859

Hays DB, Yeung EC, Pharis RP (2002) The role of gibberellins in embryo axis development. J Exp Bot 53:1747–1751

Heck GR, Perry SE, Nichols KW, Fernandez DE (1995) AGL15, a MADS domain protein expressed in developing embryos. Plant Cell 7:1271–1282

Holm PB, Knudsen S, Mouritzen P, Negri D, Olsen FL, Roue C (1994) Regeneration of fertile barley plants from mechanically isolated protoplasts of the fertilized egg cell. Plant Cell 6:531–543

Hu T, Kasha KJ (1997) Improvement of isolated microspore culture of wheat (*Triticum aestivum* L.) through ovary co-culture. Plant Cell Rep 16:520–525

Hueros G, Royo J, Maitz M, Salamini F, Thompson RD (1999) Evidence for factors regulating transfer cell-specific expression in maize endosperm. Plant Mol Biol 41:403–414

Jansen RC, Nap JP (2001) Genetical genomics: the added value from segregation. Trends Genet 17:388–391

Jofuku KD, den Boer BGW, van Montagu M, Okamuro JK (1994) Control of *Arabidopsis* flower and seed development by the homeotic gene *APETALA2*. Plant Cell 6:1211–1225

Kim HUC (1997) Characterization of three anther-specific genes isolated from Chinese cabbage. Plant Mol Biol 33:193–198

Köhler F, Wenzel G (1985) Regeneration of isolated barley microspores in conditioned media and trial to characterize the responsible factor. J Plant Physiol 121:181–191

Kreuger M, van Holst GJ (1993) Arabinogalactan proteins are essential in somatic embryogenesis of *Daucus carota* L. Planta 189:243–248

Kreuger M, van Holst GJ (1995) Arabinogalactan-protein epitopes in somatic embryogenesis of *Daucus carota* L. Planta 197:135–141

Kreuger M, van Holst GJ (1996) Arabinogalactan proteins and plant differentiation. Plant Mol Biol 30:1077–1086

Laux T (2003) The stem cell concept in plants: a matter of debate. Cell 113:281–283

Laux T, Jürgens G (1997) Embryogenesis: a new start in life. Plant Cell 9:989–1000

Laux T, Mayer KFX, Berger J, Jurgens G (1996) The WUSCHEL gene is required for shoot and floral meristem integrity in *Arabidopsis*. Development 122:87–96

Li H, Devaux P (2001) Enhancement of microspore culture efficiency of recalcitrant barley genotypes. Plant Cell Rep 20:475–481

Li Z, Thomas TL (1998) PEI1, an embryo-specific zinc finger protein gene required for heart-stage embryo formation in *Arabidopsis*. Plant Cell 10:383–398

Liu CM, Xu ZH, Chua NH (1993) Auxin polar transport is essential for the establishment of bilateral symmetry during early plant embryogenesis. Plant Cell 5:621–630

Lotan T, Ohto M, Matsudaira Yee K, West MAL, Lo R, Kwong RW, Yamagishi K, Fischer RL, Goldberg RB, Harada JJ (1998) *Arabidopsis* LEAFY COTYLEDON1 is sufficient to induce embryo development in vegetative cells. Cell 93:1195–1205

Magnard JL, Le Deunff E, Domenech J, Rogowsky PM, Testillano PS, Rougier M, Risueno MC, Vergne P, Dumas C (2000) Genes normally expressed in the endosperm are expressed at early stages of microspore embryogenesis in maize. Plant Mol Biol 44:559–574

Maluszynski M, Kasha KJ, Forster BP, Szarejko I (2003) Doubled haploid production in crop plants; a manual. Kluwer, Dordrecht

Matsubayashi Y, Takagi L, Sakagami Y (1997) Phytosulfokine-alpha, a sulfated pentapeptide, stimulates the proliferation of rice cells by means of specific high- and low-affinity binding sites. Proc Natl Acad Sci USA 94:13357–13362

McCabe PF, Valentine TA, Forsberg LS, Pennell RI (1997) Soluble signals from cells identified at the cell wall establish a developmental pathway in carrot. Plant Cell 9:2225–2241

Meinke DW, Franzmann LH, Nickle TC, Yeung EC (1994) *Leafy Cotyledon* mutants of *Arabidopsis*. Plant Cell 6:1049–1064

Mordhorst AP, Toonen MAJ, de Vries SC (1997) Plant embryogenesis. Crit Rev Plant Sci 16:535–576

Ogas J, Cheng J-C, Sung RZ, Somerville C (1997) Cellular differentiation regulated by gibberellin in the *Arabidopsis thaliana pickle* mutant. Science 277:91–94

Ohme Takagi M, Shinshi H (1995) Ethylene-inducible DNA binding proteins that interact with an ethylene-responsive element. Plant Cell 7:173–182

Opsahl-Ferstad HG, Le Deunff E, Dumas C, Rogowsky PM (1997) ZmEsr, a novel endosperm-specific gene expressed in a restricted region around the maize embryo. Plant J 12:235–246

Paire A, Devaux P, Lafitte C, Dumas C, Matthys Rochon E (2003) Proteins produced by barley microspores and their derived androgenic structures promote in vitro zygotic maize embryo formation. Plant Cell Tissue Organ Cult 73:167–176

Pechan PM, Smykal P (2001) Androgenesis: affecting the fate of the male gametophyte. Physiol Plant 111:1–8

Pechan PM, Bartels D, Brown DCW, Schell J (1991) Messenger-RNA and protein changes associated with induction of *Brassica* microspore embryogenesis. Planta 184:161–165

Perry SE, Lehti MD, Fernandez DE (1999) The MADS-domain protein AGAMOUS-like 15 accumulates in embryonic tissues with diverse origins. Plant Physiol 120:121–129

Reynolds TL (2000) Effects of calcium on embryogenic induction and the accumulation of abscisic acid, and an early cysteine-labeled metallothionein gene in androgenic microspores of *Triticum aestivum*. Plant Sci 150:201–207

Reynolds TL, Crawford RL (1996) Changes in the abundance of an abscisic acid-responsive, early cysteine-labeled metallothionein transcript during pollen embryogenesis in bread wheat (*Triticum aestivum*). Plant Mol Biol 32:823–829

Riechmann JL, Meyerowitz EM (1998) The AP2/EREBP family of plant transcription factors. Biol Chem 379:633–646

Schoof H, Lenhard M, Haecker A, Mayer KFX, Jurgens G, Laux T (2000) The stem cell population of *Arabidopsis* shoot meristems is maintained by a regulatory loop between the CLAVATA and WUSCHEL genes. Cell 100:635–644

Schulze D, Pauls KP (1998) Flow cytometric characterization of embryogenic and gametophytic development in *Brassica napus* microspore cultures. Plant Cell Physiol 39:226–234

Schulze D, Pauls KP (2002) Flow cytometric analysis of cellulose tracks development of embryogenic *Brassica* cells in microspore cultures. New Phytol 154:249–254

Sentenac H, Bonneaud N, Minet M, Lacroute F, Salmon JM, Gaymard F, Grignon C (1992) Cloning and expression in yeast of a plant potassium ion transport system. Science 256:663–665

Smykal P, Pechan PM (2000) Stress, as assessed by the appearance of sHsp transcripts, is required but not sufficient to initiate androgenesis. Physiol Plant 110:135–143

Stone SL, Kwong LW, Yee KM, Pelletier J, Lepiniec L, Fischer RL, Goldberg RB, Harada JJ (2001) LEAFY COTYLEDON2 encodes a B3 domain transcription factor that induces embryo development. Proc Natl Acad Sci USA 98:11806–11811

Takayama S, Sakagami Y (2002) Peptide signalling in plants. Curr Opin Plant Biol 5:382–387

Testillano PS, Ramirez C, Domenech J, Coronado MJ, Vergne P, Matthys Rochon E, Risueno MC (2002) Young microspore-derived maize embryos show two domains with defined features also present in zygotic embryogenesis. Int J Dev Biol 46:1035–1047

Thibaud-Nissen F, Shealy RT, Khanna A, Vodkin LO (2003) Clustering of microarray data reveals transcript patterns associated with somatic embryogenesis in soybean. Plant Physiol 132:118–136

Toonen MAJ, Schmidt EDL, vanKammen A, deVries SC (1997) Promotive and inhibitory effects of diverse arabinogalactan proteins on *Daucus carota* L. somatic embryogenesis. Planta 203:188–195

Touraev A, Ilham A, Vicente O, Heberle-Bors E (1996) Stress-induced microspore embryogenesis in tobacco: an optimized system for molecular studies. Plant Cell Rep 15:561–565

Van Hengel AJ, Guzzo F, van Kammen A, de Vries SC (1998) Expression pattern of the carrot EP3 endochitinase genes in suspension cultures and in developing seeds. Plant Physiol 117:43–53

Van Hengel AJ, Tadesse Z, Immerzeel P, Schols H, van Kammen A, de Vries SC (2001) N-acetylglucosamine and glucosamine-containing arabinogalactan proteins control somatic embryogenesis. Plant Physiol 125:1880–1890

Van Hengel Arjon J, van Kammen A, de Vries Sacco C (2002) A relationship between seed development, arabinogalactan-proteins (AGPs) and the AGP mediated promotion of somatic embryogenesis. Physiol Plant 114:637–644

Van Zyl L, Bozhkov PV, Clapham DH, Sederoff RR, von Arnold S (2003) Up, down and up again is a signature global gene expression pattern at the beginning of gymnosperm embryogenesis. Gene Expr Patterns 3:83–91

Von Arnold S, Sabala I, Bozhkov P, Dyachok J, Filonova L (2002) Developmental pathways of somatic embryogenesis. Plant Cell Tissue Organ 69:233–249

Vrinten PL, Nakamura T, Kasha KJ (1999) Characterization of cDNAs expressed in the early stages of microspore embryogenesis in barley (*Hordeum vulgare*) L. Plant Mol Biol 41:455–463

West MAL, Yee KM, Danao J, Zimmerman JL, Fischer RL, Goldberg RB, Harada JJ (1994) *LEAFY COTYLEDON1* is an essential regulator of late embryogenesis and cotyledon identity in *Arabidopsis*. Plant Cell 6:1731–1745

Wobus U, Weber H (1999) Seed maturation: genetic programmes and control signals. Curr Opin Plant Biol 2:33–38

Yeung EC (1995) Structural and developmental patterns in somatic embryogenesis. In: Thorpe TA (ed) In vitro embryogenesis in plants, vol 20. Kluwer, Dordrecht, pp 205–247

Yeung EC (2002) The canola microspore-derived embryo as a model system to study developmental processes in plants. J Plant Biol 45:119–133

Yeung EC, Rahman MH, Thorpe TA (1996) Comparative development of zygotic and microspore-derived embryos in *Brassica napus* L. cv Topas. I. Histodifferentiation. Int J Plant Sci 157(1):27–39

Zhang SB, Wong L, Meng L, Lemaux PG (2002) Similarity of expression patterns of knotted1 and ZmLEC1 during somatic and zygotic embryogenesis in maize (*Zea mays* L.). Planta 215:191–194

Zheng MY, Weng Y, Liu W, Konzak CF (2002) The effect of ovary-conditioned medium on microspore embryogenesis in common wheat (*Triticum aestivum* L.). Plant Cell Rep 20:802–807

Ziauddin A, Simion E, Kasha KJ (1990) Improved plant generation from shed microspore culture in barley (*Hordeum vulgare* L.) cultivar Igri. Plant Cell Rep 9:69–72

Zimmerman LJ (1993) Somatic embryogenesis: a model for early development in higher plants. Plant Cell 5:1411–1423

Zuo JR, Niu QW, Frugis G, Chua NH (2002) The *WUSCHEL* gene promotes vegetative-to-embryonic transition in *Arabidopsis*. Plant J 30:349–359

I.6 Storage Product Metabolism in Microspore-Derived Cultures of Brassicaceae

RANDALL J. WESELAKE[1]

1 Introduction

Major storage products that accumulate in the seeds of oleaginous crops and cereals include starch, lipid and protein. In terms of haploid systems, extensive investigations of storage product metabolism have been conducted with microspore-derived cultures of Brassicaceae. In particular, microspore-derived embryos (MDEs) of *Brassica napus* and *B. rapa* progress through similar morphogenetic and metabolic changes as zygotic embryos (Weselake and Taylor 1999). Thus, studies of storage product accumulation using MDEs may be more representative of events in the developing seed then those based on investigations with somatic embryos. The ability to maintain a continuous supply of MDEs under carefully controlled conditions has not only provided a tool to accelerate breeding programs, but has also greatly increased our knowledge of storage product metabolism in this oilseed crop. Within the last 15 years, four reviews have been published which deal primarily with storage lipid biosynthesis in MDEs (Weber et al. 1992; Taylor and Weber 1994; Weselake and Taylor 1999; Weselake 2000a). The current chapter presents a somewhat broader picture of metabolism in MDEs by also including a discussion of carbohydrate and protein metabolism. The bulk of the chapter, however, is based on lipid metabolism, which continues to represent most of the storage product research on MDEs of Brassicaceae. Indeed, efforts to understand the details of canola oil formation and improve oil yield and fatty acid (FA) composition of the oil are directly linked to the economic importance of the crop. The discussion is extended to include storage product metabolism in an MD cell suspension culture of *B. napus* L. cv. Jet Neuf which was initially used in studies of freezing tolerance (Orr et al. 1986; Johnson-Flanagan and Singh 1987; Johnson-Flanagan et al. 1991). The MD cell suspension culture was produced early in 1983 in an attempt to generate MDEs (Simmonds et al. 1991). Where appropriate, the effects of abiotic stress and external carbon source on storage product metabolism in MD cultures of *B. napus* are also examined.

[1] Department of Chemistry and Biochemistry, University of Lethbridge, 4401 University Dr, Lethbridge, Alberta, T1K 3M4, Canada. Present address: Department of Agriculture, Food and Nutritional Science, 410 Agriculture/Forestry Centre, University of Alberta, Edmonton, Alberta, T6G 2P5, Canada

Biotechnology in Agriculture and Forestry, Vol. 56
Haploids in Crop Improvement II (ed. by C.E. Palmer, W.A. Keller, and K.J. Kasha)

MD cultures are sensors of environmental change and what we learn about the changes in metabolism in response to environmental change may prove useful in the identification of genes that impart abiotic stress tolerance to crops.

2 Lipid Biosynthesis in MD Cultures

2.1 Characteristics of Lipid Accumulation in MDEs

The earliest studies of metabolism in MDEs of *Brassica* spp. focused on the characterization of lipid accumulation during embryo development. In general, the stage-dependent accumulation of total lipid and FA composition of total acyl lipid was similar for MDEs in comparison to zygotic embryos of *Brassica* spp. (Taylor et al. 1990, 1993; Chen and Beversdorf 1991; Pomeroy et al. 1991; Wiberg et al. 1991). Given the similarities in lipid accumulation in MDEs and zygotic embryos, it has been suggested that MDEs represent a model system for studies of storage lipid biosynthesis (Weselake and Taylor 1999). After 5 weeks in culture, triacylglycerol (TAG) has been shown to represent more than 90% of total acyl lipids in MDEs of *B. napus* L. cv. Topas, which was consistent with observations on mature seed (Pomeroy et al. 1991). Also, late-stage cotyledonary embryos have been shown to have similar lipid content on a dry weight basis (>40%) to zygotic embryos (Taylor et al. 1990; Chen and Beversdorf 1991).

The accumulation of seed reserves is influenced by several factors, some of which are affected by abscisic acid (ABA) (Finkelstein and Somerville 1989). ABA is known to inhibit growth and induce desiccation tolerance in MDEs of *Brassica napus* (Senaratna et al. 1991; Brown et al. 1993; Pomeroy et al. 1994). The application of exogenous ABA (5–15 μM) to MDEs of high erucic acid *B. napus* cultivars has been shown to result in an increase in the proportion of 22:1 (Holbrook et al. 1992; Albrecht et al. 1994; Pomeroy et al. 1994; Wilmer et al. 1997). ABA treatment also resulted in an elevation in embryo oil content (Holbrook et al. 1992; Wilmer et al. 1997).

In studies with MDEs, developmental rate has been shown to be lower at 15°C than at 25°C, but the FA composition of the final oil was similar for growth under both temperature regimes (Albrecht et al. 1994; Möllers et al. 1994). Wilmer et al. (1996, 1997) investigated the effects of temperature, and temperature in relation to ABA, on oil accumulation and FA composition in both developing seeds and MDEs from high and low erucic acid cultivars of *B. napus*. Temperature effects were found to be dependent on the cultivar used, and temperature and ABA had independent and additive effects on 22:1 accumulation in the oil of MDEs from high erucic acid *B. napus*.

Although lipid accumulation characteristics are similar for both MDEs and zygotic embryos, MDEs are considerably greater in fresh weight than zygotic embryos at the equivalent stage of development, a difference that may be related to restrictions imposed by the seed coat of zygotic embryos (Pomeroy et al. 1991). The relatively large size of a late cotyledonary stage MDE compared to a mature seed makes it easier to dissect the cotyledon of the MDE for FA analysis, while the remainder can be used to regenerate a plant (Albrecht et al. 1994, 1995; Möllers et al. 2000). This has obvious advantages in selecting for and breeding for FA composition.

2.2 Lipid Accumulation in MD Cell Suspension Cultures of *B. napus*

Investigations of TAG biosynthesis in MD Brassicaceae cultures have also involved studies with MD cell suspension cultures of *B. napus* L. cv. Jet Neuf. A transmission electron micrograph depicting cross sections of cells from the MD cell suspension cultures of *B. napus* is shown in Fig. 1 with a lipid body identified. The standard culture condition reported by Orr et al. (1986), with 6.5% (w/v) sucrose in the growth medium, has been shown to result in cells that contain 3–6% TAG on a dry weight basis (Weselake et al. 1993, 1998). Increasing sucrose concentration to 22% resulted in cells with about 9% TAG on a dry weight basis. The percentage of TAG in total acyl lipid of the cell suspension cultures was close to 70% for both sucrose concentrations. Although the lipid content of the MD cell suspension cultures is considerably less than that of late-stage MDEs, the cell suspension cultures offer the advantage of being able to study storage lipid biosynthesis in the absence of embryogenesis. Also, growth chamber facilities are not required to maintain developing plants as a continuous source of microspores.

2.3 FA Elongation in MDEs

In a number of oilseeds, including *B. napus*, 16- and 18-carbon FAs are synthesized in the plastid through the action of type II acetyl-coenzyme A (CoA) carboxylase and the FA synthase complex (Ohlrogge and Browse 1995). Malonyl-CoA, generated from the acetyl-CoA carboxylase-catalyzed reaction, provides two carbon units to the growing FA chain, which is linked to acyl carrier protein (ACP) as a thioester. A soluble Δ^9-stearoyl (18:0)-ACP desaturase, which is dependent on reduced ferridoxin and molecular oxygen, catalyzes the formation of 18:1-ACP from 18:0-ACP (Shanklin and Cahoon 1998). Acyl-ACP hydrolase catalyzes the release of free FAs from acyl-ACP, which are then exported into the plastidial envelope where an acyl-CoA synthetase catalyzes the formation of acyl-CoA thioesters (Ohlrogge and Browse 1995). Some of these acyl-CoAs are further elongated in the endoplasmic reticulum (ER).

Most of the investigations on FA biosynthesis in MDEs have focused on FA elongation from 18:1 to 22:1. In developing seeds of high erucic acid *B. napus*,

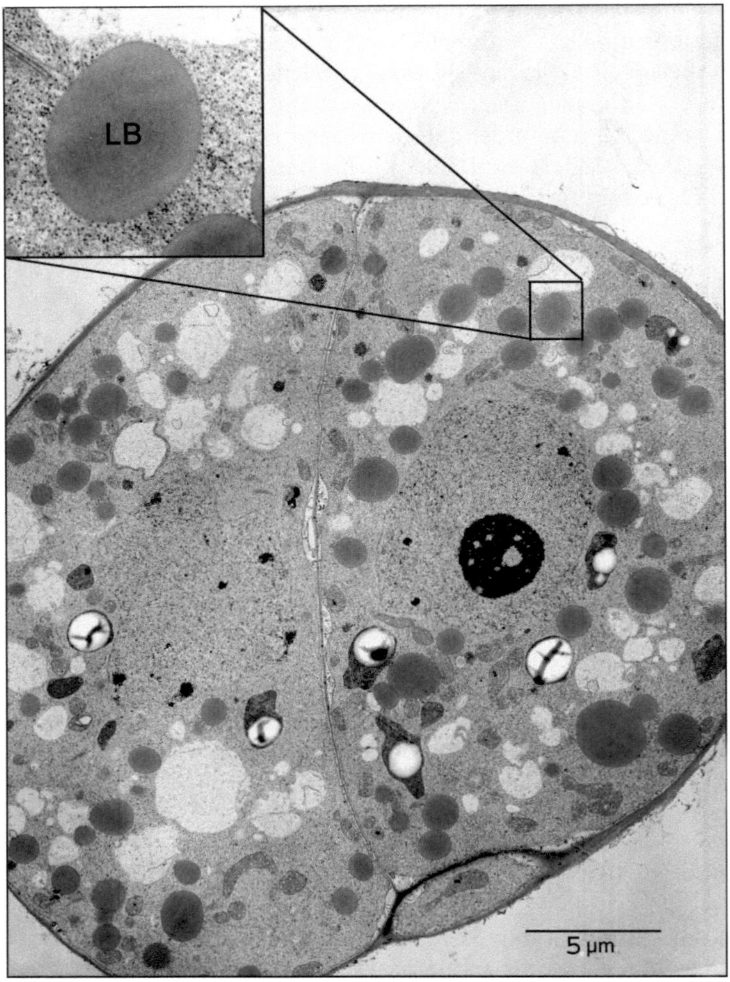

Fig. 1. Transmission electron micrograph of cells from MD cell suspension cultures of *B. napus* L. cv. Jet Neuf. *LB* Lipid body. The cells were fixed overnight in 2.5% glutaraldehyde and 4% formaldehyde in 0.1 M sodium cacodylate buffer (pH 6.8) containing 6% (w/v) sucrose. The cells were rinsed in 0.1 M sodium cacodylate buffer (pH 6.8) and then post-fixed in 1% osmium tetroxide in 0.1 M sodium cacodylate buffer (pH 6.8) for 1 h. Dehydration in an ascending series of ethanol solutions, from 30 to 50% (v/v), over a few hours was followed by stepwise infiltration in Spurrs resin (1:3; 2:2; 3:1; 100%) over a 24-h period. Samples were then placed in polyethylene embedding capsules and polymerized at 60 °C for 24 h. Silver sections were cut with a diamond knife and then stained in uranyl acetate and lead citrate. Finally, sections were photographed in a Hitachi H-500 transmission electron microscope. (Adapted from Weselake 2000b with permission from *Inform*)

the biosynthesis of very long-chain monounsaturated fatty acids (VLCMFAs) has been proposed to occur via two condensations of malonyl-CoA with oleoyl-CoA (Stumpf and Pollard 1983). Malonyl-CoA used in FA elongation is supplied through the action of a cytosolic type II ACCase (Ohlrogge and Browse 1995; Rawsthorne 2002). Downey and Craig (1964) first investigated the elongation process in studies of [^{14}C]-acetate incorporation using developing seeds of *B. napus*. Label was incorporated into 18:1, eicosenoic acid (20:1) and 22:1. Evidence for the two-carbon sequential elongations of 18:1-CoA in MDEs of *B. napus* was presented by Taylor et al. (1992a) in investigations with a 15,000 × *g* fraction which was enriched in elongase activity. Holbrook et al. (1992) showed that 10 μM ABA treatment of MDEs of *B. napus* L. cv. Reston results in an increase in elongase-specific activity, thereby accounting for an increase in levels of 20:1 and 22:1. Studies by Wilmer et al. (1998) suggested that the regulation of 22:1 production was related to total elongase activity rather than alterations in the affinity of the enzyme for its acyl-CoA substrate. Each cycle of FA elongation is believed to involve four enzyme-catalyzed reactions of an elongase complex (Fehling and Mukherjee 1991). Initially, malonyl-CoA undergoes condensation with long-chain acyl-CoA. This is followed by reduction to ß-hydroxyacyl-CoA and then dehydration to an enoyl-CoA. Finally, reduction of enoyl-CoA produces an acyl-CoA extended by two carbons. The first of the four reactions is catalyzed by 3-ketoacyl-CoA synthase, which is encoded by the *FAE1* gene (James et al. 1995; Millar and Kunst 1997). The first reaction is rate-limiting for chain extension, whereas the remaining enzymes of the elongation system are expressed ubiquitously in the plant and do not have a regulatory role in the elongation process (Millar and Kunst 1997). In studies with MDEs of high erucic acid *B. napus*, (+)-ABA has been shown to stimulate the production of VLCMFAs and transcripts encoding 3-ketoacyl-CoA synthase (Zou et al. 1995; Qi et al. 1998). Catabolism of endogenous ABA, however, may limit the extent of production of VLCMFAs (Qi et al. 1998). Recently, the low erucic acid trait of canola has been attributed to a single amino acid substitution in 3-ketoacyl-CoA synthase, which prevents the formation of 20:1 and 22:1 (Katavic et al. 2002). The mode of action of the recombinant FAE1 3-ketoacyl-CoA synthase from *Arabidopsis thaliana* has been shown to be similar to that of the soluble condensing enzyme of the FA synthase complex in the plastid (Blacklock and Jaworski 2002; Ghanevati and Jaworski 2002).

2.4 TAG Biosynthesis in MD Cultures

2.4.1 Acyl-CoA-Dependent and Acyl-CoA-Independent TAG Formation

The greatest advances in using MD cultures of Brassicaceae to investigate plant lipid metabolism have been achieved in the area of TAG biosynthesis. In developing oleaginous seeds, the process whereby fatty acyl chains are

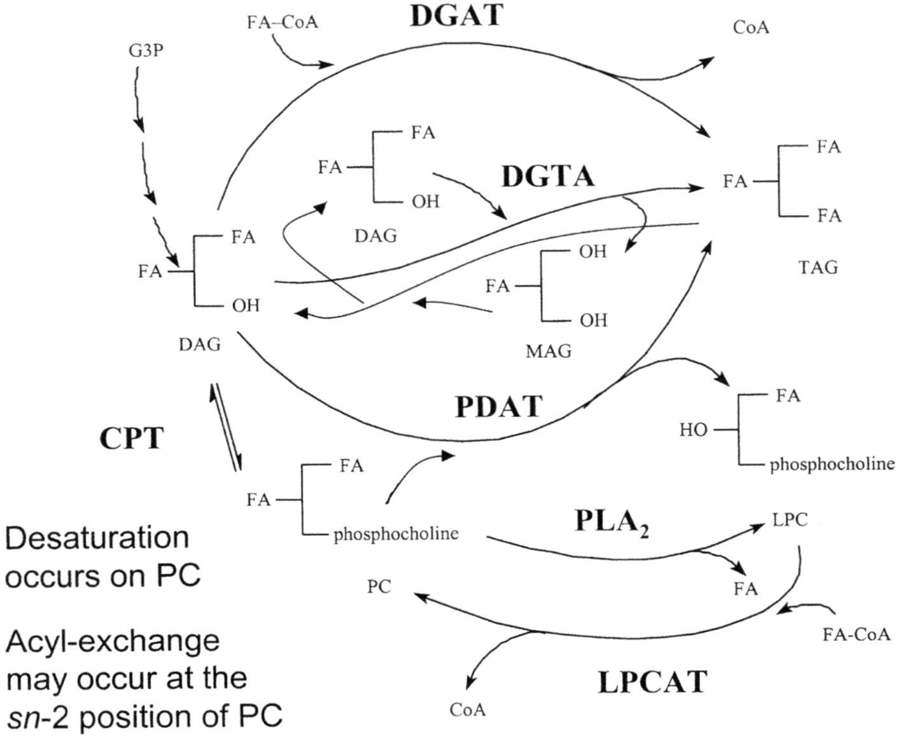

Fig. 2. Possible routes for TAG formation and remodeling in developing oleaginous seeds. *CPT* Cholinephosphotransferase; *DAG* *sn*-1,2-diacylglycerol; *DGAT* diacylglycerol acyltransferase; *DGTA* diacylglycerol transacylase; *FA* fatty acyl; *G3P* *sn*-glycerol-3-phosphate; *HO* or *OH* hydroxyl; *LPCAT* lysophosphatidylcholine acyltransferase; *MAG* monoacylglycerol; *PC* phosphatidylcholine; *PDAT* phospholipid:diacylglycerol acyltransferase; PLA_2 phospholipase A_2; *TAG* triacylglycerol (Adapted from Weselake 2000b with permission from *Inform*). Pathways presented are based on information from the following references: Stymne and Stobart (1987), Ståhl et al. (1995), Mancha and Stymne (1997), Stobart et al. (1997), Dahlqvist et al. (2000), Weselake (2002)

attached to the glycerol backbone to form TAG also occurs on the ER and involves both acyl-CoA-dependent and acyl-CoA-independent catalyzed enzyme reactions (Fig. 2). Some of these reactions are common to both TAG and membrane biosynthesis (Stymne and Stobart 1987; Kocsis and Weselake 1996; Weselake 2002). The Kennedy (1961) pathway for TAG formation involves three acyl-CoA-dependent acylations beginning with *sn*-glycerol-3-phosphate (G3P), which can be derived from the glycolytic intermediate, dihydroxyacetone phosphate, via the action of L-G3P:NAD^+ oxidoreductase (Finlayson and Dennis 1980). Prior to the final acylation catalyzed by diacylglycerol acyltransferase (DGAT), phosphate is removed from phosphatidate (PA) to form *sn*-1,2-diacylglycerol (DAG) via the action of PA phosphatase.

For a long time, DGAT was thought to be the only enzyme involved in the direct formation of TAG. Recent evidence has demonstrated that both DAG and phosphatidylcholine (PC) can serve as acyl donors to DAG in the non-acyl-CoA-dependent synthesis of TAG in reactions catalyzed, respectively, by diacylglycerol transacylase (DGTA) (Stobart et al. 1997) and phospholipid:diacylglycerol acyltransferase (PDAT) (Dahlqvist et al. 2000). DGTA might also catalyze the generation of two DAG molecules from TAG and monoacylglycerol (MAG), thereby providing metabolic opportunities for remodeling of TAG (Mancha and Stymne 1997). Non-acyl-CoA-dependent processes involved in TAG formation and remodeling, however, have only been identified in a few oilseeds. Further investigations of the relative contributions of these activities to TAG biosynthesis in other oleaginous plant systems, including *B. napus*, and at various stages of seed development are required. Other aspects of membrane metabolism that could potentially influence the FA composition of seed TAG are also depicted in Fig. 2. Lysophosphatidylcholine acyltransferase (LPCAT) catalyzes the acylation of lysophosphatidylcholine (LPC) and may also catalyze the exchange of an FA at the *sn*-2 position of PC with the acyl-CoA pool using both forward and reverse reactions (Stymne and Stobart 1987). Phospholipase A_2 catalyzes the removal of an FA from the *sn*-2 position of PC (Ståhl et al. 1995) and along with PDAT may represent a means of removing undesirable acyl chains from membranes. In addition, it is conceivable that both of these reactions can clear the way for reacylation of the *sn*-2 position of LPC catalyzed by LPCAT. The transfer of phosphocholine from cytidine diphosphate (CDP)-choline to DAG is catalyzed by CDP-choline:1,2-diacylglycerol cholinephosphotransferase (CPT). This has been shown to be a reversible process that facilitates the transfer of the DAG skeleton between the mainstream of TAG biosynthesis and phospholipid (Stymne and Stobart 1987). In developing oilseeds, the formation of polyunsaturated FAs occurs while the FAs are attached to PC (Stymne and Stobart 1987). These FA modifications can be introduced into TAG via the DAG generated through CPT-catalyzed reverse reaction.

Lipid or oil bodies of 0.2–2.5 μm represent the ultimate reservoirs of TAG in oleaginous seeds (Huang 1992; Murphy 2001) and their production during seed development may be associated with specialized regions of the ER (Lacey and Hills 1996; Vogel and Browse 1996; Lacey et al. 1999). Theories on the production of lipid bodies are controversial, with some investigators suggesting that lipid bodies pinch off of the ER and are surrounded by a half-unit membrane, whereas other researchers suggest that these subcellular structures arise initially as naked oil droplets (Murphy 2001).

2.4.2 Incorporation of Erucoyl Moieties into TAG in MD Embryos

Given our finite petrochemical reserves and the current focus on reducing the concentration of greenhouse gases, there has been a strong interest in the use of plant oils and their derivatives as substitutes for petrochemicals (Murphy 2002). One of the major goals of *Brassica* lipid biotechnology has been to produce trierucin in the seed oil of high erucic acid cultivars. High erucic acid plant oils can serve as feedstock in the production of slip-promoting/antiblocking agents, detergents, plasticizers and surface coatings, with trierucin serving as a high-temperature lubricant (Princen and Rothfus 1984). Incorporation of erucoyl moieties into the seed oil of high erucic acid *B. napus* L. (e.g. cultivar Reston), however, is limited to the *sn*-1 and *sn*-3 positions of the glycerol backbone (Norton and Harris 1983; Taylor et al. 1991). The absence of significant levels of 22:1 at the *sn*-2 position of TAG from zygotic embryos of high erucic acid *B. napus* has been shown to be attributable to the inability of the resident lysophosphatidate acyltransferase (LPAAT) to accept 22:1-CoA as a substrate (Sun et al. 1988; Bernerth and Frentzen 1990). This enzyme catalyzes the acyl-CoA-dependent acylation of the *sn*-2 position of lysophosphatidate (LPA) (Stymne and Stobart 1987). It has been also reported that 22:1-CoA is not an effective substrate for *B. napus* LPCAT, suggesting that acylation of LPC is probably not a route for incorporation of 22:1 moieties at the *sn*-2 position of the glycerol backbone (Bernerth and Frentzen 1990). The inability of *B. napus* LPAAT to utilize 22:1 has also been demonstrated in incubations of homogenate or microsome prepared from MDEs with [1-^{14}C]22:1-CoA in the presence of G3P (Taylor et al. 1991). The erucoyl moieties were rapidly incorporated at *sn*-3 position of TAG and were virtually excluded from the Kennedy pathway intermediates and complex polar lipids. In contrast, incubations with [1-^{14}C]18:1-CoA resulted in the incorporation of radiolabel into LPA, PA, DAG and membrane lipids, suggesting the action of a DGAT that was highly specific for 22:1-CoA. Also, the microsomal DGAT activity of MDEs of high erucic acid *B. napus* L. cv. Reston effectively produced radiolabeled trierucin when assayed with *sn*-1,2-dierucin and [1-^{14}C]22:1-CoA (Taylor et al. 1991, 1992b).

In contrast to *B. napus*, high-erucic acid breeding lines of *B. oleracea* have been found to contain 22:1 at the *sn*-2 position of TAG (Taylor et al. 1994). Ferrie et al. (1999) have generated MDEs from 15 of 19 *B. oleracea* accessions. Comparison of the FA distribution at the *sn*-2 position of TAG from MDEs of accession line Bo-1 and the TAG of the mature seed of the parent germplasm line indicated a significant proportion of 22:1 at the *sn*-2 position in both cases, demonstrating that this trait was stably inherited in the MDEs.

In studies aimed at investigating the relationship between 18:1-CoA elongation and TAG biosynthesis in MDEs of *B. napus*, Taylor et al. (1992a, 1993) found that radiolabeled 22:1-CoA produced through elongation of [1-^{14}C]18:1-CoA in the presence of G3P resulted in incorporation of 22:1 at both the *sn*-1 and *sn*-3 positions of TAG, although incorporation of radiolabel was about eight-fold greater at the *sn*-3 position. The studies indicated that elongation of 18:1-CoA to 22:1-CoA was somehow linked to the incorporation of 22:1 at the *sn*-1 position of TAG. Some radiolabeled VLCMFA was also found in the Kennedy pathway intermediates and membrane lipids, suggesting that newly synthesized VLCMFAs may be channeled rapidly through the intermediates and ultimately accumulate in TAG. The exclusion of unusual FAs, such as 22:1, from membrane lipids may involve a number of possible factors including phospholipase A_2 action, PDAT action, the selectivity of DGAT for molecular species of *sn*-1,2-DAG and spatial separation of TAG biosynthesis from membrane biosynthesis (Roscoe et al. 2002; Weselake 2002).

2.4.3 Diacylglycerol Acyltransferase

Microsomal DGAT activity has been the focus of numerous other investigations with both MDEs and MD cell suspension cultures of *B. napus* (Weselake and Taylor 1999; Weselake 2000a, 2002). DGAT has been implicated as a key enzyme in TAG biosynthesis in *B. napus* because the level of DGAT activity may have a substantial effect on the flow of carbon into the seed oil (Perry et al. 1999). Also, an *A. thaliana* mutant (AS11) with reduced DGAT activity has been shown to have a decreased TAG/DAG ratio compared to the wild type (Katavic et al. 1995). In addition, in developing seeds of *B. napus*, DGAT activity reaches a maximum during the active phase of oil accumulation and then declines markedly as the seed lipid level reaches a plateau, indicating that the appearance of enzyme activity is coordinated with storage lipid production (Weselake et al. 1993). The specific activity of DGAT has also been shown to increase steadily during development of MDEs of *B. napus* L. cv. Topas, reaching a maximum level at mid-cotyledonary stage (Weselake et al. 1993).

The specificity of microsomal DGAT for acyl-CoAs has been examined in MD cultures of *B. napus*. In studies with microsomes from MDEs of low erucic acid *B. napus* L. cv. Topas, Weselake et al. (1991) showed that DGAT activity was the same with either 18:1-CoA or 22:1-CoA at concentrations of 5 μM and lower in the reaction mixture, but with increased thioester concentration, the enzyme activity exhibited an increased preference for 22:1-CoA. The ability of DGAT activity from low erucic acid MDEs to effectively utilize 22:1-CoA supports the previous suggestion by Cao and Huang (1987) that the DGAT gene(s) was unaffected during the breeding process used to generate low erucic acid varieties. Solubilized DGAT from MDEs of *B. napus* L. cv. Reston was shown to have a considerably greater preference for 18:1-CoA and palmitoyl (16:0)-CoA over stearoyl (18:0)-CoA at thioester concentrations greater than 2 μM (Little et al. 1994). Investigations of specificity effects with the co-substrate DAG have proven difficult to conduct because of interference by endogenous DAG (Little et al. 1994; Weselake 2002).

A number of salts, various biomolecules and an unknown low molecular weight organic factor have been shown to stimulate microsomal DGAT activity from membrane fractions of MD cultures of *B. napus* (Little et al. 1994; Byers et al. 1999). PA has been shown to stimulate microsomal DGAT activity from MD cell suspension cultures by about twofold at a bulk concentration of 500 μM (Byers et al. 1999). This effect may have been attributable to production of DAG that was more easily used by DGAT than exogenous DAG, thus suggesting that substrate channeling may be involved in TAG biosynthesis in the ER. Attempts to demonstrate that DGAT was downregulated through phosphorylation by cytosolic DGAT kinase revealed a low molecular weight stimulatory factor that has yet to be purified and identified (Byers et al. 1999). Bovine serum albumin (BSA) and human acylation stimulating protein (ASP) have also been shown to stimulate DGAT activity (Little et al. 1994; Weselake et al. 2000a). It has been suggested that BSA might stimulate acyltransferase activity by preventing the formation of acyl-CoA micelles that could inhibit enzyme activity (Stobart and Stymne 1990). ASP is a small molecular mass (9 kDa) basic protein in the bloodstream that is produced through cleavage of the complement C3, which originates in the adipocyte (Baldo et al. 1995). Although human ASP has been shown to stimulate storage lipid accumulation in adipocytes via a protein kinase C signaling pathway (Baldo et al. 1995), Yasruel et al. (1991) showed that the protein may also interact directly with DGAT in microsomes from human adipose tissue. The degree of activation of plant DGAT by human ASP remained constant regardless of the concentration of acyl-CoA, suggesting that ASP did not improve substrate delivery to the active site of the enzyme.

A cDNA encoding plant DGAT1 isoform was first cloned from *A. thaliana* (Hobbs et al. 1999; Routaboul et al. 1999; Zou et al. 1999; Bouvier-Navé et al. 2000). The cDNA encoded a 520 amino acid protein with a predicted molecular mass of 59 kDa sharing about 38 % identity with the predicted amino acid sequence of mouse DGAT1 (Hobbs et al. 1999). The first cDNAs encoding BnDGAT1 and BnDGAT2 from *B. napus* were obtained from MD cell suspension cultures of *B. napus* L. cv. Jet Neuf (Nykiforuk et al. 1999a,b, 2002). Both mRNAs were produced in cell suspension cultures and the cDNAs were functionally expressed in *Pichia pastoris*. *BnDGAT1* cDNA encoded a 503 amino acid protein with predicted molecular mass of 56.9 kDa whereas *BnDGAT2* cDNA encoded a 341 amino acid protein with molecular mass of 39.5 kDa. BnDGAT1 is 96 % homologous to BnDGAT2 across 341 overlapping amino acids, and shares 85 % identity with *A. thaliana* DGAT. The hydrophilic N-terminal region of BnDGAT1 may have a role in binding cellular acyl-CoA (Weselake et al. 2000b). More recently, a second DGAT gene family was identified in the oleaginous fungus *Mortierella ramanniana* with members in numerous other species including *A. thaliana* and *B. napus* (Lardizabal et al. 2001). In *M. ramanniana*, two cDNAs were identified that corresponded to related proteins (54 % homology) with molecular masses of 36 and 36.5 kDa. The proteins were unrelated to the previously identified gene family and were designated DGAT2. It is important, however, not to confuse the DGAT2 gene family with the BnDGAT2 from *B. napus* MD cell suspension cultures, which represents a truncated form of BnDGAT1.

Propagation of MD cell suspension cultures of *B. napus* for 2 weeks at progressively higher concentrations of sucrose ranging from 2 to 22 % (w/v) has been shown to result in increased TAG accumulation (Weselake et al. 1998).

This effect may be due to both an increase in carbon supply and change in osmotic potential. The relative amount of BnDGAT1 transcript was positively associated with DGAT activity as the sucrose concentration was raised from 2 to 6%, but the level of transcript decreased markedly at 14% sucrose (Weselake et al. 1997; Davoren et al. 2002). Cells cultured in 14% sucrose may have synthesized TAG earlier in the culture period than cells cultured in 6% sucrose, or the temporal regulation of the transcript may have been different under the two sucrose concentrations. Also, some of the DGAT activity may have been attributable to the DGAT2 gene family identified by Lardizabal et al. (2001). Differential display of mRNA from cells cultured in 2, 6 and 14% sucrose also revealed the induction of transcript encoding a putative mitochondrial acyl-carrier protein, suggesting a possible relationship between mitochondrial FA biosynthesis and TAG biosynthesis in the ER (Davoren et al. 2002). In another study, sucrose-mediated changes in TAG metabolism and *BnDGAT1* expression over time were examined in cells transferred from 6 to 14% sucrose (Nykiforuk et al. 2002). TAG content was maximal 6 h after transfer but increased again from 24 to 72 h. DGAT activity per gram of fresh weight and the relative amount of BnDGAT1 polypeptide as detected by Western blotting generally followed the TAG content profile. The level of BnDGAT1 transcript, however, was highest at 12 h, suggesting some control of expression at the post-transcriptional level. TAG lipase displayed a marked increase at 12 h at the time when TAG content was lowest, thereby metabolically accounting for the reduction in the level of storage lipid. Thus, the MD cell suspension system appears to display attributes associated with both lipid deposition and mobilization, and may be useful in assessing the mechanisms that regulate these processes.

2.4.4 Phosphatidate Phosphatase

Studies of PA phosphatase have been conducted using both MDEs and MD cell suspension cultures of *B. napus*. The first PA phosphatase solubilized from *B. napus* was from MD embryos of the cv. Topas (Kocsis et al. 1996). The enzyme was solubilized with 0.4% (w/v) Tween 20 at a detergent to protein ratio of 1:1. The solubilized PA phosphatase was active with various molecular species of PA and a number of other phosphate-containing compounds. The *B. napus* PA phosphatase may have broad substrate specificity for PAs with various acyl chains and/or the solubilized preparation may consist of PA phosphatase isoforms differing in substrate specificity. The substrate preference of PA phosphatase in the plant cell, however, may not be that critical if the enzyme is limited to accepting endogenous PA generated via the LPAAT-catalyzed reaction through substrate-channeling. Wilmer et al. (1998) reported substantial production of PA, with DAG being the next most abundant product, when a particulate fraction from MDEs of *B. napus* L. cv. Reston was incubated with [U-^{14}C]G3P and 18:1-CoA. The results suggested that PA phosphatase activity was limiting in TAG formation, possibly sharing some control with DGAT. Furukawa-Stoffer et al. (1998) identified both *N*-ethylmaleimide-sensitive and -insensitive PA phosphatase activity in microsomal fractions from both MDEs and MD cell suspension cultures of *B. napus*, suggesting that plant cells, as in mammalian systems (Kocsis and Weselake 1996), may contain different isoforms of PA phosphatase involved in both glycerolipid biosynthesis and intracellular signaling.

2.4.5 Lysophosphatidylcholine Acyltransferase

Recently, some characteristics of LPCAT activity have been examined in MD cell suspension cultures of *B. napus* L. cv. Jet Neuf (Furukawa-Stoffer et al. 2003). Microsomal LPCAT activity utilized 16:0-, 18:0- and 18:1-CoA in the direction of PC formation, but was not active with 12:0-CoA, suggesting that the specificity of this enzyme may be important in attempts to alter the FA composition of the *sn*-2 position of TAG from *B. napus* seed. PC produced through LPCAT activity can potentially serve as a source of *sn*-1,2-DAG via the reverse reaction of CPT, and the forward/backward reactions catalyzed by LPCAT may alter the FA composition of the acyl-CoA pool, which in turn can affect TAG composition (Stymne and Stobart 1987). Metabolic engineering studies aimed at altering the FA composition of the acyl moiety at the *sn*-2 position of TAG have only focused on the modification of the LPAAT-catalyzed step (Weselake and Taylor 1999; Weselake 2002). The availability of a LPCAT cDNA clone could lead to new and complementary approaches for modifying the FA composition of the *sn*-2 position of TAG.

3 Carbohydrate Metabolism in MD Cultures of *B. napus*

In developing zygotic embryos of *B. napus*, the formations of oil, starch and amino acids are linked through reactions in carbohydrate metabolism, which begin with the uptake of sucrose, hexoses and amino acids from the liquid endosperm (Rawsthorne 2002; Schwender and Ohlrogge 2002; Hill et al. 2003). In working with MDEs of *B. napus*, the liquid endosperm is replaced by a more chemically defined medium (Pomeroy et al. 1991). A schematic representation of some aspects of carbon metabolism in the cytosol, plastid and ER of developing zygotic embryos of *B. napus* is depicted in Fig. 3. In addition to being the site of FA biosynthesis, the transient accumulation of starch also occurs in these organelles, with deposition of the carbohydrate polymer beginning prior to oil accumulation (Rawsthorne 2002). At relatively early stages of development, about half of the carbon from glucose-6-phosphate (Glc6P) is used for starch synthesis, with the remaining portion partitioned between FA synthesis and the oxidative pentose phosphate path-

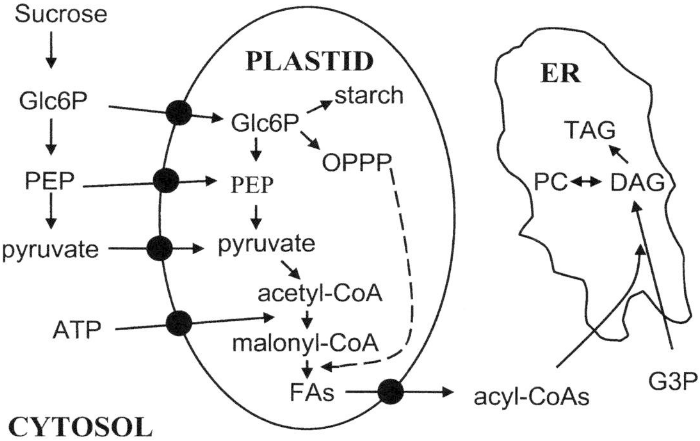

Fig. 3. Generalization of some aspects of carbon metabolism in the cytosol, plastid and endoplasmic reticulum (*ER*) of developing zygotic embryos of *B. napus*. The oxidative pentose phosphate pathway (*OPPP*) may provide reducing power for fatty acid (*FA*) biosynthesis. Transporters in the plastidial envelope are indicated as *black circles*. *ATP* Adenosine triphosphate; *DAG* *sn*-1,2-diacylglycerol; *Glc6P* glucose-6-phospate; *G3P* *sn*-glycerol-3-phosphate; *OPPP* oxidative pentose phosphate pathway; *PC* phosphatidylcholine; *PEP* phospho*enol*pyruvate; *TAG* triacylglycerol. (Adapted from Rawsthorne 2002; White et al. 2000, with permission from *Progress in Lipid Research* and *Plant Physiology*, respectively. The American Society of Plant Biologists holds the copyrights for the original illustration from *Plant Physiology*)

way (OPPP) (Rawsthorne 2002). As the embryo becomes more active in oil accumulation at later stages of development, the flow of Glc6P to starch and FA synthesis decreases while OPPP activity increases. At later stages of embryo development, however, substantial increases in the activity of a plastidial pyruvate transporter have been shown to be concomitant with a major increase in the utilization of pyruvate for FA synthesis by plastids (Eastmond and Rawsthorne 2000). Thus, at later stages of embryo development, there is a metabolic shift, with FA biosynthesis and oil accumulation relying on a different precursor. The metabolic role of transient starch accumulation during development of zygotic embryos, however, remains unclear (Rawsthorne 2002).

Ilić-Grubor et al. (1998) demonstrated that MDEs of *B. napus* L. cv. Topas could be effectively induced in a medium where sucrose was replaced to a very large extent with high-molecular-weight polyethylene glycol (PEG). Sucrose was present in the induction medium at about 0.1 % (w/v), which was considerably lower than the concentrations normally used (8–17 %). The PEG embryos resembled immature zygotic embryos dissected from ovules. The investigators, however, did not examine the lipid accumulation characteristics of the PEG embryos. They suggested that alterations in the carbohydrate composition of MDE growth medium might lead to more synchronous populations of the embryos. Previous investigations have indicated that cultures of MDEs do not exhibit synchronous growth, with many of the embryos only proceeding through early stages of embryogenesis and not continuing onto the cotyledonary stage of development (Pomeroy et al. 1991; Weselake and Taylor 1999).

From Fig. 3, it can be seen that both cytosolic and plastidial glycolysis are operative in plants. Cytosolic and plastidial forms of pyruvate kinase (PK) have been investigated in both MDEs (Sangwan et al. 1992) and MD cell suspension cultures (Smith et al. 2000; Plaxton et al. 2002) of *B. napus*. PK is a regulatory enzyme that catalyzes the irreversible phosphorylation of ADP using phospho*enol*pyruvate (PEP) as a phosphate donor, generating pyruvate and ATP. Dennis and Miernyk (1982) have proposed that the main role of plastidial PK (PK_p) of developing oil seeds is to generate pyruvate and ATP to support FA biosynthesis. Sangwan et al. (1992) investigated the levels of cytosolic PK (PK_c) and PK_p during development of MDEs of *B. napus* L. cv. Topas using an immunochemical approach. The amount of PK_p reached a maximum at about 17 days in culture (at the late cotyledonary stage), followed by a marked decrease at 21 days. The amount of PK_c, which was less abundant than PK_p, remained relatively constant over the 35-day culture period. Also, total PK_p activity was closely associated with the level of PK_p protein throughout development, and the ratio of pyruvate/PEP in 14-day-old MDEs was about four-fold higher than in 7-day-old MDEs. The developmental period during which PK activity, PK_p protein and pyruvate/PEP ratio showed maximal increases corresponded to the active phase of TAG accumulation in MDEs reported by Pomeroy et al. (1991).

The cytosolic form of PK represents the major site for the control of glycolytic flux to pyruvate as precursor to acetyl-CoA production for the mitochondrial tricarboxylic acid (TCA) cycle (Plaxton 1996). Smith et al. (2000) purified PK_c 200-fold to electrophoretic homogeneity from the *B. napus* cell suspension cultures. The purified native enzyme was a 220-kDa homotetramer consisting of 56-kDa subunits. Phospho*enol*pyruvate carboxylase (PEPC) of the cytosol catalyzes the irreversible β-carboxylation of PEP in the presence of the bicarbonate ion and Mg^{2+} to generate oxaloacetate (OAA) and Pi (Plaxton 1996). The purified native enzyme from *B. napus* cell suspension cultures was a homotetramer consisting of identical 104-kDa subunits (Moraes and Plaxton 2000). In C_3 plants, such as *B. napus*, the proposed functions of PEPC are diverse (Chollet et al. 1996). In addition to having a role in the anaplerotic replenishment of OAA for the TCA intermediates consumed in biosyntheses, PEPC has an additional metabolic role in supplying OAA for L-aspartate biosynthesis by aspartate aminotransferase during nitrogen assimilation (Plaxton 1996). The coordination of the regulation of PK_c and PEPC, which both utilize PEP, is essential in the control of glycolytic flux in terms of integrating carbon partitioning with the generation of 2-oxoglutarate needed for nitrogen assimilation by the glutamine synthetase/glutamine 2-oxoglutarate aminotransferase system. Based on studies of the effects of a number of metabolic effectors on PK_c and PEPC activity from *B. napus* cultures, a model was developed that highlighted the central role of L-aspartate and L-glutamate in coordinating allosteric regulation of the PEP branchpoint in vascular plants (Moraes and Plaxton 2000; Smith et al. 2000). The fact that both PK_c and PEPC were subject to feedback inhibition by L-glutamate provided a rationale for the known activation of the two cytosolic enzymes which has been shown to take place in vivo during periods of increased nitrogen assimilation when cellular levels of L-glutamate were low (Huppe and Turpin 1994). L-Aspartate was also an inhibitor of PEPC (Moraes and Plaxton 2000), but served as an activator of PK_c and completely reversed the inhibition of PK_c by L-glutamate (Smith et al. 2000). The differential effects of L-glutamate and L-aspartate on PK_c are interesting considering that the side chains of the two amino acids only differ by one carbon atom. The reciprocal control of PK_c and PEPC by L-aspartate provided a mechanism for reduced flux from PEP to L-aspartate via PEPC and aspartate aminotransferase while promoting PK_c activity when cytosolic L-aspartate levels became elevated. When the cell's requirement for nitrogen was satisfied, protein biosynthesis would become more dependent on the availability of ATP rather than the supply of amino acids. Under these cellular conditions, PK_c would then have a more prominent role in facilitating ATP generation instead of the production of biosynthetic precursors.

Pi plays an important role in cellular metabolism and energetics, especially with regard to carbohydrate metabolism where a number of intermediates are phosphorylated. Pi starvation can lead to substantial decreases in the ATP pools (Plaxton 1999). Inorganic pyrophosphate (PPi) accumulates, however, as a by-product of various biosyntheses and in some cases can replace ATP in some biochemical reactions during abiotic stresses such as anoxia or Pi starvation. For example, the vacuolar H^+-translocating pyrophosphatase (H^+-PPiase) allows plants to utilize PPi to circumvent the ATP-consuming vacuolar H^+-ATPase (Palma et al. 2000). A recent study with MD cell suspension cultures of *B. napus* L. cv. Jet Neuf demonstrated that Pi starvation led to a substantial increase in H^+-PPiase protein levels that was associated with an increase in PPi-dependent tonoplast H^+ transport and PPi-dependent hydrolytic activity (Palma et al. 2000). It was proposed that the H^+-PPiase facilitates the conservation of ATP pools and recycling of Pi during Pi starvation.

Pi starvation has also been shown to cause an increase in PEPC activity in the MD cell suspension cultures of *B. napus* (Moraes and Plaxton 2000). In cells deprived of Pi, increased PEPC, with malic dehydrogenase and NAD-malic enzyme, may facilitate a sustained supply of pyruvate to the TCA cycle while contributing Pi to cellular metabolism (Plaxton and Carswell 1999). Moraes and Plaxton (2000) demonstrated that PEPC existed mainly in the dephosphorylated L-malate sensitive form in both +Pi and −Pi *B. napus* cells. The increased PEPC activity in −Pi cells appeared to be attributable to increased biosynthesis and/or reduced proteolytic turnover of PEPC polypeptide.

4 Storage Protein Biosynthesis in MD Cultures of *B. napus*

Cruciferin, napin and oleosin represent major storage proteins that accumulate during seed development in *B. napus* (Murphy and Cummins 1989; Murphy et al. 1989). Cruciferin is the major storage protein in zygotic embryos, representing about 50% of the total cellular protein (Murphy et al. 1989; Hoglund et al. 1992). Oleosins, which are embedded on the surface of oil bodies, can make up about 10% of the cellular protein (Murphy 2001). Oleosins prevent lipid bodies from coalescing and may serve as receptor sites for TAG lipase action during germination (Huang 1992; Murphy 2001). Gene expression in terms of transcript and polypeptide levels for these storage proteins have been examined during seed development (DeLisle and Crouch 1989; Murphy and Cummins 1989; Murphy et al. 1989; Hoglund et al. 1992; Cummins et al. 1993; Tzen et al. 1993).

In an early study of storage protein accumulation in *B. napus*, Crouch (1982) compared the storage protein content of zygotic embryos with anther-cultured MDEs. Storage proteins accumulated earlier in the development of MDEs than for zygotic embryos, but the final storage protein levels in the MDEs were lower. Treatment of microspores at high temperature (e.g., 32 °C) results in a reprogramming of the microspores to form haploid embryos rather than pollen (Keller and Armstrong 1979; Pechan et al. 1991). Numerous mRNAs and proteins have been shown to appear in microspores during the first 8 h of high-temperature treatment (Pechan et al. 1991). Cordewener et al. (1994) conducted in situ [^{35}S]methionine-labeling of microspores coupled with two-dimensional polyacrylamide gel electrophoresis to assess the effect of high temperature on protein expression. Eighteen polypeptides were shown to incorporate radiolabeled methionine at statistically significant higher levels at 32 °C than control microspores at 18 °C. In the analysis of napin and cruciferin gene expression in MDEs of *B. napus* L. cv. Topas, Boutilier et al. (1994) demonstrated that napin transcripts correlated with the induction of embryogenesis and suggested that napin transcripts might serve as an early marker of the process. In addition to monitoring napin transcript levels, the investigators induced embryogenesis in transgenic microspores containing a β-glucuronidase marker gene fused to a promoter of a napin gene. The promoter was strongly induced in the microspores that had morphogenetic characteristics associated with embryogenesis. Cruciferin gene expression was initiated at a later stage of development and cruciferin mRNA accumulated to a lower level than napin mRNA. Napin and cruciferin transcript levels remained high for the remainder of MDE development. In contrast, mRNA levels for these proteins in zygotic embryos have been shown to decline at later stages of development (DeLisle and Crouch 1989).

Treatment of isolated zygotic embryos of *B. napus* with exogenous ABA or increased osmoticum has been shown to lead to increases in transcript levels for cruciferin and napin (Finkelstein et al. 1985; Finkelstein and Crouch

1986). Also, the application of ABA (1–10 μM) to MDEs of *B. napus* increased transcript levels for these storage proteins (Taylor et al. 1990; Wilen et al. 1990). The effect of exogenous ABA (10 μM) on stimulating the production of cruciferin protein, however, was considerably less marked than the transcript accumulation for this protein (Taylor et al. 1990). Treatment of MDEs of *B. napus* with high osmoticum (due to 12.5% sorbitol) also led to increased transcript levels for cruciferin and napin, but the ABA effect occurred more rapidly, suggesting that the osmotic effect could be mediated by ABA that was formed in response to the lower water potential (Wilen et al. 1990). ABA levels were increased about six-fold within 2 h of culture in high osmoticum. Application of jasmonic acid, a growth regulator derived from linolenic acid (Meyer et al. 1984; Vick and Zimmerman 1984), has also been shown to cause increased napin transcript levels in MDEs of *B. napus* (Hays et al. 1999). Jasmonic acid is known to be a natural product in *Brassica* embryos (Wilen et al. 1991).

Treatment with ABA or high osmoticum has also been shown to affect the expression of lipid body proteins in both MDEs (Taylor et al. 1990; Holbrook et al. 1991; van Rooijen et al. 1992). The structure, biosynthesis and incorporation of the 20-kDa oleosin into lipid bodies has been the subject of a number of investigations (Huang 1992; Murphy 2001). In Western blotting studies, treatment of MDEs with ABA, jasmonic acid or 12.5% sorbitol resulted in an increase in oleosins in the 20- to 23-kDa range (Holbrook et al. 1991). The response of oleosin synthesis to ABA was found at all developmental stages (Holbrook et al. 1991) and was in agreement with observations on transcript levels (van Rooijen et al. 1992). In contrast, the synthesis of a 32-kDa lipid body protein was inhibited by treatment with ABA or sorbitol, but synthesis of this protein was unaffected by jasmonic acid. Hays et al. (1999) suggested that the effects of ABA on expression of oleosin and napin might be modulated by jasmonic acid.

Increases in transcript levels for the 20-kDa oleosin transcript have also been noted in MD cell suspension cultures of *B. napus* treated with ABA or 13% sorbitol (Crowe et al. 2000) or increasing sucrose concentrations (2, 6 and 14%) (Weselake et al. 1998). Transient expression of the seed-specific activator ABI3, using a biolistic approach, has been shown to promote oleosin gene expression in the MD cell suspension cultures (Crowe et al. 2000). The effect was demonstrated by coexpressing an oleosin promoter-GUS fusion and a full-length ABI3 gene from *Arabidopsis*.

The appearance of late embryogenesis abundant (Lea) proteins during seed development suggests that these proteins may have a role in conferring desiccation tolerance to the seed (Skriver and Mundy 1990). In Section 2.1, it was indicated that treatment of MDEs with ABA induced desiccation tolerance. In an early study with MDEs of *B. napus*, Johnson-Flanagan et al. (1992) reported that increases in desiccation tolerance did not correspond to an increase in the Lea.76 transcript. In contrast, a more recent study by Wakui and Takahata (2002) reported that Lea proteins and transcripts do appear to be strongly associated with desiccation tolerance in MDEs of *B. napus*.

5 Glucosinolate Biosynthesis in MDEs of *B. napus*

Glucosinolates are thioglucosides that are common to a number of families of dicotyledonous plants including Brassicaceae such as *B. napus* (Gijzen et al. 1994; Wittstock and Halkier 2002). Glucosinolates are often concentrated in seeds, constituting as high as 10% of the seed mass (van Etten et al. 1974). When tissue that contains glucosinolates is disrupted, the thioglucosides come in contact with myrosinase (thioglucoside glucohydrolase), which catalyzes their hydrolysis, leading to a number of degradation products including glucose, isothiocyanates, thiocyanates and nitriles. In the case of high-glucosinolate *B. napus*, the presence of glucosinolate degradation products in the meal can be toxic to livestock. It should be noted, however, that the presence of glucosinolates in plant tissue could also impart anti-microbial properties to the plant and serve a nutraceutical function as anti-cancer agents (Wittstock and Halkier 2002).

The accumulation of glucosinolates in the embryo stops when the embryo is separated from the maturing seed (Gijzen et al. 1989). Based on studies with zygotic embryos of *B. napus*, Gijzen et al. (1994) reported that reduced glucosinolate content in seeds was related to the reduced glucosinolate content of the maternal plant and to differences in the rate of glucosinolate uptake by the embryo. MDEs of *B. napus* are useful in examining glucosinolate accumulation because metabolic studies can be conducted without the influence of the maternal tissue. Two studies have shown that transferring MDEs to a germination medium containing a lower concentration of sucrose that favors root and shoot growth leads to an induction of glucosinolate biosynthesis (McClellan et al. 1993; Iqbal et al. 1995). Substitution of sucrose with an equimolar amount of non-metabolizable osmoticum such as mannitol substantially suppressed glucosinolate accumulation, indicating that the drop in osmotic potential was a key factor in altering secondary metabolism in favor of glucosinolate production (McClellan et al. 1993). Iqbal et al. (1995) further demonstrated that plantlets generated from MDEs of high glucosinolate genotypes produce increased amounts of alkenyl glucosinolates when compared to plantlets derived from MDEs of low glucosinolate genotypes. Glucosinolate analysis of in vitro plantlets could be used to select for low glucosinolate tissue in segregating doubled haploid populations at early stages of culture in breeding programs (McClellan et al. 1993; Iqbal et al. 1995). Recently, Iqbal and Möllers (2003) reported that MDEs of high and low glucosinolate cultivars were equally effective in the uptake of the alkenyl glucosinolate sinigrin. The investigators suggested that monitoring the extent of sinigrin uptake might also be useful in breeding programs aimed at reducing glucosinolate content.

In recent years, a number of advances in understanding glucosinolate biosynthesis have come through the use of functional genomics approaches with *Arabidopsis* (Wittstock and Halkier 2002). These studies have set the foundation for metabolic engineering of glucosinolate biosynthesis to enhance both nutritional quality and crop protection.

6 Conclusions and Future Directions

MD cultures of Brassicaceae, particularly *B. napus*, have contributed to increasing our understanding of storage product metabolism in developing oilseeds. The greatest advances have been in the enzymology of TAG biosynthesis where research has been conducted with both MDEs and MD cell suspension cultures of *B. napus*. Working with MDEs is more convenient than using zygotic embryos because limited growth chamber facilities are required to produce plants for a source of microspores, thereby eliminating the need for extensive greenhouse space. From a practical point of view, MDEs will continue to be valuable in accelerating *Brassica* spp. breeding programs for obtaining desired seed characteristics. By using MDEs for metabolic work, we avoid having to collect seeds at different stages of development, followed by dissections that can be potentially damaging to zygotic embryos. MDEs have been shown to have lipid accumulation characteristics that are similar to zygotic embryos and investigations of nutrient/precursor uptake and osmotic effects can be examined without the influence of the maternal liquid endosperm. MDEs have been particularly useful in examining FA elongation, and biosynthesis of lipid body proteins and glucosinolates. In the future, proteomics-based approaches should contribute to developing insight into the reprogramming of microspores to produce embryos, oil biosynthesis in specialized membrane fractions and the onset of desiccation tolerance. Also, MDEs produced from transgenic plants should continue to prove useful in more rapid investigations of the effects of modifying the expression of specific genes on seed metabolism. MD cell suspension cultures feature the added advantage of only needing a temperature-controlled incubator(s) with lighting for culture maintenance. MD cell suspension cultures of *B. napus* L. cv. Jet Neuf have been shown to synthesize and degrade TAG in the absence of the complexities of gene expression associated with tissue undergoing embryogenesis, and should prove useful in future investigations of the factors that regulate these processes. Further attempts should be made to develop MD cell suspension cultures from other cultivars of *B. napus* and *B. rapa*. MD cell suspension cultures have proven particularly useful in examining the regulation of oilseed carbohydrate metabolism as it relates to nitrogen metabolism and phosphate starvation. Currently, we are using the MD cell suspension cultures to examine the effects of low temperature and osmotic stress on gene expression associated with lipid metabolism.

Acknowledgements I would like to thank the following agencies for supporting our research on plant lipid biochemistry and molecular biology: the Alberta Crop Industry Development Fund, Alberta Agricultural Research Institute, Alberta Network for Proteomics Innovation, Flax Council of Canada, Genome Prairie and Genome Canada, Imperial Oil Ltd., Natural Sciences and Engineering Research Council of Canada, Saskatchewan Agriculture Development Fund, University of Lethbridge and Western Economic Diversification. Also, I thank Nancy Fineberg and Tara Furukawa-Stoffer for their critical evaluation of the manuscript. In addition, I thank Doug Bray and Keri Colwell for imaging of cell suspension cultures by transmission electron microscopy.

References

Albrecht S, Möllers C, Röbbelen G (1994) Selection for fatty acid composition in microspore-derived embryoids (MDE) of rapeseed, *Brassica napus* (L.). J Plant Physiol 143:526–529

Albrecht S, Möllers C, Röbbelen G (1995) Selection in vitro for erucic-acid content in segregating populations of microspore-derived embryoids of *Brassica napus*. Plant Breed 114:210–214

Baldo A, Sniderman AD, St. Luce S, Zhang X-J, Cianflone K (1995) Signal transduction pathway of acylation stimulating protein: involvement of protein kinase C. J Lipid Res 36:1415–1426

Bernerth R, Frentzen M (1990) Utilization of erucoyl-CoA by acyltransferases from developing seeds of *Brassica napus* (L.) involved in triacylglycerol biosynthesis. Plant Sci 67:21–28

Blacklock BJ, Jaworski JG (2002) Studies into factors contributing to substrate specificity of membrane-bound 3-ketoacyl-CoA synthases. Eur J Biochem 269:4789–4798

Boutilier KM, Giné M-J, DeMoor JM, Huang B, Baszczynski CL, Iyer VN, Miki BL (1994) Expression of the BnmNAP subfamily of napin genes coincides with the induction of *Brassica* microspore embryogenesis. Plant Mol Biol 26:1711–1723

Bouvier-Navé P, Benveniste P, Oelkers P, Sturley SL, Schaller H (2000) Expression in yeast and tobacco of plant cDNAs encoding acyl CoA:diacylglycerol acyltransferase. Eur J Biochem 267:85–96

Brown DCW, Watson EM, Pechan PM (1993) Induction of dessication tolerance in microspore-derived embryos of *Brassica napus*. In Vitro Cell Dev Biol 29P:113–118

Byers SD, Laroche A, Smith KC, Weselake RJ (1999) Factors enhancing diacylglycerol acyltransferase activity in microsomes from cell-suspension cultures of oilseed rape. Lipids 34:1143–1149

Cao YZ, Huang AHC (1987) Acyl coenzyme A preference of diacylglycerol acyltransferase from the maturing seeds of *Cuphea*, maize, rapeseed, and canola. Plant Physiol 84:762–765

Chen JL, Beversdorf WD (1991) Evaluation of microspore-derived embryos as models for studying lipid biosynthesis in seed of rapeseed (*Brassica napus* L.). Euphytica 58:145–155

Chollet R, Vidal J, O'Leary MH (1996) Phospho*enol*pyruvate carboxylase: a ubiquitous, highly regulated enzyme in plants. Annu Rev Plant Physiol Plant Mol Biol 47:273–298

Cordewener JHG, Busink R, Traas JA, Custers JBM, Dons HJM, van Lookeren Campagne MM (1994) Induction of microspore embryogenesis in *Brassica napus* L. is accompanied by specific changes in protein synthesis. Planta 195:50–56

Crowe AJ, Abenes M, Plant A, Moloney MM (2000) The seed-specific transactivator, ABI3, induces oleosin gene expression. Plant Sci 151:171–181

Crouch ML (1982) Non-zygotic embryos of *Brassica napus* L. contain embryo-specific storage proteins. Planta 156:520–524

Cummins I, Hills MJ, Ross JHE, Hobbs DH, Watson MD, Murphy DJ (1993) Differential, temporal and spatial expression of genes involved in storage oil and oleosin accumulation in developing rapeseed embryos: implications for the role of oleosins and the mechanisms of oil-body formation. Plant Mol Biol 23:1015–1027

Dahlqvist A, Ståhl U, Lenman MA, Lee M, Sandager L, Ronne H, Stymne S (2000) Phospholipid:diacylglycerol acyltransferase: an enzyme that catalyzes the acyl-CoA-independent formation of triacylglycerol in yeast and plants. Proc Natl Acad Sci USA 97:6487–6492

Davoren JD, Nykiforuk CL, Laroche A, Weselake RJ (2002) Sucrose-induced changes in the transcriptome of cell suspension cultures of oilseed rape reveal genes associated with lipid biosynthesis. Plant Physiol Biochem 40:719–725

DeLisle AJ, Crouch ML (1989) Seed storage protein transcription and mRNA levels in *Brassica napus* during development and in response to exogenous abscisic acid. Plant Physiol 91:617–623

Dennis DT, Miernyk JA (1982) Compartmentation of nonphotosynthetic carbohydrate metabolism. Annu Rev Plant Physiol 33:27–50

Downey RK, Craig BM (1964) Genetic control of fatty acid biosynthesis in rapeseed (*Brassica napus* L.). J Am Oil Chem 41:475–478

Eastmond PJ, Rawsthorne S (2000) Coordinate changes in carbon partitioning and plastidial metabolism during the development of oilseed rape embryos. Plant Physiol 122:767–774

Fehling E, Mukherjee KD (1991) Acyl-CoA elongase from a higher plant (*Lunaria annua*): metabolic intermediates of very-long-chain acyl-CoA products and substrate specificity. Biochim Biophys Acta 1082:239–246

Ferrie AMR, Taylor DC, MacKenzie SL, Keller WA (1999) Microspore embryogenesis of high *sn*-2 erucic acid *Brassica oleracea* germplasm. Plant Cell Tiss Org Cult 57:79–84

Finkelstein RR, Crouch ML (1986) Rapeseed embryo development in culture on high osmoticum is similar to that in seeds. Plant Physiol 81:907–912

Finkelstein R, Somerville C (1989) Abscisic acid or high osmoticum promote accumulation of long-chain fatty acids in developing embryos of *Brassica napus*. Plant Sci 61:213–217

Finkelstein RR, Tenbarge KM, Shumway JE, Crouch ML (1985) Role of ABA in maturation of rapeseed embryos. Plant Physiol 78:630–636

Finlayson SA, Dennis DT (1980) NAD^+-specific glycerol 3-phosphate dehydrogenase from developing castor bean endosperm. Arch Biochem Biophys 199:179–185

Furukawa-Stoffer TL, Byers SD, Hodges DM, Laroche A, Weselake RJ (1998) Identification of *N*-ethylmaleimide-sensitive and -insensitive phosphatidate phosphatase activity in microspore-derived cultures of oilseed rape. Plant Sci 131:139–147

Furukawa-Stoffer TL, Boyle RM, Thomson AL, Sarna MA, Weselake RJ (2003) Properties of lysophosphatidylcholine acyltransferase from *Brassica napus* cultures. Lipids 38:651–656

Ghanevati M, Jaworski JG (2002) Engineering and mechanistic studies of the *Arabidopsis* FAE1 β-ketoacyl-CoA synthase, FAE1 KCS. Eur J Biochem 269:3531–3539

Gijzen M, McGregor I, Ségun-Swartz G (1989) Glucosinolate uptake by developing rapeseed embryos. Plant Physiol 86:260–263

Gijzen M, Ségun-Swartz G, McGregor I (1994) Glucosinolate metabolism in rapeseed embryos: effect of feeding glucosinolate precursors and uptake of glucosinolate by different plant cultivars. J Plant Physiol 144:17–21

Hays DB, Wilen RW, Sheng C, Moloney MM, Pharis RP (1999) Embryo-specific gene expression in microspore-derived embryos of *Brassica napus*. An interaction between abscisic acid and jasmonic acid. Plant Physiol 119:1065–1072

Hill LM, Morley-Smith ER, Rawsthorne S (2003) Metabolism of sugars in the endosperm of developing seeds of oilseed rape. Plant Physiol 131:228–236

Hobbs DH, Lu C, Hills MJ (1999) Cloning of a cDNA encoding diacylglycerol acyltransferase from *Arabidopsis thaliana* and its functional expression. FEBS Lett 452:145–149

Hoglund AS, Rodin J, Larsson E, Rask L (1992) Distribution of napin and cruciferin in developing rape seed embryos. Plant Physiol 98:509–515

Holbrook LA, van Rooijen GJH, Wilen RW, Moloney MM (1991) Oilbody proteins in microspore-derived embryos of *Brassica napus*. Plant Physiol 97:1051–1058

Holbrook LA, Magus JR, Taylor DC (1992) Abscisic acid induction of elongase activity, biosynthesis and accumulation of very long chain monounsaturated fatty acids and oil body proteins in microspore-derived embryos of *Brassica napus* L. cv. Reston. Plant Sci 84:99–115

Huang AHC (1992) Oil bodies and oleosins in seeds. Annu Rev Plant Physiol Plant Mol Biol 43:177–200

Huppe H, Turpin DH (1994) Integration of carbon and nitrogen metabolism in plant and algal cells. Annu Rev Plant Physiol Plant Mol Biol 45:577–607

Ilić-Grubor K, Attree SM, Fowke LC (1998) Induction of microspore-derived embryos of *Brassica napus* L. with polyethylene glycol (PEG) as osmoticum in a low sucrose medium. Plant Cell Rep 17:329–333

Iqbal MC, Möllers C (2003) Uptake and distribution of sinigrin in microspore derived embryos of *Brassica napus* L. J Plant Physiol 160:961–966

Iqbal MCM, Röbbelen R, Möllers C (1995) Biosynthesis of glucosinolates by microspore derived embryoids and plantlets in vitro of *Brassica napus* L. Plant Sci 112:107–115

James Jr DW, Lim E, Keller J, Plooy I, Ralston E, Dooner HK (1995) Directed tagging of the *Arabidopsis FATTY ACID ELONGATION1* (*FAE1*) gene with maize transposon *activator*. Plant Cell 7:309–319

Johnson-Flanagan AM, Singh J (1987) Alteration of gene expression during the induction of freezing tolerance in *Brassica napus* suspension cultures. Plant Physiol 85:699–705

Johnson-Flanagan AM, Huiwen Z, Thiagarajah MR, Saini HS (1991) Role of abscisic acid in the induction of freezing tolerance in *Brassica napus* suspension-cultured cells. Plant Physiol 95:1044–1048

Johnson-Flanagan AM, Huiwen Z, Geng X-M, Brown DC, Nykiforuk CL, Singh J (1992) Frost, abscisic acid, and desiccation hasten embryo development in *Brassica napus*. Plant Physiol 99:700–706

Katavic V, Reed DW, Taylor DC, Giblin EM, Barton DL, Zou J, MacKenzie SL, Covello PS, Kunst L (1995) Alteration of seed fatty acid composition by an ethyl methanesulfonate-induced mutation in *Arabidopsis thaliana* affecting diacylglycerol acyltransferase activity. Plant Physiol 108:399–409

Katavic V, Mietkiewska E, Barton DL, Giblin EM, Reed DW, Taylor DC (2002) Restoring enzyme activity in nonfunctional low erucic acid *Brassica napus* fatty acid elongase 1 by a single amino acid substitution. Eur J Biochem 269:5625–5631

Keller WA, Armstrong KC (1979) Stimulation of embryogenesis and haploid production in *Brassica campestris* anther cultures by elevated temperature treatments. Theor Appl Genet 55:65–67

Kennedy EP (1961) Biosynthesis of complex lipids. Fed Proc Am Soc Exp Biol 20:934–940

Kocsis MG, Weselake RJ (1996) Phosphatidate phosphatases of mammals, yeast, and higher plants. Lipids 31:785–802

Kocsis MG, Weselake RJ, Eng JA, Furukawa-Stoffer TL, Pomeroy MK (1996) Phosphatidate phosphatase from developing seeds and microspore-derived cultures of *Brassica napus*. Phytochemistry 41:353–363

Lacey DJ, Hills MJ (1996) Heterogeneity of the endoplasmic reticulum with respect to lipid synthesis in developing seeds of *Brassica napus* L. Planta 199:545–551

Lacey DJ, Beaudoin F, Dempsey CE, Shewry PR, Napier JA (1999). The accumulation of triacylglycerols within the endoplasmic reticulum of developing seeds of *Helianthus annuus*. Plant J 17:397–405

Lardizabal KD, Mai JT, Wagner NW, Wyrick A, Voelker T, Hawkins DJ (2001) DGAT2 is a new diacylglycerol acyltransferase gene family. J Biol Chem 276:38862–38869

Little D, Weselake R, Pomeroy K, Furukawa-Stoffer T, Bagu J (1994) Solubilization and characterization of diacylglycerol acyltransferase from microspore-derived cultures of oilseed rape. Biochem J 304:951–958

Mancha M, Stymne S (1997) Remodelling of triacylglycerols in microsomal preparations from developing castor bean (*Ricinus communis* L.) endosperm. Planta 203:51–57

McClellan D, Kott L, Beversdorf W, Ellis BE (1993) Glucosinolate metabolism in zygotic and microspore-derived embryos of *Brassica napus* L. J Plant Physiol 141:153–159

Meyer A, Miersch O, Buttner C, Dathe W, Sembdner G (1984) Occurrence of plant growth regulator jasmonic acid in plants. J Plant Growth Regul 3:1–8

Millar AA, Kunst L (1997) Very-long-chain fatty acid biosynthesis is controlled through the expression and specificity of the condensing enzyme. Plant J 12:121–131

Möllers C, Albrecht S, Röbbelen GJ (1994) Effect of in vitro culture conditions on fatty acid desaturation in microspore-derived embryoids of *Brassica napus*. Plant Physiol 143:530–533

Möllers C, Rücker B, Stelling D, Schierholt A (2000) In vitro selection for oleic and linoleic acid content in segregating populations of microspore derived embryos of *Brassica napus*. Euphytica 112:195–201

Moraes TF, Plaxton WC (2000) Purification and characterization of phospho*enol*pyruvate carboxylase from *Brassica napus* (rapeseed) suspension cell cultures. Implications for phosphoenolpyruvate carboxylase regulation during phosphate starvation, and the integration of glycolysis with nitrogen assimilation. Eur J Biochem 267:4465–4476

Murphy DJ (2001) The biogenesis and functions of lipid bodies in animals, plants and microorganisms. Prog Lipid Res 40:325–438

Murphy DJ (2002) Biotechnology and the improvement of oil crops – genes, dreams and realities. Phytochem Rev 1:67–77

Murphy DJ, Cummins I (1989) Biosynthesis of seed storage products during embryogenesis in rapeseed, *Brassica napus*. J Plant Physiol 135:63–69

Murphy DJ, Cummins I, Kang AS (1989) Synthesis of the major oil-body membrane protein in developing rapeseed (*Brassica napus*) embryos. Integration with storage-lipid and storage-protein synthesis and implications for the mechanism of oil-body formation. Biochem J 258:285–293

Norton G, Harris JF (1983) Triacylglycerols in oilseed rape during seed development. Phytochemistry 22:2703–2707

Nykiforuk CL, Laroche A, Weselake RJ (1999a) Isolation and sequence analysis of a novel cDNA encoding a putative diacylglycerol acyltransferase from a microspore-derived cell suspension culture of *Brassica napus* L. cv. Jet Neuf (accession no AF155224) (PGR99–123). Plant Physiol 120:1207

Nykiforuk CL, Laroche A, Weselake RJ (1999b) Isolation and characterization of a cDNA encoding a second putative diacylglycerol acyltransferase from a microspore-derived cell suspension culture of *Brassica napus* L. cv. Jet Neuf (accession no AF164434) (PGR99–158). Plant Physiol 121:1957

Nykiforuk CL, Furukawa-Stoffer TL, Huff PW, Sarna M, Laroche A, Moloney MM, Weselake RJ (2002) Characterization of cDNAs encoding diacylglycerol acyltransferase from cultures of *Brassica napus* and sucrose-mediated induction of enzyme biosynthesis. Biochim Biophys Acta 1580:95–109

Ohlrogge JB, Browse J (1995) Lipid biosynthesis. Plant Cell 7:957–970

Orr W, Keller WA, Singh J (1986) Induction of freezing tolerance in an embryogenic cell suspension culture of *Brassica napus* by abscisic acid at room temperature. J Plant Physiol 126:23–32

Palma DA, Blumwald E, Plaxton WC (2000) Upregulation of vacuolar H^+-translocating pyrophosphatase by phosphate starvation of *Brassica napus* (rapeseed) suspension cell cultures. FEBS Lett 486:155–158

Pechan PM, Bartels D, Brown DCW, Schell J (1991) Messenger-RNA and protein changes associated with induction of *Brassica* microspore embryogenesis. Planta 184:161–165

Perry HJ, Bligny R, Gout E, Harwood JL (1999) Changes in Kennedy pathway intermediates associated with increased triacylglycerol synthesis in oilseed rape. Phytochemistry 52:799–804

Plaxton WC (1996) The organization and regulation of plant glycolysis. Annu Rev Plant Physiol Plant Mol Biol 47:185–214

Plaxton WC (1999) Metabolic aspects of the phosphate starvation in plants. In: Deikman J, Lynch J (eds) Phosphorus in plant biology: regulatory roles in molecular, cellular, organismic, and ecosystem processes. American Society of Plant Physiologists, Rockville, Maryland, pp 164–176

Plaxton WC, Carswell MC (1999) Metabolic aspects of the phosphate starvation response in plants. In: Lehner HR (ed) Plant responses to environmental stresses: from phytohormones to genome reorganization. Dekker, New York, pp 349–372

Plaxton WC, Smith CR, Knowles VL (2002) Molecular and regulatory properties of leucoplast pyruvate kinase from *Brassica napus* (rapeseed) suspension cells. Arch Biochem Biophys 400:54–62

Pomeroy MK, Kramer JKG, Hunt DJ, Keller WA (1991) Fatty acid changes during development of zygotic and microspore-derived embryos of *Brassica napus*. Physiol Plant 81:447–454

Pomeroy K, Brown DCW, Takahata Y (1994) Response of *Brassica napus* L. microspore-derived embryos to exogenous abscisic acid and dessication. In Vitro Cell Dev Biol 30P:196–203

Princen LH, Rothfus JA (1984) Development of new crops for industrial raw materials. J Am Oil Chem Soc 61:281–289

Qi Q, Rose PA, Abrams GD, Taylor DC, Abrams SR, Cutler AJ (1998) (+)-Abscisic acid metabolism, 3-ketoacyl-coenzyme A synthase gene expression, and very-long-chain monounsaturated fatty acid biosynthesis in *Brassica napus* embryos. Plant Physiol 117:979–987

Rawsthorne S (2002) Carbon flux and fatty acid synthesis in plants. Prog Lipid Res 41:182–196

Roscoe T, Maisonneuve S, Delseny M (2002) Production of unusual fatty acids in rapeseed. Oleagineux Corps Gras Lipides 9:24–30

Routaboul J-M, Benning C, Bechtold N, Caboche M, Lepiniec L (1999) The TAG1 locus of *Arabidopsis* encodes for a diacylglycerol acyltransferase. Plant Physiol Biochem 37:831–840

Sangwan RS, Gauthier DA, Turpin DH, Pomeroy MK, Plaxton WC (1992) Pyruvate-kinase isoenzymes from zygotic and microspore-derived embryos of *Brassica napus*. Developmental profiles and subunit composition. Planta 187:198–202

Schwender J, Ohlrogge J (2002) Probing in vivo metabolism by stable isotope labeling of storage lipids and proteins in developing *Brassica napus* embryos. Plant Physiol 130:347–361

Senaratna T, Kott L, Beversdorf WD, McKersie BD (1991) Dessication of microspore derived embryos of oilseed rape (*Brassica napus* L.). Plant Cell Rep 10:342–344

Shanklin J, Cahoon EB (1998) Desaturation and related modifications of fatty acids. Annu Rev Plant Physiol Plant Mol Biol 49:611–641

Simmonds DH, Long NE, Keller WA (1991) High plating efficiency and plant regeneration frequency in low density protoplast cultures derived from an embryogenic *Brassica napus* cell suspension. Plant Cell Tissue Organ Cult 27:231–241

Skriver K, Mundy J (1990) Gene expression in response to abscisic acid and osmotic stress. Plant Cell 2:503–512

Smith CR, Knowles VL, Plaxton WC (2000) Purification and characterization of cytosolic pyruvate kinase from *Brassica napus* (rapeseed) suspension cell cultures. Implications for the integration of glycolysis with nitrogen assimilation. Eur J Biochem 267:4477–4485

Ståhl U, Bánas ÑA, Stymne S (1995) Plant microsomal phospholipid acyl hydrolases have selectivities for uncommon fatty acids. Plant Physiol 107:953–962

Stobart AK, Stymne S (1990) Triacylglycerol biosynthesis. Methods Plant Biochem 4:19–46

Stobart K, Mancha M, Lenman M, Dahlqvist A, Stymne S (1997) Triacylglycerols are synthesized and utilized by transacylation reactions in microsomal preparations of developing safflower (*Carthamus tinctorius* L.) seeds. Planta 203:58–66

Stumpf PK, Pollard MR (1983) Pathways of fatty acid biosynthesis in higher plants with particular reference to developing rapeseed. In: Kramer KG, Sauer FD, Pidgen WJ (eds) High and low erucic acid rapeseed oils. Academic Press, New York, pp 131–141

Stymne S, Stobart AK (1987) Triacylglycerol biosynthesis. In: Stumpf PK (ed) Triacylglycerol biosynthesis. The biochemistry of plants, vol 9. Lipids: structure and function. Academic Press, New York, pp 175–214

Sun C, Cao Y-Z, Huang AHC (1988) Acyl coenzyme A preference of the glycerol phosphate pathway in the microsomes from the maturing seeds of palm, maize, and rapeseed. Plant Physiol 88:56–60

Taylor DC, Weber N (1994) Microspore-derived embryos of the Brassicaceae – model system for studies of storage lipid bioassembly and its regulation. Fat Sci Technol 96:228–235

Taylor DC, Weber N, Underhill EW, Pomeroy MK, Keller WA, Scowcroft WR, Wilen RW, Moloney MM, Holbrook LA (1990) Storage-protein regulation and lipid accumulation in microspore embryos of *Brassica napus* L. Planta 181:18–26

Taylor DC, Weber N, Barton DL, Underhill EW, Hogge LR, Weselake RJ, Pomeroy MK (1991) Triacylglycerol bioassembly in microspore-derived embryos of *Brassica napus* L. cv. Reston. Plant Physiol 97:65–79

Taylor DC, Barton DL, Rioux KP, MacKenzie SL, Reed DW, Underhill EW, Pomeroy KM, Weber N (1992a) Biosynthesis of acyl lipids containing very-long chain fatty acids in microspore-derived and zygotic embryos of *Brassica napus* L. cv. Reston. Plant Physiol 99:1609–1618

Taylor DC, Weber N, Hogge LR, Underhill EW, Pomeroy MK (1992b) Formation of trierucoylglycerol (trierucin) from 1,2-dierucoylglycerol by a homogenate of microspore-derived embryos of *Brassica napus* L. J Am Oil Chem Soc 69:355–358

Taylor DC, Ferrie AMR, Keller WA, Gibline EM, Pass EW, MacKenzie SL (1993) Bioassembly of acyl lipids in microspore-derived embryos of *Brassica campestris* L. Plant Cell Rep 12:375–384

Taylor DC, MacKenzie SL, McCurdy AR, McVetty PBE, Giblin EM, Pass EW, Stone SJ, Scarth R, Rimmer SR, Pickard MD (1994) Stereospecific analyses of seed triacylglycerols from high-erucic acid Brassicaceae: detection of erucic acid at the *sn*-2 position in *Brassica oleracea* L. genotypes. J Am Oil Chem Soc 71:163–167

Tzen JTC, Cao Y-Z, Laurent P, Ratnayake C, Huang AHC (1993) Lipids, proteins, and structure of seed oil bodies from diverse species. Plant Physiol 101:267–276

Van Etten CH, McGrew CE, Daxenbichler ME (1974) Glucosinolate determination in cruciferous seeds and meals by measurement of enzymatically released glucose. J Agric Food Chem 22:483–487

Van Rooijen GJH, Wilen RW, Holbrook LA, Moloney MM (1992) Regulation of accumulation of mRNAs encoding a 20-kDa oil-body protein in microspore-derived embryos of *Brassica napus*. Can J Bot 70:503–508

Vick BA, Zimmerman DC (1984) Biosynthesis of jasmonic acid by several plant species. Plant Physiol 75:458–461

Vogel G, Browse J (1996) Cholinephosphotransferase and diacylglycerol acyltransferase. Substrate specificities at a key branch point in seed lipid metabolism. Plant Physiol 110:923–931

Wakui K, Takahata Y (2002) Isolation and expression of *Lea* gene in desiccation-tolerant microspore-derived embryos of *Brassica* spp. Physiol Plant 116:223–230

Weber N, Taylor DC, Underhill EW (1992) Biosynthesis of storage lipids in plant cell and embryo cultures. In: Fiechter A (ed) Advances in biochemical engineering/biotechnology, vol 45. Springer, Berlin Heidelberg New York, pp 99–131

Weselake RJ (2000a) Lipid biosynthesis in cultures of oilseed rape. In Vitro Cell Dev Biol Plant 36:338–348

Weselake RJ (2000b) The slippery roads to oil and fat. Inform 11:1281–1286

Weselake RJ (2002) Biochemistry and biotechnology of plant triacylglycerol glycerol biosynthesis. In: Kuo TM, Gardner HW (eds) Lipid biotechnology. Dekker, New York, pp 27–56

Weselake RJ, Taylor DC (1999) The study of storage lipid biosynthesis using microspore-derived cultures of oilseed rape. Prog Lipid Res 38:401–460

Weselake RJ, Taylor DC, Pomeroy MK, Lawson SL, Underhill EW (1991) Properties of diacylglycerol acyltransferase from microspore-derived embryos of *Brassica napus*. Phytochemistry 30:3533–3538

Weselake RJ, Pomeroy MK, Furukawa TL, Golden JL, Little DB, Laroche A (1993) Developmental profile of diacylglycerol acyltransferase in maturing seeds of oilseed rape. Plant Physiol 102:565–571

Weselake RJ, Davoren JM, Byers SD, Laroche A, Hodges DM, Pomeroy MK (1997) Differential display of mRNA from oil-forming cell suspension cultures of *Brassica napus*. In: Williams JP, Khan MU, Lem NW (eds) Physiology, biochemistry and molecular biology of plant lipids. Kluwer, Boston, pp 357–359

Weselake RJ, Byers SD, Davoren JM, Laroche A, Hodges DM, Pomeroy MK, Furukawa-Stoffer TL (1998) Triacylglycerol biosynthesis and gene expression in microspore-derived cell suspension cultures of oilseed rape. J Exp Bot 49:33–39

Weselake RJ, Kazala EC, Cianflone K, Boehr DD, Middleton CK, Renni CD, Laroche A, Recnik I (2000a) Human acylation stimulating protein enhances triacylglycerol biosynthesis in plant microsomes. FEBS Lett 481:189–192

Weselake RJ, Nykiforuk CL, Laroche A, Patterson NA, Wiehler WB, Szarka SJ, Moloney MM, Tari LW, Derekh U (2000b) Expression and properties of diacylglycerol acyltransferase from cell-suspension cultures of oilseed rape. Biochem Soc Trans 28:684–686

White JA, Todd J, Newman T, Focks N, Girke T, de Ilárduya OM, Jaworski JG, Ohlrogge JB, Benning C (2000) A new set of *Arabidopsis* expressed sequence tags from developing seeds. The metabolic pathway from carbohydrates to seed oil. Plant Physiol 124:1582–1594

Wiberg E, Råhlen L, Hellman M, Tillberg E, Glimelius K, Stymne S (1991) The microspore-derived embryo of *Brassica napus* L. as a tool for studying embryo-specific lipid biogenesis and regulation of oil quality. Theor Appl Genet 82:515–520

Wilen RW, Mandel RM, Pharis RP, Holbrook LA, Moloney MM (1990) Effects of abscisic acid and high osmoticum on storage protein gene expression in microspore embryos of *Brassica napus*. Plant Physiol 94:875–881

Wilen RW, van Rooijen GJH, Pearce DW, Pharis RP, Holbrook LA, Moloney MM (1991) Effects of jasmonic acid on embryo-specific processes in *Brassica* and *Linum* oilseeds. Plant Physiol 95:399–405

Wilmer JA, Helsper JPFG, van der Plas LHW (1996) Effect of growth temperature on erucic acid levels in seeds and microspore-derived embryos of oilseed rape, *Brassica napus* L. J Plant Physiol 147:486–492

Wilmer JA, Helsper JPFG, van der Plas LHW (1997) Effects of abscisic acid and temperature on erucic acid accumulation in oilseed rape (*Brassica napus* L.). J Plant Physiol 150:414–419

Wilmer JA, Lessire R, Helsper JPFG, van der Plas LHW (1998) Regulation of elongase activity by abscisic acid and temperature in microspore-derived embryos of oilseed rape (*Brassica napus*). Physiol Plant 102:185–191

Wittstock U, Halkier BA (2002) Glucosinolate research in the *Arabidopsis* era. Trends Plant Sci 7:263–270

Yasruel Z, Cianflone K, Sniderman AD, Rosenbloom M, Walsh M, Rodriguez MA (1991) Effect of acylation stimulating protein on the triacylglycerol synthetic pathway of human adipose tissue. Lipids 26:495–499

Zou J, Abrams GD, Barton DL, Taylor DC, Pomeroy MK, Abrams SR (1995) Induction of lipid and oleosin biosynthesis by (+)-abscisic acid and its metabolites in microspore-derived embryos of *Brassica napus* L. cv. Reston. Biological responses in the presence of 8'-hydroxy abscisic acid. Plant Physiol 108:563–571

Zou J, Wei Y, Jako C, Kumar A, Selvaraj G, Taylor DC (1999) The *Arabidopsis thaliana TAG1* mutant has a mutation in a diacylglycerol acyltransferase gene. Plant J 19:645–653

I.7 Chromosome Doubling and Recovery of Doubled Haploid Plants

KEN J. KASHA[1]

1 Introduction

Having the ability to double the chromosome number of haploid plants has been of major concern in their utilization for breeding, as well as for genetic, mutation and transformation studies. The ideal objective is to double the chromosome number of the original haploid single cell and then regenerate a plant from it. This could provide regenerated plants that will be completely fertile doubled haploids in the case of diploid and allopolyploid species, or that will have better seed recovery from autopolyploids. It would also be an asset for mutation and transformation studies to be able to double the single haploid cell in the G1 cell cycle phase after the mutation or gene insertion had occurred in order to obtain plants homozygous for the new gene or allele. Alternatively, haploidy procedures would still be beneficial in obtaining homozygosity if used on plants heterozygous for the mutants or transgenes, although it would take another generation of plants.

The approach to doubling the chromosomes also depends upon the methods used to produce the haploids. With anther or isolated microspore culture, doubling may be feasible at the first pollen mitotic division (PMI) of the microspore, leading to plants that are completely doubled and fertile. Whereas, with wide hybridization and other gynogenetic methods, unlike the microspore, the egg cell is difficult to access and often the embryo is the earliest stage that can be treated. However, most often in gynogenetic systems it is the haploid seedling that is treated, resulting in ploidy chimeras. The ideal objectives of any haploid method that is to be used widely are to be able to produce large numbers of haploids from any genotype within the species and to achieve high success in doubling their chromosome number.

There have been two excellent reviews and analyses of the literature on chromosome doubling (Jensen 1974; Rao and Suprasanna 1996) as well as many briefer accounts. Jensen (1974) extensively reviewed the literature up to that date and provided a sound analysis for further research in the area. Rao and Suprasanna (1996) have built upon the basis developed by Jensen and provide a thorough analysis of developments since the review of Jensen. While it is necessary to outline the background on chromosome doubling,

[1] Department of Plant Agriculture, University of Guelph, Guelph, Ontario, N1G 2W1, Canada

Biotechnology in Agriculture and Forestry, Vol. 56
Haploids in Crop Improvement II (ed. by C.E. Palmer, W.A. Keller, and K.J. Kasha)

please refer to these earlier reviews for details on the earlier literature. Because of increased knowledge of gametophytic development and the genes involved, this chapter will focus on relevant concepts in cell development and chromosome behavior relative to chromosome doubling and provide tables summarizing selected recent findings for illustrative purposes.

2 Terminology

In the literature there has been some continuing confusion concerning the terminology used for doubled haploids. The term **haploid plant** is a general term used to describe a plant (sporophyte) with the gametic chromosome number, whether the parent plant is diploid or polyploid. The term **dihaploid** (2×) was used originally to denote the haploid from an autotetraploid (4×) and, thus, it is not likely to be genetically homozygous like a doubled haploid. It was agreed at the 1st International Symposium on Haploid Plants (de Fossard 1974) that the term 'dihaploid' should not be used to describe a 'doubled haploid' (DH) plant. The preferred term to use when the chromosome number of a haploid is doubled is **doubled haploid**.

Chromosome doubling is thought to occur by one or more of four mechanisms, namely endomitosis, endoreduplication, C-mitosis or nuclear fusion as summarized by Jensen (1974). **Endomitosis** is described as chromosome multiplication and separation, but failure of the spindle leads to one restitution nucleus with double the chromosome number. It has also been called 'nuclear restitution' (Sunderland and Evans 1980). **Endoreduplication** is duplication of the chromatids without their separation and leads to diplochromosomes or to polytene chromosomes if many replications occur. Endoreduplication is a common feature in specialized plant cells where cells become differentiated or enlarged or in cells that are very active in metabolite production. Rao and Suprasanna (1996) concluded that the spontaneous chromosome doubling was probably due to endoreduplication based on earlier evidence from microspores. **C-mitosis** is a specific form of endomitosis where, under the influence of colchicine, the centromeres do not initially separate in metaphase, while chromosome arms or chromatids do separate. **Nuclear fusion** was described by Sunderland et al. (1974) as occurring when two or more nuclei divide synchronously and develop a common spindle. Thus, two or more nuclei could result with doubled, polyploidy or aneuploid chromosome numbers. Sun (1978) suggested that in barley and rye the interphase nuclei following the first symmetric microspore division appeared to fuse when no cell wall was formed. Chen et al. (1984) observed the coalescence of such nuclei using electron microscopy. This type of nuclear fusion should result in a single nucleus with doubled chromosome number. Kasha et al. (2001) provided evidence of nuclear fusion resulting in a single nucleus during the pretreatment of barley microspore with mannitol at room temperature. The binucleate microspores became uninucleate microspores with the appropriate increased DNA levels. Nuclear fusion is a common phenomenon in such areas as double fertilization in plants and mating systems in yeast. Some consider it to be the most likely route of 'spontaneous' doubling of chromosomes during anther and microspore culture (Sunderland and Evans 1980; Kasha et al. 2001). However, all four methods can have a role in doubling in specific situations.

By definition, the **microspore stage** occurs between the tetrad stage and the first mitotic division. Subsequently, it is referred to as a gametophyte or pollen. For clarity and continuity in covering the early stages of microspore induction leading to embryogenesis and chromosome doubling, this chapter will refer to the structures as microspores or microspore-derived structures where gametophyte development is switched to sporophytic development.

One of the difficulties with comparing results of haploid research and chromosome doubling is the absence of a common method of expressing results. Results have been expressed as haploids per 100 anthers, per 10^4 microspores, per inflorescence, per spike, per plant, etc. Species differ in numbers of anthers per floret, numbers of microspores per anther, numbers of microspores isolated per anther and in numbers of florets per spike or type of inflorescence. In androgenesis, the number of microspores cultured would probably be the best common denominator to use, while in gynogenesis the ovary is probably the best common base. For chromosome doubling, the percentage of haploids doubled by a method or treatment is usually used, but utility of a system is also dependent upon the numbers of haploids that can be produced and the survival of the treated plants.

The methods of producing doubled haploids dictate to some extent the approach to chromosome doubling and these methods are summarized in Table 1. Also, Horlow and Raquin (1998) have produced a good figure summary of the methods of producing haploids. These methods can be grouped broadly as either **androgenetic** (male gamete origin) or **gynogenetic** (female gamete). Some methods, for example the semigamy system in cotton (Tur-

Table 1. Methods of producing doubled haploids in plants, related to androgenesis and gynogenesis

Haploid procedure	Category of origin	Examples of crop species
Anther culture	Androgenetic	*Oryza sativa* (rice)
Isolated microspore culture	Androgenetic	*Brassica napus* (canola)
Ovary culture	Gynogenetic	*Beta vulgaris* (sugar beet)
Spontaneous (marker selected)	Gynogenetic	*Zea mays* (corn, maize)
Gene induced	Gynogenetic	*Zea mays*, *Hordeum vulgare*
Wide hybridization	Gynogenetic	*Solanum tuberosum* (potato)
Chromosome elimination with wide hybridization	Gynogenetic	*Hordeum vulgare* (barley) *Triticum aestivum* (wheat)
Semigamy	Andro- and gynogenetic	*Gossypium hirsutum* (cotton)

cotte and Feaster 1974) or the indeterminate gametophyte (*ig*) gene system in maize (Kermicle 1969), can produce haploids of both male and female gametes. However, for chromosome doubling, they require systems similar to those used with gynogenetic systems. The discussion on doubling will be covered under the topics androgenesis and gynogenesis, with additional sections on aspects such as meiotic doubling and genetic stability.

3 Chromosome Doubling of Androgenetic Haploids

The systems of anther culture and isolated microspore culture are the most widely used systems for haploid production in plants. The large numbers of microspores per anther and the multiple anthers per floret provide the potential for the largest number of haploid plants per floret. In contrast, most other systems of haploid production in angiosperm species are gynogenic where there is a single egg cell, or else a small number of ovules per floret, such as in legume species. Successful haploid plant production has been recorded in hundreds of species by these methods. However, the ability to produce sufficient numbers of haploids from microspores to be suitable for breeding programs and research has been difficult to achieve. There are some exceptions such as canola, tobacco, barley, wheat and rice where much more extensive research has been applied. Furthermore, strong genotype differences in response tend to be the rule. Nevertheless, research on all species has been important for making progress with this method of haploid production and chromosome doubling.

A key feature of androgenetic haploid production systems is the potential to double the chromosomes during the very early stages of embryogenesis, leading to plants being completely doubled and fertile. The mechanism of doubling the chromosome number is unclear in many instances, with wide differences in response among and within species. In order to gain some insight into the process of doubling, knowledge of the potential pathways to embryogenesis and factors that influence their success is necessary. Induction of androgenesis and chromosome doubling both appear to involve changes in microfilaments and microtubules in the mitotic divisions in the microspore. Five potential pathways to embryogenesis are outlined and discussed in Chapter I.2 and illustrated in Chapter I.2, Fig. 2. Some of the induction factors are also covered in Chapters I.3 and I.4. Recent reviews also provide information on the stages of gametophyte development and induction of embryogenesis, such as McCormick (1993), Reynolds (1997), Heberle-Bors (1998), Twell and Howden (1998), Twell et al. (1998), Pechan and Smykal (2001) and Touraev et al. (2001). In particular, the paper of Twell et al. (1998) has provided a good illustrative account of the asymmetric first mitotic division in the microspore and cell-fate determination in developing pollen. They suggest that microfilaments and microtubules may be responsible for the nuclear migration around the uninucleate microspore wall and are involved in the establishment of polarization of the microspore cell prior to PMI. If a pretreatment system for inducing embryogenesis also disrupts microtubules, it might also lead to chromosome doubling. It is also possible that pretreatment might suspend the progression of the nucleus through the first mitotic division. This is one possible interpretation of the results of Shim and Kasha (2003) who observed that a combination of mannitol and cold for pretreatment induction in barley appeared to stall cell cycle progression.

During the migration of the nucleus in barley from the pore to a position opposite to the germ pore, it enters into cell cycle change and moves from G1 into DNA synthesis (S stage). By the time it reaches its destination it is in the G2 stage (Shim and Kasha 2003). These migration stages are used for staging uninucleate microspores for culturing (Wheatley et al. 1986). Within the polarized cell the nucleus enters into the M stage of PMI and an asymmetrical spindle is formed perpendicular to the cell wall. The blunt end of the spindle is at the generative cell (GC) pole at the wall and the sharp spindle end is at the vegetative cell (VC) pole pointing towards the centre of the cell (Twell and Howden 1998). The phragmoplast and cell plate form in an asymmetrical position closer to the GC pole. When the asymmetric division is complete the generative nucleus (GN) is enclosed in a small cell with very little cytoplasm or few organelles. Its wall is usually attached to the intine of the microspore wall and the nucleus remains highly condensed and stains darkly. In most species, the GN immediately enters into a second cell cycle. The large and diffuse vegetative nucleus (VN) remains in the large area of the microspore with most of the organelles and proteins. It remains at the G1 stage and becomes metabolically very active; filling the vegetative cell with cytoplasm that is rich in organelles, proteins, lipids and eventually starch. It remains for a time, often for 5–8 days in barley or 6–12 days in tobacco, at the G1 stage during pollen development (Pechan and Keller 1988).

Disruption of the asymmetric spindle leads to the division being symmetrical, as is commonly observed after uninucleate microspores have been induced towards embryogenesis. The subsequent failure of cell wall formation as observed in the cereals (Sun 1978; Hu and Kasha 1999; Magnard et al. 2000) could allow the nuclei to fuse, resulting in chromosome doubling as observed by Kasha et al. (2001). However, in *Brassica*, Pechan and Keller (1988) concluded that only those 30–40 % of microspores that formed a preprophase band and cell wall after PMI continued towards embryogenesis. The difference in wall formation between the cereals and *Brassica* species may be related to differences in spontaneous chromosome doubling which is low in *Brassica* and high in cereals such as barley (up to 80 %). However, Testillano et al. (2002) observed two domains in young 5- to 7-day-old maize microspore-derived embryos with features similar to zygotic embryogenesis. The denser section with initial straight wall formation appeared more like an embryo with smaller cells and nuclei. The larger and less dense domain was usually without cell walls at this stage and when cell walls formed they were very wavy or irregular shaped. The latter is similar to endosperm development that forms in conjunction with zygotic embryos. Various in situ and biochemical studies that were performed supported this conclusion (Testillano et al. 2002). The ploidy level of these domains was not reported. In addition, the stage of pretreatment or the type of pretreatment could also be factors in these species differences.

3.1 Microspore Culture Stage Influence on Chromosome Doubling

One widely adopted theory is that haploid induction during anther or microspore culture begins with some form of stress applied at a critical stage before or during the culture of the microspores (Nitsch 1974; Touraev et al. 1996, 1997; Zoriniants et al., Chap. I.3, this Vol.). Simmonds and Keller (1999) proposed that microtubule reorganization was the key event in the induction of embryogenesis. When microspores are induced by a stress pretreatment or culture conditions, the gametophytic development is blocked and the microspore nuclear behavior changes. It leads to a rapid release (usually within the first 8 h of treatment) of the VN from G1 into the S phase and subsequently nuclear divisions (Pechan and Keller 1988). Twell et al. (1998) proposed that this switch to VC development is a default program that is normally repressed by the GC as a result of PMI asymmetry. When this asymmetry is disrupted, the VN is released to enter DNA synthesis and may continue divisions to develop in a sporophytic manner. It has been reported in the vast majority of cases that the embryo or callus formation arises from the vegetative nucleus (Aionesei et al., Chap. I.2, this Vol.), although it has been reported to be derived from the generative nucleus in some instances (Raghavan 1976). In view of the recent observations of Testillano et al. (2002), this concept may need to be re-examined. There could have been some confusion in earlier studies with the loose and dense regions observed by Testillano et al. (2002). They could have been considered to be evidence of both VN and GN divisions respectively. Conversely, are the dense and less dense coenocytic regions observed by Testillano et al. (2002) derived from the GN and VN respectively?

Sunderland (1974) divided the development of the microspore and pollen into six stages relative to *Nicotianna tabacum* and *Datura innoxia*. These six stages for *Datura* are described in Table 2 along with the chromosome numbers of progeny plants obtained when microspore cultures were initiated at those stages. Included in Table 2 are results from an earlier similar study by Engvild et al. (1972) on *Datura*. It should be cautioned that although the stages of the two studies are grouped as the same in Table 2, they might not be exactly the same. It is notable in both studies that the ploidy levels increased when later stages of microspores were cultured, emphasizing the importance of microspore stage for induction of chromosome doubling. Culturing of early uninucleate stages gave predominantly haploid progeny, while

Table 2. Chromosome number of structures from anther culture of *Datura innoxia* relative to microspore stage at time of anther culture. *E* (adapted from Engvild et al. 1972), *S* (adapted from Sunderland 1974)

Microspore stage and (cell cycle)	No. of haploids		No. of diploids		No. of triploids		No. of tetraploids		Percentage of anther response	
	E	S	E	S	E	S	E	S	E	S
Tetrad	20	21	–	1	–	–	–	–	14	–
Early uninucleate (G1)	20	6	–	13	–	–	–	–	22	–
Late uninucleate (G2)	10	20	10	63	–	2	–	1	54	–
Mitosis (M)	8	9	12	15	–	5	–	1	88	–
Early binucleate	4	5	16	16	–	12	–	1	76	–
Late binucleate	–	3	16	5	4	25	1	3	24	–

culturing the binucleate stages gave more doubled haploids and polyploids. The ranges of ploidy levels of the derived plants also need to be explainable by the pathways to chromosome doubling. The anthers were cultured at 30 °C, which today would be described as a stress condition. One possible explanation of the triploid plants is that the GN of binucleates had undergone DNA synthesis prior to pretreatment or during in vitro culture and that the fusion of the VN and GN would provide a triploid. Such a fusion might produce the cell with both diplo and normal chromosomes as observed in Fig. 8 of Sunderland (1974). Presumably, the diplo-chromosomes would later separate, resulting in an increase in the triploid chromosome number. The tetraploids would likely result from a second round of doubling, either by fusion or endomitosis.

Pechan and Keller (1988) working with *Brassica napus* also described six stages of microspore development that are slightly more condensed in time than those of Sunderland (1974). Pechan and Keller (1988) found that the best microspore stages for highest embryo induction were the late uninucleate to early binucleate stages. They proposed that there was a fairly narrow window of about 8 h for best induction. However, embryo induction is still possible over a broad range of stages, as shown in Table 2, and when using a more severe stress of 41 °C (Binarova et al. 1997) on more mature pollen. While the microspores within an anther show a high degree of synchrony (Shim and Kasha 2003), they are not completely synchronized, so that the stages in Table 2 should be considered as the modal group of microspores. Aionesei et al. (Chap. II.2, Fig. 2, this Vol.) proposed five potential pathways for multicellular formation from microspores. They are based on the A, B and C pathways of Sunderland (1974) and the observations of generative nuclear development (Raghavan 1976). The microspore stage at the time of pretreatment or culture would appear to be one of the many factors that might influence which pathway would develop. The key two stages are the uninucleate stage and the binucleate stage. A symmetrical PMI (pathway B) would most likely arise from treatment (stress) at the uninucleate stage of the microspore. The pathway appears to be related to the disruption of the asymmetric spindle formation in gametophyte development and the subsequent failure of cell wall formation which could permit fusion of the two nuclei and thus chromosome doubling. This type of development has been observed quite frequently in cereals such as barley (Sun 1978), wheat (Hu and Kasha 1999) and maize (Magnard et al. 2000). Given the right conditions, this can readily lead to nuclear fusion to produce a single doubled haploid cell (Sunderland and Evans 1980; Chen et al. 1984; Kasha et al. 2001) that continues to divide and produce embryo structures. Most of the progeny are doubled haploids (Kasha et al. (2001) while the remaining plants are haploid, but up to 8–10% polyploids may occur.

The failure of cell wall formation usually persists for 5–10 days in these species, resulting in a coenocytic structure. If cell wall formation does not subsequently occur the coenocytic structures will abort (pathway D). Sunderland and Evans (1980) suggest that the structures developing with highly polyploid chromosome numbers would tend to abort or to be less competitive in cultures.

In other species such as *Brassica napus*, Pechan and Keller (1988) observed that a cell wall was formed in 30–40% of the symmetrical PMI and concluded that those in which the wall formed were the ones that developed into embryos. Simmonds and Keller (1999) showed that a preprophase band of microtubules was necessary for the cell wall to form and proposed that it could be used to select the embryogenic microspores. However, if the cell wall is formed, the potential for nuclear fusion may be reduced, resulting in a lower frequency of spontaneously doubled haploids. This concept may have some merit in that the frequency of spontaneously doubled haploids in *Brassica* is low, while in barley it can be as high as 80% but somewhat lower in other cereals.

When pretreatments are applied at the early to mid binucleate stage of the microspore, the PMI is usually asymmetric (pathway A). In most instances, it is the VC that will continue to divide to form the embryogenic structure (Pechan and Smykal 2001; Zoriniants et al., Chap. I.3, this Vol.). When the cell wall formation between the VN and GN is incomplete, it is possible to have fusion of these two nuclei (pathway C; Sunderland 1974). It is possible that there are two stages of doubling by fusion. The second is predicted from the observations that a high frequency of chromosome doubling occurs in barley after a long cold pretreatment (Kasha et al. 2001). An asymmetric PMI is predominantly formed after such a pretreatment, which suggests that the GN cell wall remains largely intact. Since the vegetative cell continues to divide, perhaps the cell wall between the daughter vegetative nuclei fails, permitting nuclear fusion. Alternatively, there is also good evidence for the fusion of asymmetric nuclei as follows: first, there is the failure of the cell wall development around the generative nucleus (Chen et al. 1984; Pechan and Keller

1988), followed by the multinucleate cells in cereal anther/microspore cultures such as barley (Sun 1978; Wilson et al. 1978), rye (Sun 1978), wheat (Hu and Kasha 1999) and maize (Magnard et al. 2000); second, the beautiful photographs of triploid nuclei in Sunderland (1974) show the more condensed chromosomes from the GN associated with the less condensed chromosomes from the VN (pathway C); third, there was a higher frequency of fused nuclei than of symmetrically dividing microspores (Kasha et al. 2001) following mannitol pretreatment at 25 °C for 4 days. In multinucleate cells, higher polyploid cells may result from further nuclear fusions. In some instances, both the VN and the GN will start to divide separately as identified by the dense chromatin of the GN. In these instances, GN nuclear division usually stops by the third or fourth division and degeneration begins, while the VN continues to divide and develop the multinucleate structure for embryogenesis. On rare occasions, the GN nucleus is the one that continues to divide and develop into the embryo (pathway E; Raghavan 1976). Alternatively, these observations may be confused with the two types of developing structures observed by Testillano et al.(2002) as mentioned previously and which may be species-specific.

Another possible route to chromosome doubling in some of the above instances is endomitosis or C-mitosis. It is known that some pretreatments such as cold and colchicine will block spindle formation, resulting in a restitution nucleus with double the chromosome number. Thus, it would appear that the type of PMI (symmetrical or asymmetrical) may depend to some extent upon the microspore stage (uninucleate or binucleate) when pretreatment or culture stress is applied. The degree of synchrony of the microspore stages is also important in such research as well as the type of treatment used for induction.

3.2 Role of Pretreatments in 'Spontaneous' Chromosome Doubling

Based on Section 3.1 which discussed male gametophyte stages and development, it appears that the process of embryo induction could also have a major influence on chromosome doubling. Many of the pretreatments used for microspore induction of embryogenesis also improve the frequency of chromosome doubling. Such processes appear to be related to microtubule and microfilament elements that form the cytoskeleton and are critical for an asymmetric PMI.

Recent studies have been directed at finding inducing agents that also give high chromosome doubling frequencies in androgenesis. This work was initiated in *Brassica napus* by Zaki and Dickinson (1991) using colchicine in the initial stages of culture and followed by Zhao et al. (1996) and Zhou et al. (2002a,b). Colchicine and a number of other antimicrotubule agents have been used to improve chromosome doubling in anther/microspore cultures of many other species (Table 3). Zhao et al. (1996) observed that while heat

Table 3. Some recent results on the efficiency of chromosome doubling in haploid microspore/anther cultures. *colc.* Colchicine; *oryz.* oryzalin; *APM* amiprophos methyl; *trifl.* trifluralin; *man.* mannitol; *2HNA* 2-hydroxy nicotinic acid; *DMSO* dimethyl sulfoxide; *c+m* cold plus mannitol

Species	Stage treated	Treatment	Best results	Reference
Avena sativa	Anther culture	Colc. 0.5% + 10% DMSO for 5 h	91%	Kiviharja et al. (2000)
Brassica napus	Microspore culture	Colc. in media vs. heat shock	90% Colc. 6% Heat	Zhao et al. (1996)
B. napus	Microspore culture	Colc. 500 mg/l, first 15 h		
Winter			88%	Zhou et al. (2002a)
Spring			91%	Zhou et al. (2002b)
Coffea arabica	Microspores	Colc. 100 mg/l	<5%	Herrera et al. (2002)
Hordeum vulgare	Microspores	4 °C + 0.3 M man., 4 days or cold 4 °C 21 days	79% c+m 83% Cold	Kasha et al. (2001)
	Spikes and microspores	Cold 21 days on spikes + man. 7 days on microspores	79% Cold 94% Combined	Li and Devaux (2003)
Lillium longiflorum	Anther culture	Cold + colc. in media	60%	Antoine-Michard and Beckert (1997)
Oryza sativa	Panicles and plants	Cold 10 °C, 10 days	8–78%	Bishonoi et al. (2000)
		Colc. 1%, 16–18 h	+18%	
Phleum pratense	Microspore culture	Heat shock + 0.1% colc. + 0.1% DMSO 3 h	66%	Guo and Pulli (2000a)
Secale cereale	Microspore culture	Mannitol pretreatment	90%	Guo and Pulli (2000b)
Triticale	Anther culture	Colc. in media + hydroponic recovery	98 vs. 15% Spontaneous	Arzani and Darvey (2001)
Triticum aestivum	Anthers	Colc. 0.03%, 3 days in culture media	100%	Zamani et al. (2000)
		Colc. 100 mg/l to media, 1–3 days	86% at 3 days	Redha et al. (2000)
	Microspore culture	Colc., trifl., APM, oryz.	94% Colc. Others 65%	Hansen and Andersen (1996)
		Colc. 300–1,000 μM, 24 or 48 h	53%	Hansen and Andersen (1998a)
		APM or trifl. 0.1–10 μM first 48 h	74% APM 66% Trifl. 15% Continuous	Hansen and Andersen (1998b)
	Microspores	Spike cold + 0.4 M man., 4 °C, 7 days	80%	Hu and Kasha (1997)
	Donor spikes	Cold 4 °C	44%	Stober and Hess (1997)

Table 3. (Continue)

Species	Stage treated	Treatment	Best results	Reference
	Tillers at microspore stage	Pretreatment in 0.01% 2HNA	64%	Liu et al. (2002)
Zea mays	Anther culture	Cold Colc. + cold in media	40% 60% Combined	Antoine-Michard and Beckert (1997)
	Anther culture	GA3, cold Oryz., pronamide	+52% GA3 =Reduced	Martin and Widholm (1996)
		Colc. 5–1,000 mg/l 1–7 days at 14°C	250 mg 7 days 56%	Saisington et al. (1996)
		Colc. 0.3%, 3 days in media	100%	Barnabas et al. (1999)

pretreatment caused a symmetric PMI, it did not lead to high frequencies of chromosome doubling as was observed with a colchicine treatment during the initial stages of culture at room temperature. The response to some of the pretreatments appears to be species specific. The work in *Brassica* and tobacco (as reviewed by Pechan and Smykal 2001) points to a role of high temperature and small heat shock proteins being associated with induction of microspore embryogenesis. In this situation, cell wall formation after division of the VN appears to be essential for embryogenesis to continue and form plants. This is in contrast to the work in cereals where most often induction by low temperatures and/or mannitol is used and failure of the cell wall formation leads to coenocytic structures (Sun 1978; Hu and Kasha 1999; Magnard et al. 2000; Testillano et al. 2002). Wall formation occurs after 5–10 days and is likened to the development of the endosperm in these species (namely barley, rye, wheat and maize). This difference could be responsible for the great differences in spontaneous doubled haploids between cereals and the dicots and relate to the different pretreatments mentioned.

However, when spindle inhibitors or antimicrotubule agents are used, they appear to be effective in improving chromosome doubling across most of the species so far reported (Table 3). When colchicine was used in place of heat stress in *Brassica*, the frequencies of doubled haploids was greatly improved (Zhao et al. 1996; Zhou et al. 2002b). This indicates that the pretreatments that affect microtubules can influence the frequencies of 'spontaneously' doubled haploids and that they really are '**induced**' doubled haploids. The fairly broad application of the antimicrotubule agents supports the induction theories mentioned earlier by Twell et al. (1998) and Simmonds and Keller (1999) that a reorganization of the microtubules or microfilaments and the disruption of the asymmetric PMI were key triggers for the induction of embryogenesis from microspores.

Most researchers consider the B pathway as evidence of successful androgenic induction of microspore cultures (Fan et al. 1988; Reynolds 1997;

Zoriniants et al., Chap. I.3, this Vol.). Kasha et al. (2001) provided good evidence that nuclear fusion could occur at high rates in barley microspores during a pretreatment in 0.3 M mannitol at 25 °C for 4 days. The microspores were predominantly at the mid-uninucleate stage when pretreatments were initiated. When sampling cultures each day of the pretreatment they observed that the majority (55–60 %) of the first divisions were type B (symmetrical) and that by as early as day 2, most cells had started to become uninucleate again with DNA levels typical of fused nuclei. The high frequencies, up to 80 %, of microspores with a large fused nucleus would indicate that fusion most likely occurred after both symmetric and asymmetric PMI, since the initial frequency of symmetric divisions was only about 55–60 %. Often this wall around the GN is only partially formed (Simmonds and Keller 1999) which may be sufficient to constrict the GN size but not prevent its fusion with the VN. The frequency of 'spontaneous' doubled haploids from isolated microspore culture of barley is roughly 80 % and is consistent with the proportion of fused nuclei observed.

Earlier results of Keller and Melchers (1973) showed that 0.4 M mannitol with 0.05 M $CaCl_2$ at a pH of 10.5 greatly improved protoplast fusion in plants. Thus, the mannitol pretreatment used for induction of embryogenesis in barley may be responsible to some extent for the nuclear fusion that appears to lead to high frequencies of chromosome doubling observed in isolated microspore cultures (Kasha et al. 2001). Since the cold pretreatment alone also gave good induction and chromosome doubling, it is likely that any microtubule disruption agent exhibiting symmetric PMI would also lead to both embryo induction and spontaneous chromosome doubling. It would be interesting to try more combinations of antimicrotubule agents with mannitol for chromosome doubling experiments. Li and Devaux (2003) used the combination of 21 days of cold followed by 7 days in mannitol which improved doubling over the cold pretreatment alone.

Sunderland (1974) showed beautiful photos of microspore cells at metaphase with both normal and endoreduplicated (diplo) chromosomes; the latter are more condensed and most likely from the GN that had gone through the S phase prior to the stress treatment. They suggested that the diplo chromosomes were formed from two cycles of DNA replication (endoreduplication) in the GN. The photos could also be interpreted, based upon the chromosomes being mixed together, as being from a single fused nucleus of generative and vegetative nuclei that had entered into mitosis. An alternative explanation of fusion was that the two nuclei formed a common spindle so that two nuclei with doubled chromosome numbers would result (Sunderland 1974). However, this form of fusion does not explain the results of Kasha et al. (2001) in which coalescence of membranes of adjacent nuclei does and has been observed (Chen et al. 1984).

The degree of stress applied may also be important as Eady et al. (1995) reported that high levels of colchicine blocked PMI whereas lower levels resulted in a symmetric division at PMI. Colchicine is considered to cause doubling through endomitosis (C-mitosis) when plants have been treated. This would result in a blocked PMI. However, in microspore cultures, the symmetric PMI observed at lower concentrations could lead to nuclear fusion.

The increased awareness of nuclear fusion for doubling chromosomes would lead one to also speculate about the best stages of microspore induction relative to the frequency of chromosome doubling. As illustrated in Table 2, treatment at the late uninucleate stage or at mitosis in *Datura* resulted in a higher proportion of doubled haploids when compared to induction treatment at the early binucleate stage. The latter stage for fusion of GN and VN could give rise to triploids, or larger ploidy levels due to additional cycles of nuclear fusion and possibly endoreduplication. However, it is also possible that the fusion of generative and vegetative nuclei may not be frequent in some species so that the VN would divide to give rise to the embryos and subsequent haploid or doubled haploid plants. If the VN divisions are occurring during treatment for doubling, fusion of vegetative nuclei could lead to chromosome doubling. This could explain the results of Li and Devaux (2003) who used mannitol after the cold pretreatment to improve chromosome doubling in barley.

The best response in microspore culture in some species is at mitosis or early binucleate stages and the best time for doubling may have to be tempered by the best stage for response in culture. To get high frequencies of completely fertile doubled haploids the pretreatment should be carried out at the uninucleate microspore stage. Stress at the early binucleate stage could lead to more triploids or other polyploids.

One of the effects of stress pretreatment would appear to be disruption of the formation of the wall around the GN so that the cell becomes binucleate following PMI. This wall appears to be temporary and easily disrupted. In normal pollen development the wall around the GN tends to become separated from the intine wall and to disappear by the time the sperm nuclei enter the pollen tube and function independently in double fertilization.

3.3 Molecular Analysis of Androgenesis and Chromosome Doubling

Boutilier et al. (Chap. I.5, this Vol.) present a thorough analysis of the efforts to characterize the genes and pathways to induction and development of microspore-derived embryos and only a brief mention of aspects related to chromosome doubling will be covered here. There are also relevant background reviews that are helpful in understanding microspore development and induction. Dewitte and Murray (2003) review the plant cell cycle and its gene controls while Vantard et al. (2000) review the cell cycle regulation of the microtubule cytoskeleton.

In recent years, the use of molecular tools such as gene isolation and gene tagging through transformations has greatly enhanced our understanding of current and earlier results of anther and microspore culture, but this work is only beginning. The sequencing of the *Arabidopsis* genome has permitted the isolation of genes related to embryo induction in microspores and their subsequent tagging and isolation in crop species. Genes affecting the nuclear migration in microspores, the asymmetric PMI and proteins relating to the fusion of nuclei during fertilization have been identified (Cordewener et al. 1998; Twell and Howden 1998; Twell et al. 1998; Pechan and Smykal 2001; Boutilier et al., Chap. I.5, this Vol.).

Cordewener et al. (1998) and Pechan and Smykal (2001) discussed the close relationship in both tobacco and *Brassica napus* between induction of embryogenesis and the expression of heat shock proteins (HSP). There are many small HSPs produced by stress, with HSP70 being the predominant one in treated microspores. Cordewener et al. (1998) and Zoriniants et al. (Chap. I.3, this Vol.) concluded that stress seems to be a common factor in induction of the sporophytic pathway in microspores and further research is needed on the HSPs.

The degree of doubling through nuclear fusion could also be species specific, depending upon the extent of gene action related to proteins that may enhance nuclear fusion. The review of Twell et al. (1998) describes the various mutants produced in *Arabidopsis* that affect polarity within the microspore and the formation of the asymmetric PMI. They discuss the controls over microspore development and the similarity between pretreatment effects and known gene mutations. Unfortunately, microspore culture has not been successful yet in *Arabidopsis* so that such mutations could not be studied.

Nuclear fusion in plants is probably related to the fusion during mating in yeast where genes and their functions have been investigated more thoroughly. Rose (1996) in his review on yeast stated that there are two main steps in fusion. First, there is microtubular involvement in nuclear aggregation, and second, fusion of nuclear membranes. These steps could be similar to what we are seeing in chromosome doubling in plant microspore culture. The first sign of disruption may be blocking of nuclear migration in the uninucleate microspore (Twell et al. 1998), and then disruption of the asymmetric wall initiation that normally develops around the generative nucleus. This results in a symmetric nuclear division with the two nuclei lying close together. The second step of nuclear membrane coalescence has also been observed in plants (Chen et al. 1984).

Simmonds and Keller (1999) studied cell wall formation in *Brassica napus* and found that the wall around the GN was often disrupted. They concluded that microspore embryogenesis only continued in those microspores where the GC wall was complete. This is in contrast to the cereals where the first divisions often occur without wall formation. Magnard et al. (2000) isolated two types of genes from induction stages in microspore embryos in maize and found that they were not expressed in zygotic embryos but were expressed in the early stages of endosperm development which was coenocytic like the early microspore structures. This supports the concept that induction and chromosome doubling in cereals may be different from that found in dicot species. However, the pathway may also depend upon the induction treatments that were quite different in *Brassica* and maize. Zoriniants et al. (Chap. I.3, this Vol.) discuss the influence of colchicine on blocking the microtubule assembly during induction of microspore embryogenesis. The use of such microtubule inhibitors (antimicrotubule agents) can improve chromosome doubling over temperature treatments in most species and are also useful agents for the induction of microspore embryogenesis (Hansen and Andersen 1996; Martin and Widholm 1996; Zhao et al. 1996). Most of these inhibitors were tried earlier on haploid seedlings produced from androgenesis, wide hybridization or gynogenesis and were found not to be as effective as colchicine for chromosome doubling (Rao and Suprasanna 1996). However, in microspore cultures, there is easier access for such chemicals, and the disruption of the temporary wall formation around the generative nucleus may be more sensitive to these microtubule inhibitors. They could also be less toxic to cells than colchicine.

3.4 Recent Literature on Chromosome Doubling in Microspores

The more recent literature on microspore culture and pretreatments has concentrated on chemicals that might influence the cell microtubules that are involved in cell plate formation as well as in spindle development. This has involved a wide range of species as different chemicals may work more effectively in a particular species (Table 3). Zaki and Dickenson (1991) and Zhao et al. (1996) have shown that adding colchicine to the initial culture medium of *Brassica napus* microspores could cause induction of embryogenesis and resulted in much higher frequencies of doubled haploids. Subsequently, Zhou et al. (2002a,b) greatly improved chromosome doubling in isolated microspores of *Brassica napus* (Table 3) by using colchicine in the induction culture medium. The use of mannitol for pretreatment of microspores in cereals has provided consistently high doubling frequencies in barley (Kasha et al. 2001; Li and Devaux 2003), rye (Guo and Pulli 2000b) and wheat (Hu and Kasha 1997). Thus, it appears from the studies of Kasha et al. (2001) that mannitol may be acting as an anti-microtubule agent disrupting the asymmetrical PMI in addition to possible induction through starvation as proposed by Heberle-Bors (1989) or as an osmotic stress (Hoekstra et al. 1997). Alternatively, Keller and Melchers (1973) found that mannitol assisted in protoplast fusion in plants. Thus, the mannitol pretreatment used for induction in barley could be enhancing induction as well as nuclear fusion, leading to high frequencies of chromosome doubling in barley isolated microspore cultures (Kasha et al. 2001; Li and Devaux 2003). Since the cold pretreatment alone in barley gives good embryo induction and chromosome doubling at frequencies similar to those of mannitol, it is possible that any microtubule disruption agent could cause both embryo induction and spontaneous chromosome doubling.

Many of the recent studies are now using antimicrotubule agents in induction culture media for the first few hours (Table 3). Again, this would lead one to question whether starvation or osmotic stresses are operative in such situations and this needs to be explored further. Colchicine is the main standard against which other antimicrotubule agents are compared and it is quite effective in most instances. However, other antimicrotubule agents such as oryzalin, trifluralin and amiprophos-methyl (APM) as well as 2-hydroxynicotinic acid (2HNA) are showing good doubling frequencies (Table 3) with perhaps less toxic effects than colchicine and with less concern for handler safety. 2HNA has been used on tillers or spikes of wheat prior to microspore isolation (Liu et al. 2002), while the others are being used at the initiation of in vitro microspore culture. The use of DMSO (dimethyl sulfoxide) as a carrier to assist in the chemical penetration of tissues and cells is quite common with these antimicrotubule agents. This work with antimicrotubule agents is one area where we can expect to see further improvements in induction of both embryogenesis and chromosome doubling.

A number of the studies with colchicine have reported improved induction of sporophytic development with lower doses of such chemicals (Saisingtong et al. 1996; Zhao et al. 1996). It would be interesting to know how often colchicine induces C-mitosis (endomitosis) versus disruption of the asymmetric PMI and subsequent doubling by nuclear fusion. Eady et al. (1995) observed that low dosages of colchicine produced a symmetric PMI while higher dosages blocked PMI. Temporary blockage of PMI could result in endomitosis and restitution nuclei, so the degree of stress might influence the doubling procedure. The recent study of Li and Devaux (2003) compared 21 days of cold pretreatment with the same pretreatment followed by 7 days in mannitol and showed significant improvement in chromosome doubling (predominantly doubled haploids with no haploid progeny observed) when the mannitol was added. Since the microspores would most likely have passed through the first division by the end of this treatment (Shim and Kasha 2003), it is possible that the subsequent VN divisions were influenced by mannitol to block cell wall formation and to enhance nuclear fusion. It would be interesting to follow the use of other antimicrotubule agents in combination with mannitol to see if such combinations could enhance the recovery of doubled haploids. Kao and Michayluk (1974) also found that polyethylene glycol (PEG) enhanced protoplast fusion and Ilic-Grubor et al. (1998) reported enhanced haploid production from microspore culture of *Brassica napus* when using this chemical. The improvement had been interpreted to be osmotic effects in that most of the sugar was replaced by PEG. However, the development of the embryos was also greatly improved and compared very favorably with zygotic embryos. They did not report on the effect of PEG on chromosome doubling, but its role in fusion of nuclei should also be examined.

Both the length of the pretreatment and the strength of pretreatment need to be determined for each particular species. The strength could either be the concentration of a chemical used or the temperature used for the stress. The degree of synchronization of the microspore population would also influence the duration of the microspore pretreatment. The duration of treatment may be shorter when the microspores are highly synchronized and probably should be longer for microspores collected from species that have multiple florets varying in microspore stage. Microspores tend to be a fairly highly synchronized population, and could be further synchronized through density gradient centrifugation or the use of cell cycle blocker chemicals. Alternatively, one could spend more time in selecting florets or anthers to provide a more uniform population of microspores.

The duration of pretreatment stress may also be important in species where the best response is obtained at the early binucleate stage. In this case, one may need to ensure that the stress is present when the VN undergoes division to possibly prevent cell wall formation and enable subsequent nuclear fusion of vegetative nuclei. As mentioned earlier, by delaying the pretreatment until after the PMI in the microspore, the generative nucleus may have undergone its normal DNA synthesis or a second round of synthesis (endoreduplication). Thus, when fusion is involved, triploids or higher ploidy levels in progeny plants may arise.

While more research is required to clarify the process of chromosome doubling in microspores, new molecular tools will provide interpretations of the process of doubling and opportunities to improve the systems. Being able to follow specific microspores through induction stages by the method reported by Indrianto et al. (2001) would be valuable for clarifying the doubling process.

4 Gynogenetic Haploid Production and Chromosome Doubling

A number of methods of haploid production fall under this category of producing haploid plants from the egg cell. The three main ones are parthenogenesis, wide hybridization/chromosome elimination and ovule culture. The problem for chromosome doubling is the difficulty in isolating the egg cell from the ovule or attempting to rapidly penetrate the ovule to reach the egg with chromosome doubling agents or stresses. There are continuing reports on improved chromosome doubling (Table 4) with such systems, but no major breakthroughs have appeared since the review of Rao and Suprasanna (1996). The percentage of doubling of young plants is quite high in some species so that these systems of haploid production are useful in *Brassica* and cereals (Devaux and Pickering, Chap. II.3, this Vol.)

Table 4. Some recent results on chromosome doubling of haploids from ovary or embryo culture and seedling treatments. Chemical abbreviations: *colc.* colchicine; *oryz.* oryzalin; *APM* amiprophos methyl; *trifl.* trifluralin; *man.* mannitol; *N_2O* nitrous oxide; *DMSO* dimethyl sulfoxide

Species and method	Stage treated	Treatment	Best results	Reference
Allium cepa (ovary culture)	Clones and plantlets 24-h treatment	Colc. 0.625–125 mM Oryz. 10–200 μM	65% 57%	Geoffriau et al. (1997)
Beta vulgaris (ovary culture)	Ovaries after 10 days culture	APM 150 μM, 5 h	64%	Hansen et al. (2000)
	Ovaries cold pretreatment	Colc. 50–60 mg/l, 48 h Trifl. 1.7–5 mg/l	25%	Guerel et al. (2000)
Brassica juncea (microspores)	Plants	Colc.	18%, 48 h	Lionneton et al. (2001)
Gerbera sp. (ovary culture)	Plants	Colc. or oryz. + 1% DMSO	68%, 60 μM colc. 61% oryz.	Tosca et al. (1996)
Populus nigra (anther culture)	Plants	4 °C, 8–14 days	85%	Kiss et al. (2001)
Pyrus communis	Plants	Oryz. 200–300 μM in 7.5% DMSO, 48 h	39%	Bouvier et al. (2002)
Triticum durum (× maize)	Seedlings	Colc. 0.25% for 4–5 h	100% of survivors	Saidi et al. (1998)
(Ovary culture)	Seedlings	Colc. 0.5% + 10% DMSO + PO_4KH_2, 4 h	100% colc. + spontaneous	Sibi et al. (2001)
Triticum aestivum (× maize)	1–2 Leaf seedling in media	Colc. in media	50%	Sharma et al. (2002)
(Anther culture)	Embryos and plantlets	Colc. 500 mg/l, 1–3 days	100% Embryos	Mentewab and Sarrafi (1997)
Triticale (× maize)	3–4 Tiller stage	Colc .1%, 4% DMSO, GA3, Tween 20	37%	Wedzony et al. (1998)
Zea mays (haploid inducer)	Seedlings	Colc. 0.6% + 0.5% DMSO, 5 h	27%	Eder and Chalyk (2002)
	6 Leaf stage (flower primordia)	N_2O 2 days at 600 kPa	44%	Kato (2002)
(Marker selected)	Seedlings	Colc. 1.5 mg/l, 3 h	60%	Bordes et al. (1997)

4.1 Parthenogenesis

The first haploid system utilized in crop breeding was the work of S.S. Chase (1949) in maize. He summarized his work at the 1st International Symposium on Haploid Higher Plants (Chase 1974), and that paper provides an interesting and thorough historical review of haploids produced for breeding up to that time. Chase used a dominant purple marker gene in the male parent to identify the low frequency (ca. 1 in 1,000 seeds) of parthenogenetic (colorless) haploid plants from the crosses. Chromosome doubling in maize turned out to be more difficult than selecting the haploid plants, but by 1949, he had produced 34 doubled haploid inbreds from 298 original haploids selected (Chase 1949). These were doubled by colchicine injection into the scutellar node using a syringe. He used 0.5 ml of an aqueous solution containing 0.05 % colchicine and 10 % glycerin. Many more such inbreds were produced by Chase and a few of the DH lines were used as inbreds to produce hybrids, some of which went into commercial production by DeKalb where Chase worked as a breeder for 12 years. Coe (1959) obtained a frequency of over 3 % parthenogenesis when he used a selected line called 'stock 6' as a pollinator in maize. Other such inducer stocks have been developed in maize (Bordes et al. 1997; Eder and Chalyk 2002). It would appear that after a lull of many years, the combination of inducer genes and marker genes is again being utilized to establish inbred lines for commercial maize breeding or homozygous transgenic lines.

4.2 Chromosome Elimination/Bulbosum Method

The next method developed for commercial utilization was wide hybridization in barley (Kasha and Kao 1970). When barley lines (*Hordeum vulgare*) were crossed with diploid *Hordeum bulbosum* pollen, fertilization occurred, but the chromosomes of *H. bulbosum* were preferentially eliminated in early embryo development so that haploid embryos were produced. These were rescued by embryo culture and grown into haploid sporophytes with very little spontaneous chromosome doubling (up to 3 %; Subrahmanyam and Kasha 1975). This was the first haploid system that worked well across genotypes and could produce a random array of gametes as plants in sufficient numbers to be used as a complete breeding program. At last count, at least 58 registered barley cultivars had been produced around the world by this method (Thomas et al. 2003). For a more thorough account of the development of this haploid system (bulbosum method) see Kasha (2003) or Devaux and Pickering (Chap. II.3, this Vol.).

The chromosome elimination method also worked on wheat using *H. bulbosum* pollen (Barclay 1975), but there were strong genotype effects due to crossability genes in wheat reacting to pollen of rye and *H. bulbosum*. Later it was found that maize (*Zea mays* L.) pollen would work well on wheat without a strong genotype effect (Laurie and Bennett 1986) and this system is now widely used in wheat breeding and genetics. Other pollinators such as sorghum and millets have also worked on wheat and to a lesser extent on oats (see Kasha 2003). Mochida and Tsujimoto (2001) found that pollen from Job's-tears (*Coix lachryma-jobi*) works as well on wheat as maize pollen and, being a perennial plant, it can produce pollen throughout the year.

Since nearly all the barley plants produced by wide hybridization/chromosome elimination were haploid, there was a need to develop an efficient method to double the chromosomes. Before the bulbosum method, doubling procedures were mainly with colchicine using long (often days), laborious procedures that resulted in stunting and quite high plant mortality. Thus, both nitrous oxide and colchicine methods were investigated (Jensen 1974; Subrahmanyam and Kasha 1975; Thiebaut and Kasha 1978; Thiebaut et al. 1979) and a system for a short colchicine treatment evolved that gave doubled sectors on a high proportion of the haploids produced. This system has been improved and widely used and adapted for chromosome doubling of plants in many species. The key features of the doubling system that evolved was a 5-h treatment under bright lights at room temperature with 0.1% aqueous colchicine solution containing 2% DMSO (dimethyl sulfoxide) plus a wetting agent. It was used on actively growing two- to three-leaf stage seedlings on medium in culture vials (Thiebaut and Kasha 1978) or on potted seedlings at the three-tiller stage. The concept was to maximize the uptake of the chemical through treating actively growing plants at a time of maximum cell division and then to enhance the recovery of the treated plants. In the tiller system the roots were cut back to 2 or 3 cm before placing the plants in sufficient colchicine to cover the roots and crown. After the treatment, the tops were cut back and the seedlings were washed in running water and treated with GA3 to stimulate recovery and growth. For smaller plants in culture vials, the GA3 was added to the colchicine treatment solution. The colchicine and DMSO concentrations vary with the species and the laboratory procedures used. In the cereals this often results in 80–90% of the plants having doubled sectors.

There have been some recent reports on chromosome doubling procedures for haploid seedlings, as summarized in Table 4. The trends are the treatment of younger plants or clones, shortening of the length of time in treatment to about 5 h or less and treatment under good growing conditions. Most often the plants are set in the solution of colchicine or other microtubule inhibiting agents along with DMSO to improve penetration. Other microtubule inhibiting agents tend to be less toxic than colchicine, but the doubling rate is usually lower. For specific species, there is some information on doubling in the chapters on those species in the volumes on haploidy edited by Jain et al. (1996/1997), in addition to the review by Rao and Suprassana (1996).

4.3 Ovary/Ovule Culture

Ovary culture is used for haploid production in crops such as sugar beets and onions. The problem is again the penetration of doubling agents to induce doubling of the early parthenogenically developing egg cell. As shown in Table 4, much of the recent effort has been on comparing colchicine with other antimicrotubule agents, such as oryzalin, APM or trifluralin (Geoffriau et al. 1997; Guerel et al. 2000; Hansen et al. 2000). Again, the objective has been to find agents that are not as toxic to deal with as colchicine, even though they are usually not quite as effective in inducing chromosome doubling at the present time.

5 Other Avenues for Chromosome Doubling

A recent study by Kato (2002) tried a different approach to chromosome doubling in maize. Plants were treated at the six-leaf stage with nitrous oxide. This was the stage determined to be the time of floral primordial development in maize and would have been expected to have one or only a few cells that needed to be doubled to obtain fertile florets. The results were quite good for maize in that Kato obtained seed from self-pollination on 44% of the plants compared to 11% on non-treated plants.

Chen and Beversdorf (1992) observed that the cooling step of isolated microspores prior to placement in liquid nitrogen improved the frequency of doubled haploids in *Brassica napus*. The lowest cooling temperature of −40 °C resulted in about 50% doubled haploids. They recommended that a cold pretreatment prior to treatment with an antimicrotubule agent might greatly enhance the frequency of doubled haploids in *Brassica*.

5.1 Meiotic Irregularities and Homozygosity of Doubled Haploids

Jensen (1974), Sunderland (1974), Veilleux (1985) and Rao and Suprasanna (1996) reviewed irregularities at meiosis prior to microspore development as a source of progeny plants having doubled haploid numbers of chromosomes or with aneuploid or polyploid chromosome numbers. While chromosomally doubled, such progeny could also be heterozygous, leading to genetic variation among their progeny. The source of this meiotic disruption is not always clear, but stress to pollen donor plants prior to or during meiosis could be one cause of the irregular meiotic products observed. One example is the dyads produced that appeared to have both generative and vegetative type nuclei that could divide (Sunderland 1974). Doubling from this process leads to infrequent seed set on the haploid plants and these seeds have the potential to have some genetic heterozygosity. It is suggested that these occasional seeds could be described as spontaneously doubled haploids as contrasted with the completely fertile homozygous plants arising from anther or isolated microspore culture which are induced by pretreatment or treatments applied to the cultures.

Various studies of the variability from doubled haploids have been conducted. For example, Muranty et al. (2002) compared the homozygosity of spontaneous with colchicine-induced doubled haploids in hexaploid wheat and spontaneously doubled hexaploid triticale plants using microsatellite markers. They found heterozygosity for one to three loci in 12% of spontaneous hexaploid wheat plants, in 11% of induced hexaploid wheat plants and in 4% of spontaneously doubled triticale plants. They proposed that first or second division restitution during meiosis leads to unreduced gametes or that possibly transposition events could explain these irregularities and the occasional aneuploids. Less extensive studies with similar results were found for rice (Afza et al. 2001). Lim et al. (2001) examined the meiotic nuclear restitution mechanism in lily interspecific hybrids using FISH and GISH technology and observed that the formation of stainable 2*n*-pollen ranged from 3–30%. Meiotic irregularities could also give rise to segregation, aneuploidy or polyploidy in gynogenetically haploid systems.

5.2 Unreduced Gametes

Unreduced gametes are related to meiotic irregularities covered in Section 5.1 above and are utilized in breeding autotetraploid crops. First division restitution (FDR) or second division restitution (SDR) during meiosis are a source of unreduced gametes (2*n* gametes) that could produce plants with double the gametic chromosome number (Mok and Peloquin 1975; Rao and Suprasanna 1996). Their discovery has led to many breeding schemes for their use in producing heterozygous 4× progeny from dihaploids or crosses with other diploid potato species (Peloquin et al. 1989; Bingham et al. 1994). While FDR should produce more heterozygosity than SDR at the 4× level, it has been difficult to show improved plant production (Douches and Mass 1998).

Lough et al. (2001) used flow cytometry to compare potato monoploids ($2n=1\times=12$) produced by anther or microspore culture with those produced from wide hybridization. The wide hybridization in this situation is thought to produce dihaploids via parthenogenesis rather than chromosome elimination. They observed that gynogenic monoploid plants contained 5% more haploid cells while the androgenic monoploids exhibited 27% more endopolyploidy cells at the tetraploid level and were larger in size and produced more tubers. Their suggested explanation was that selection was inherent in the monoploid derivation mechanism for potato.

In reality, the culture of 2*n* microspores would lead to plants that would appear to be completely fertile doubled haploids but would likely have some heterozygosity.

6 Chromosomal Variation Associated with Chromosome Doubling

Chromosomal variation has been known to occur in tissue cultures for many years and has often been described as somaclonal variation where it is unknown whether the causes are chromosomal changes or mutations in nuclear or cytoplasmic genomes. In the production of haploids through culture, this variation has been termed **gametoclonal variation.** In anther cultures, early work was quite extensive in Chinese laboratories and has been reviewed by Sun (1978) and Hu (1983). More recently, Ziauddin and Kasha (1990) and Logue (1996) also reviewed the genetic stability of haploid cell cultures and derived doubled haploids, and readers are referred to their tabular lists of species and reported variations in chromosomes. The review of Logue (1996) is extensive and covers topics such as DNA amplification, structural and molecular changes and cytoplasmic changes, and readers are referred to this publication for details rather than repeating those variations here. The report of initiation of doubling at the globular stage in *Brassica* microspore cultures (XuHan et al. 1999) is most likely the start of tissue differentiation which is often accompanied by endopolyploidy (Logue 1996).

Hu (1996) reviewed the details of the chromosomal variations observed in wheat (*Triticum aestivum* L.) haploid production, with emphasis on androgenetic methods. In earlier studies of anther and microspore culture, plants were most often regenerated from callus produced from the culture of the anthers containing microspores and it was soon established that chromosome variations occurred during callus growth and these variations were reflected in the progeny plants (Hu 1983). This relationship was closer in polyploid species such as wheat than in the diploid species because survival of plants with aneuploid chromosome numbers or chromosome changes was not as frequent in diploids. Thus, one of the objectives in androgenesis was to develop culture systems that would provide more direct embryogenesis from microspores, rather than callus. Such improvements have been a long tedious process because of the many factors in culture environment, culture media and pretreatments, most of which are species specific (Ziauddin and Kasha 1990; Logue 1996).

Early literature proposed that the chromosome doubling agent colchicine caused morphological mutations (Jensen 1974; Logue 1996). It was, however, difficult to determine whether the changes were temporary due to long treatments with colchicine or in frequencies greater than observed in non-colchicine-treated plants. Foisset et al. (1997) compared segregation distortion in androgenesis in *Brassica napus* when the doubled haploids were produced spontaneously or by colchicine. They found no differences in distortion around the parental genes that differed for culturability. At this time there is no clear consensus that colchicine is mutagenic, but it can be toxic to researchers and care should be taken when handling chromosome-doubling agents.

Doubled haploids produced by anther culture and chromosome elimination have been compared for the production of random samples of the gametes as haploids (Devaux et al. 1995). This is important in molecular genetic map construction using populations of doubled haploid plants. If there is a difference in recombination rate or survival between male and female gametes it would affect the maps produced. Numerous studies have found segregation distortion from the anther culture and this appears to be related to the location of parental genes that cause differences in anther culture productivity. The populations produced from chromosome elimination procedures usually tend to be a more random sample of gametes, but some genotype differences in response also occurs (see Devaux and Pickering, Chap. II.3, this Vol.).

7 Conclusions

The accumulating evidence indicates that induction of embryogenesis and of chromosome doubling in androgenetic systems are closely related. Some of the stresses leading to induction appear to affect the organization of the microtubule system that initiates cell wall formation. Whether one uses microtubule inhibitors or other forms of stress such as high or low temperatures for the induction of embryogenesis, the same results will be observed, namely the formation of a symmetric PMI in the microspore. It is apparent based on a number of studies that doubling will also occur during the next one or two nuclear divisions in the microspore-derived structure, most often in the VN, but the mechanism of doubling is not always clear. It is possible that some induction agents such as mannitol may delay cell wall formation and lead to nuclear fusion, while antimicrotubule agents may induce endomitosis. The concentration of the chemicals may determine whether symmetric divisions and fusion or blocking of divisions (endomitosis) may occur. Both the stage of pretreatment and the type of pretreatment can influence doubling and the particular method employed may vary with the species. The use of microtubule inhibitors at normal growing temperatures can also produce good induction of embryogenesis in microspores and higher frequencies of chromosome doubling than heat treatment alone. A number of reports indicate that antimicrotubule agents enhance the production of embryos from material that had already been pretreated by a different stress.

Research on improving doubled haploid production through androgenesis is now focused on two areas, one being the testing of antimicrotubule agents in the initial stages of microspore culture, and the other the molecular analysis of genes associated with the induction of androgenesis. Many genes have already been identified that influence tubulins and their organization and are thus related to the asymmetric PMI. Other genes may influence nuclear fusion and gene actions as a result of treatments such as heat shock proteins. The length of treatment is likely dependent upon the synchrony of the microspores in culture and this can vary with the species and the methods of preparing the microspores for culture. What is needed to clarify the origins of doubling in microspores is to be able to follow them through cultures as conducted by Indrianto et al. (2001).

For gynogenetic systems of producing haploids, the treatment at the single haploid cell stage is very difficult or often impossible. However, doubling treatments should be carried out as soon as feasible in order to get a higher proportion of the plant doubled, i.e. approaching the levels feasible with androgenetic systems. In order to reduce the toxic effects of colchicine, treatments should be as short as possible and carried out on healthy and rapidly growing embryos or seedlings. Agents such as DMSO and wetting agents are beneficial in improving the tissue uptake of the inducing agent. Washing and subsequent GA treatments may enhance the recovery of treated materials, reducing the mortality from the treatment. Doubling of a high proportion of treated plants has been achieved in crop plants such as *Brassica* and the cereals. Future efforts should be directed towards doubling the initial haploid cells or critical tissues of the reproductive systems. Molecular analyses of genes involved in embryogenesis and chromosome doubling should provide additional clues to the direction of future research on chromosome doubling.

References

Afza R, Xie J, Shen M, Zapata-Arias FJ, Fundi HK, Lee K-S, Bobadilla-Mucino E, Kodym A (2001) Detection of androclonal variation in anther-cultured rice lines using RAPDs. In Vitro Cell Dev Biol Plant 37:644–647

Antoine-Michard S, Beckert M (1997) Spontaneous versus colchicine-induced chromosome doubling in maize anther culture. Plant Cell Tissue Organ Cult 48:203–207

Arzani A, Darvey NL (2001) The effect of colchicine on triticale anther-derived plants: microspore pretreatment and haploid plant treatment using a hydroponic recovery system. Euphytica 122:235–241

Barclay IR (1975) High frequencies of haploid production in wheat (*Triticum aestivum* L.) by chromosome elimination. Nature 256:410–411

Barnabas B, Obert B, Kovacs G (1999) Colchicine, an efficient genome doubling agent for maize (*Zea mays* L.) microspores cultured in anthers. Plant Cell Rep 18:858–862

Binarova P, Hause G, Cenklova V, Cordewener JHG, Van Lookeren Campagne MM (1997) A short severe heat shock is required to induce embryogenesis in late bicellular pollen of *Brassica napus* L. Sex Plant Reprod 10:200–208

Bingham E, Goose R, Woodfield D, Kidwell K (1994) Complementary gene interactions in alfalfa are greater in autotetraploids than in diploids. Crop Sci 34:823–829

Bishonoi U, Jain R, Rohilla J, Chowdhury V, Gupta K, Chowdhury J (2000) Anther culture of recalcitrant indica × basmati rice hybrids. Euphytica 114:93–101

Bordes J, Dumas de Vaulx R, Lapierre A, Pollacsek M (1997) Haplodiploidization of maize (*Zea mays* L.) through induced gynogenesis assisted by glossy markers and its use in breeding. Agronomie 17:291–297

Bouvier L, Guerif P, Djulbic M, Durel D, Chevreau E, Lespinasse Y (2002) Chromosome doubling of pear haploid plants and homozygosity assessment using isozyme and microsatellite markers. Euphytica 123:255–262

Chase SS (1949) The reproductive success of monoploid maize. Am J Bot 36:795–796

Chase SS (1974) Utilization of haploids in plant breeding: breeding diploid species. In: Kasha KJ (ed) Haploids in higher plants, advances and potential. University of Guelph, Guelph, pp 211–230

Chen CC, Howarth MJ, Peterson RL, Kasha KJ (1984) Ultrastructure of androgenic microspores of barley during the early stages of anther cultures. Can J Genet Cytol 26:484–491

Chen JL, Beversdorf WD (1992) Production of spontaneous diploid lines from isolated microspores following cryopreservation in spring rapeseed (*Brassica napus* L). Plant Breed 108:324–327

Coe EH Jr (1959) A line of maize with high haploid frequency. Am Nat 93:381–382

Cordewener JHG, Custers JBM, Van Lookeran Campagne MM (1998) Microspore culture. A model for investigating the role of stress in the induction of embryogenesis. In: Chupeau Y, Caboche M, Henry Y (eds) Androgenesis and haploid plants. INRA edn. Springer, Berlin Heidelberg New York, pp 54–68

De Fossard RA (1974) Terminology in 'haploid' work. In: Kasha KJ (ed) Haploids in higher plants: advances and potential. University of Guelph, Guelph, pp 403–410

Devaux P, Kilian A, Kleinhofs A (1995) Comparative mapping of the barley genome with male and female-derived doubled haploid populations. Mol Gen Genet 249:600–608

Dewitte W, Murray JAH (2003) The plant cell cycle. Annu Rev Plant Biol 54:235–264

Douches DS, Mass DL (1998) Comparison of FDR and SDR-derived tetraploid progeny from using haploids of *Solanum tuberosum* L. that produce mixed modes of 2n eggs. Theor Appl Genet 97:1307–1313

Eady C, Lindsay K, Twell D (1995) The significance of microspore division and division symmetry for vegetative cell-specific transcription and generative cell differentiation. Plant Cell 7:65–74

Eder J, Chalyk S (2002) In vitro haploid induction in maize. Theor Appl Genet 104:703–708

Engvild KC, Linde-Laursen I, Lundqvist A (1972) Anther cultures of *Datura innoxia*: floral bud stage and embryoid level of ploidy. Hereditas 72:331–332

Fan Z, Armstrong K, Keller W (1988) Development of microspores in vivo and in vitro in *Brassica napus*. Protoplasma 147:191–199

Foisset N, Delourme R, Lucas MO, Renard M (1997) In vitro androgenesis and segregation distortion in *Brassica napus* L: spontaneous versus colchicine-doubled lines. Plant Cell Rep 16:464–468

Geoffriau E, Kahne R, Bellamy C, Rancillac M (1997) Ploidy stability and in vitro chromosome doubling in gynogenic clones of onion (*Allium cepa* L.). Plant Sci 122:201–208

Guerel S, Guerel E, Kaya Z (2000) Doubled haploid plant production from unpollinated ovules of sugar beet (*Beta vulgaris* L.). Plant Cell Rep 19:1155–1159

Guo YD, Pulli S (2000a) An efficient androgenic embryogenesis and plant regeneration method through isolated microspore culture in timothy (*Phleum pratense* L.). Plant Cell Rep 19:761–767

Guo YD, Pulli S (2000b) Isolated microspore culture and plant regeneration in rye (*Secale cereale* L). Plant Cell Rep 19:875–880

Guha S, Maheshwari SC (1967) Development of embryoids from pollen grains of *Datura* in vitro. Phytomorphology 17:454–461

Hansen NJP, Andersen SB (1996) In vitro chromosome doubling potential of colchicine, oryzalin, trifluralin and APM in *Brassica napus* microspore culture. Euphytica 88:159–164

Hansen NJP, Andersen SB (1998a) Efficient production of doubled haploid wheat plants by in vitro treatment of microspores with trifluralin or APM. Plant Breed 117:401–405

Hansen NJP, Andersen SB (1998b) In vitro chromosome doubling with colchicine during microspore culture in wheat (*Triticum aestivum* L.). Euphytica 102:101–108

Hansen AL, Gertz A, Joersbo M, Andersen SB (2000) Chromosome doubling in vitro with amiprophos-methyl in *Beta vulgaris* ovule culture. Acta Agri Scand 50:89–95

Heberle-Bors E (1989) Isolated pollen culture in tobacco: plant reproductive development in a nutshell. Sex Plant Reprod 2:1–10

Heberle-Bors E (1998) Experimental control of pollen development. In: Chupeau Y, Caboche M, Henry Y (eds) Androgenesis and haploid plants. INRA edn. Springer, Berlin Heidelberg New York, pp 38–53

Herrera JC, Moreno LG, Acuna JR, De Pena M, Osorio D (2002) Colchicine-induced microspore embryogenesis in coffee. Plant Cell Tissue Organ Cult 71:89–92

Hoekstra S, van Bergen S, van Brouwershaven IR, Schilperoot RA, Wang M (1997) Androgenesis in *Hordeum vulgare* L.: effects of mannitol, calcium and abscisic acid on anther pretreatment. Plant Sci 126:211–218

Horlow C, Raquin C (1998) A critical analysis of existing haploidization techniques. In: Chupeau Y, Caboche M, Henry Y (eds) Androgenesis and haploid plants. INRA edn. Springer, Berlin Heidelberg New York, pp 7–23

Hu H (1983) Genetic stability and variability of pollen-derived plants. In: Sen SK, Giles KL (eds) Plant cell culture in crop improvement. Basic life science, vol 22. Plenum Press, New York, pp 145–157

Hu H (1996) Chromosome engineering in the Triticeae using pollen-derived plants (CETPP). In: Jain SM, Sopory SK, Veilleux RE (eds) In vitro haploid production in higher plants, vol 2. Kluwer, Dordrecht, pp 203–223

Hu T, Kasha KJ (1997) Improved embryogenesis from isolated microspore culture of wheat (*Triticum aestivum* L.) through ovary co-culture. Plant Cell Rep 16:520–525

Hu TC, Kasha KJ (1999) A cytological study of pretreatments used to improve isolated microspore cultures of wheat (*Triticum aestivum* L.) cv. Chris. Genome 42:432–441

Ilic-Grubor K, Attree SM, Fowke LC (1998) Induction of microspore-derived embryos of *Brassica napus* L with polyethylene glycol (PEG) as osmoticum in a low sucrose medium. Plant Cell Rep 17:329–333

Indrianto A, Barinova I, Touraev A, Heberle-Bors E (2001) Tracking individual wheat microspores in vitro: identification of embryogenic microspores and body axis formation in the embryo. Planta 212:163–174

Jain SM, Sopory SK, Veilleux RE (eds) (1996/1997) In vitro haploid production in higher plants, vol 1–5. Kluwer, Dordrecht

Jensen CJ (1974) Chromosome doubling techniques in haploids. In: Kasha KJ (ed) Haploids in higher plants, advances and potential. University of Guelph, Guelph, pp 153–190

Kao KN, Michayluk MR (1974) A method for high frequency intergeneric fusion of plant protoplasts. Planta 115:355–367

Kasha KJ (2003) Plants from haploid cells. In: Saxena PK (ed) Single cells to plants: concepts and applications. Oxford and IBH Publ Co Pvt Ltd, New Delhi

Kasha KJ, Kao KN (1970) High frequency haploid production in barley (*Hordeum vulgare* L.). Nature 225:874–876

Kasha KJ, Hu TC, Oro R, Simion E, Shim YS (2001) Nuclear fusion leads to chromosome doubling during mannitol pretreatment of barley (*Hordeum vulgare* L.) microspores. J Exp Bot 52:1227–1238

Kato A (2002) Chromosome doubling of haploid maize seedlings using nitrous oxide gas at the flower primordial stage. Plant Breed 121:370–377

Keller WA, Melchers G (1973) The effect of high pH and calcium on tobacco leaf protoplast fusion. Z Naturforsch 28c:737–741

Kermicle JL (1969) Androgenesis conditioned by a mutation in maize. Science 166:1422–1424

Kiss J, Kondrak M, Toerjek O, Kiss E, Gyulai G, Mazik-Toekei K, Heszky L (2001) Morphological and RAPD analysis of poplar trees of anther culture origin. Euphytica 188:213–221

Kiviharja E, Puolimatka M, Saastamoinen M, Pehu E (2000) Extension of anther culture to several genotypes of cultivated oats. Plant Cell Rep 19:674–679

Laurie DA, Bennett MD (1986) Wheat by maize hybridization. Can J Genet Cytol 28:313–316

Li H, Devaux P (2003) High frequency regeneration of barley doubled haploid plants from isolated microspore culture. Plant Sci 164:379–386

Lim KB, Ramanna MS, de Jong JH, Jacobsen E, van Tuyl JM (2001) Indeterminate meiotic restitution (IMR): a novel type of meiotic nuclear restitution mechanism detected in interspecific lily hybrids by GISH. Theor Appl Genet 103:219–230

Lionneton E, Beuret W, Delaitre C, Ochatt S, Rancillac M (2001) Improved microspore culture and doubled-haploid plant regeneration in the brown condiment mustard (*Brassica juncea*). Plant Cell Rep 20:126–130

Liu W, Zheng MY, Konzak CF (2002) Improving green plant production via isolated microspore culture in bread wheat (*Triticum aestivum* L.). Plant Cell Rep 20:821–824

Logue SJ (1996) Genetic stability in microspore-derived doubled haploids. In: Jain SM, Sopory SK, Veilleux R (eds) In vitro haploid production in higher plants, vol 2. Kluwer, Dordrecht, pp 1–51

Lough RC, Varrieur JM, Veilleux RE (2001) Selection inherent in monoploid derivation mechanisms for potato. Theor Appl Genet 103:178–184

Magnard J-L, Le Deunff E, Domenech J, Rogowsky PM, Testillano PS, Rougier M, Risueño MC, Vergne P, Dumas C (2000) Genes normally expressed in the endosperm are expressed at early stages of microspore embryogenesis in maize. Plant Mol Biol 44:559–574

Martin B, Widholm JM (1996) Ploidy of small individual embryo-like structures from maize anther cultures treated with chromosome doubling agents and calli derived from them. Plant Cell Rep 15:781–785

McCormick S (1993) Male gametophyte development. Plant Cell 5:1265–1275

Mentewab A, Sarrafi A (1997) Androgenic ability and chromosome doubling by different colchicine treatments in anther culture of hexaploid wheat genotypes (*Triticum aestivum* L.). Cereal Res Commun 25:897–903

Mochida K, Tsujimoto H (2001) Production of wheat haploids by pollination with Job's tears (*Coix lachryma-jobi* L.). J Hered 92:81–83

Mok D, Peloquin SJ (1975) Three mechanisms of *2n* pollen formation in diploid potatoes. Can J Genet Cytol 17:217–225

Muranty H, Sourdille P, Bernard S, Bernard M (2002) Genetic characterization of spontaneous diploid androgenetic wheat and triticale plants. Plant Breed 121:470–474

Nitsch C (1974) Pollen culture – a new technique for mass production of haploid and homozygous plants. In: Kasha KJ (ed) Haploids in higher plants, advances and potential. University of Guelph, Guelph, pp 123–135

Pechan PM, Keller WA (1988) Identification of potentially embryogenic microspores in *Brassica napus*. Physiol Plant 74:377–384

Pechan PM, Smykal P (2001) Androgenesis: affecting the fate of the male gametophyte. Physiol Plant 111:1–8

Peloquin S, Yerk G, Werner J, Darmo E (1989) Potato breeding with haploids and *2n* gametes. Genome 31:1000–1004

Raghavan V (1976) Role of the generative cell in androgenesis in henbane. Science 191:338–339

Rao PS, Suprasanna P (1996) Methods to double haploid chromosome numbers. In: Jain SM, Sopory SK, Veilleux RE (eds) In vitro haploid production in higher plants, vol 1. Kluwer, Dordrecht, pp 317–339

Redha A, Islam SMS, Büter B, Stamp P, Schmid JE (2000) The improvement in regenerated doubled haploids from anther culture of wheat by anther transfer. Plant Cell Tissue Organ Cult 63:167–172

Reynolds TL (1997) Pollen embryogenesis. Plant Mol Biol 33:1–10

Rose MD (1996) Nuclear fusion in the yeast *Saccharomyces cerevisiae*. Annu Rev Cell Dev Biol 12:663–695

Saidi N, Chlyah O, Chlyah H (1998) Production of green haploid durum wheat plants by pollination of wheat by maize. Can J Bot 76:652–656

Saisingtong S, Schmid JE, Stamp P, Büter B (1996) Colchicine-mediated chromosome doubling during anther culture of maize (*Zea mays* L.). Theor Appl Genet 92:1017–1023

Sharma H, Yang Y, Ohm H (2002) An assessment of doubled haploid production in soft red winter wheat by wheat × corn wide crosses. Cereal Res Commun 30:269–275

Shim YS, Kasha KJ (2003) The influence of pretreatment on cell stage progression and the time of DNA synthesis in barley (*Hordeum vulgare* L.) uninucleate microspores. Plant Cell Rep 21:1065–1071

Sibi ML, Kobaissi A, Shekafandeh A (2001) Green haploid plants from unpollinated ovary culture in tetraploid wheat (*Triticum durum* Defs.). Euphytica 122:351–359
Simmonds DH, Keller WA (1999) Significance of preprophase bands of microtubules in the induction of microspore embryogenesis of *Brassica napus*. Planta 208:383–391
Stober A, Hess D (1997) Spike pretreatments, anther culture conditions, and anther culture response of 17 German varieties of spring wheat (*Triticum aestivum* L.). Plant Breed 116:443–447
Subrahmanyam NC, Kasha KJ (1975) Chromosome doubling of barley haploids by nitrous oxide and colchicine treatments. Can J Genet Cytol 17:573–583
Sun C-S (1978) Androgenesis of cereal crops. In: Proc Symp on Plant Tissue Culture. Science Press, Peking, pp 117–124
Sunderland N (1974) Anther culture as a means of haploid production. In: Kasha KJ (ed) Haploids in higher plants, advances and potential. University of Guelph, Guelph, pp 91–122
Sunderland N, Evans LJ (1980) Multicellular pollen formation in cultured barley anthers. J Exp Bot 31:501–514
Sunderland N, Collins GB, Dunwell JM (1974) The role of nuclear fusion in pollen embryogenesis of *Datura innoxia* Mill. Planta (Berl) 117:227–241
Testillano PS, Ramirez C, Domenech J, Coronado MJ, Vergne P, Matthys Rochon E, Risueno MC (2002) Young microspore-derived maize embryos show two domains with defined features also present in zygotic embryogenesis. Int J Dev Biol 46:1035–1047
Thiebaut J, Kasha KJ (1978) Modification of the colchicine technique for chromosome doubling of barley haploids. Can J Genet Cytol 20:513–521
Thiebaut J, Kasha KJ, Tsai A (1979) Influence of plant development stage, temperature and plant hormones on chromosome doubling of barley haploids using colchicine. Can J Bot 57:480–483
Thomas WTB, Forster BP, Gertsson B (2003) Doubled haploids in breeding. In: Maluszynski M, Kasha KJ, Forster BP, Szarejko I (eds) Doubled haploid production in crop plants, a manual. Kluwer, Dordrecht, pp 337–350
Tosca A, Pandolfi R, Citterio S, Fasoli A, Sgorbati S (1996) Determination by flow cytometry of the chromosome doubling capacity of colchicine and oryzalin in gynogenetic haploids of *Gerbera*. Plant Cell Rep 14:455–458
Touraev A, Pfosser M, Vincente O, Heberle-Bors E (1996) Stress as the major signal controlling the developmental fate of tobacco microspores: towards a unified model of induction of microspore/pollen embryogenesis. Planta 200:144–152
Touraev A, Vincente O, Heberle-Bors E (1997) Initiation of microspore embryogenesis by stress. Trends Plant Sci 2:297–302
Touraev A, Pfosser M, Heberle-Bors E (2001) The microspore: a haploid multipurpose cell. Adv Bot Res 35:53–109
Turcotte EL, Feaster CV (1974) Semigametic production of cotton haploids. In: Kasha KJ (ed) Haploids in higher plants, advances and potential. University of Guelph, Guelph, pp 53–64
Twell D, Howden R (1998) Mechanisms of asymmetric division and cell fate determination in developing pollen. In: Chupeau Y, Caboche M, Henry Y (eds) Androgenesis and haploid plants. INRA edn. Springer, Berlin Heidelberg New York, pp 69–103
Twell D, Park SK, Lalanne E (1998) Asymmetric division and cell-fate determination in developing pollen. Trends Plant Sci 3:305–310
Vantard M, Cowling R, Delichère C (2000) Cell cycle regulation of the microtubular cytoskeleton. Plant Mol Biol 43:691–703
Veilleux R (1985) Diploid and polyploid gametes in crop plants: mechanisms of formation and utilization in plant breeding. Plant Breed Rev 3:253–288
Wedzony M, Marcinska I, Pontika A, Slusarkiewicz-Jarzina A, Wozna J (1998) Production of doubled haploids in triticale (× *Triticosecale* Wittin.) by means of crosses with maize (*Zea mays* L.) using picloram and dicamba. Plant Breed 117:211–215
Wheatley WG, Marsolais AA, Kasha KJ (1986) Microspore growth and anther staging in barley anther culture. Plant Cell Rep 5:47–49

Wilson HM, Mix G, Foroughi-Wehr B (1978) Early microspore divisions and subsequent formation of microspore calluses at high frequency in anthers of *Hordeum vulgare* L. J Exp Bot 108:227–238

XuHan X, Jing H-C, Cheng X-F, Iwanowska A, Kieft H, Bergervost JHW, Groot SPC, Bino RJ, van Lammeren AMM (1999) Polyploidization in embryogenic microspore cultures of *Brassica napus* L cv. Topas enables the generation of doubled haploid clones by somatic embryogenesis. Protoplasma 208:240–247

Zaki MAM, Dickenson HG (1991) Microspore-derived embryos in *Brassica*: the influence of division symmetry in pollen mitosis 1 to embryogenic development. Sex Plant Reprod 44:48–55

Zamani I, Kovács G, Gouli-Vavdinoudi E, Roupakias DG, Barnabás B (2000) Regeneration of fertile doubled haploid plants from colchicine-supplemented media in wheat anther culture. Plant Breed 119:461–465

Zhao J-P, Simmonds DH, Newcombe W (1996) High frequency production of doubled haploid plants of *Brassica napus* cv. Topas derived from colchicine-induced microspore embryogenesis without heat shock. Plant Cell Rep 15:668–671

Zhou WJ, Tang GX, Hagberg P (2002a) Efficient production of doubled haploid plants by immediate colchicine treatment of isolated microspores in winter *Brassica napus*. Plant Growth Regul 37:185–192

Zhou WJ, Hagberg P, Tang GX (2002b) Increasing embryogenesis and doubling efficiency by immediate colchicine treatment of isolated microspores in spring *Brassica napus*. Euphytica 128:27–34

Ziauddin A, Kasha KJ (1990) Genetic stability in haploid cell cultures. In: Bajaj YPS (ed) Biotechnology in agriculture and forestry, vol 12. Haploids in crop improvement I. Springer, Berlin Heidelberg New York, pp 83–89

I.8 Utilization of Microspore-Derived Embryos

Y. TAKAHATA[1], H. Fukuoka[2], and K. WAKUI[3]

1 Introduction

Embryos derived from microspores in anther culture and isolated microspore culture are of great importance in plant breeding and basic science. In plant breeding, homozygous lines are utilized as final varieties in self-pollinating crops and as parents in F1 hybrid cultivars in cross-pollinating crops. The advantages of haploids and doubled haploids produced from male gametic cells for the production of homozygous lines have been described by many researchers (Keller et al. 1987; Bajaj 1990; Khush and Virmani 1997; Takahata 1997; other chapters in this book).

In addition to practical plant breeding, a microspore culture system, which produces a large number of embryos, has several advantages over anther culture in basic science and applied genetic manipulation because of the haploid and/or doubled haploid production system from a single cell. An effective microspore culture system can be utilized as a model system of embryogenesis and a target of genetic manipulation such as mutation, in vitro selection, transformation and artificial seed (Fig. 1).

In this chapter, we describe utilization of genetic manipulation of microspore-derived embryos. Though microspore culture has been reported in many species, an effective culture system is limited to species of a few genera such as *Brassica*, *Nicotiana* and *Hordeum*. Of these genera, many studies have concentrated on *Brassica*. Although we mainly deal with the data concerning *Brassica* in this chapter, we feel that the data can be utilized universally.

2 Mutation

A microspore culture system of *Brassica* has several advantages in terms of mutagenesis and selection of mutants, namely the use of a large single cell population, low frequency of somaclonal variation, efficient and uniform

[1] Faculty of Agriculture, Iwate University, Morioka 020-8550, Japan
[2] National Institute of Vegetable and Tea Science, Mie 514-2392, Japan
[3] Junior College, Tokyo University of Agriculture, Setagaya 156-8502, Japan

Biotechnology in Agriculture and Forestry, Vol. 56
Haploids in Crop Improvement II (ed. by C.E. Palmer, W.A. Keller, and K.J. Kasha)

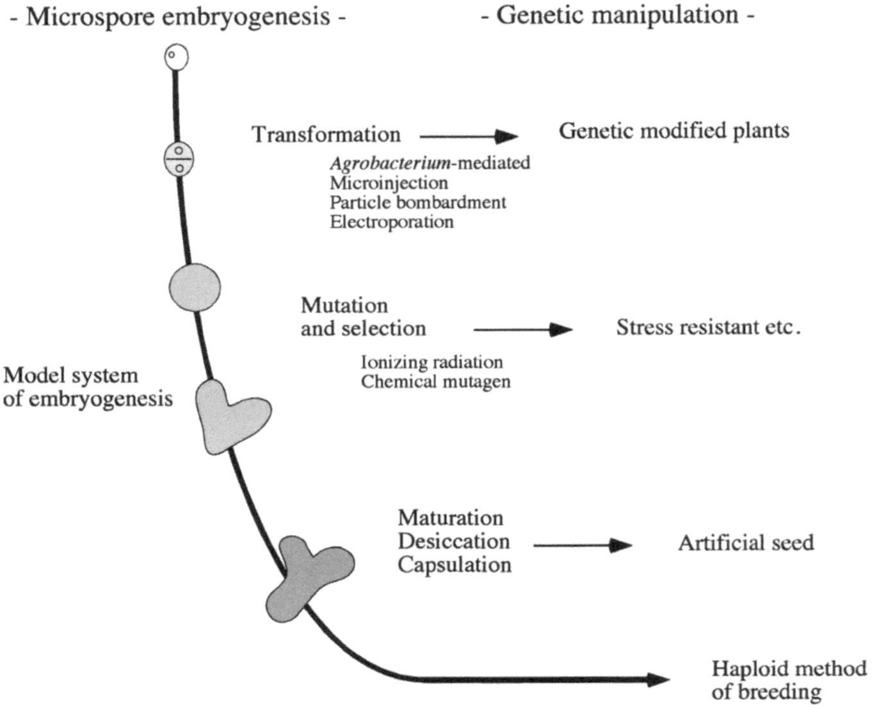

Fig. 1. Utilization of microspore culture system for genetic manipulation

application of mutagen, efficient isolation of recessive mutants and easy in vitro selection (Swanson et al. 1989; Huang 1992). For inducing variation, ionizing radiation, such as ultraviolet (UV), X-ray and gamma-ray, and chemical mutagens have been used. Huang (1992) indicated that the optimum time for applying the mutagen is 16–24 h after culture when microspores first start cell division. The results of Zhang and Takahata (1999), who reported that Chinese cabbage (*B. campestris* syn. *rapa*) microspores UV-irradiated after 4–24 h of culture were more sensitive, especially those irradiated after 8 h of culture, are consistent with this idea. The effect of different mutagens on microspore culture has been assayed using a decrease in embryo yield as an index. The LD50 of *B. napus* microspore embryogenesis was reported to correspond to a dose of 13 and 10.5 Gy of gamma- and X-radiation, respectively (MacDonald et al. 1991; Maluszynski et al. 1996). UV light, which is more convenient and shows less decrease in regeneration ability (Maluszynski et al. 1996), has been used with several species. The LD50 for *B. napus* was estimated to be exposed for 20 and 40 s when irradiated with UV at an incident dose of 3.3 and 2.0 $W\,m^{-2}\,s^{-1}$, respectively (Ahmad et al. 1991; T. Mikami and Y. Takahata, unpubl.). Similar results were obtained with Chinese cabbage whose LD50 was 12 s with a dose rate of 2.0 $W\,m^{-2}\,s^{-1}$

(Zhang and Takahata 1999). In contrast, a longer exposure time of 8 min with a dose rate of 3.0 J m^{-2} s^{-1} was reported in *B. carinata* (Barro et al. 2002). When the effects of ethyl methanesulfonate (EMS) on embryogenesis were estimated in *B. napus*, treatment with 0.1% EMS for 1 h decreased the embryo yield by 45% (Mikami et al. 1993).

Many mutants showing herbicide tolerance, disease tolerance or alteration of seed storage substances have been obtained in *Brassica* spp. Herbicide (chlorsulfuron and imidazolinone)-tolerant mutants of *B. napus* were obtained after in vitro selection of microspores treated with 20 μM ethyl nitrosourea and 0.5 Krad gamma-irradiation (Swanson et al. 1988, 1989). Ahmad et al. (1991) obtained herbicide (chlorsulfuron)-resistant mutants and *Alternaria brassicicola*-resistant ones in in vitro and in planta selection, respectively, after UV irradiation on *B. napus* microspores. *Erwinia carotovora*-resistant mutants of Chinese cabbage were selected in UV-irradiated microspore-derived embryos by in vitro selection using the culture filtrates of the pathogen after confirming the effect of the culture filtrates on embryogenesis (Zhang and Takahata 1999).

The microspore-derived embryo system of *Brassica* is applicable in modification of seed storage substances such as fatty acid composition and glucosinolates content. Seed storage composition is usually estimated in the seeds produced in M1 plants. However, since fatty acid composition of the microspore-derived embryos is similar to that of zygotic embryos (Taylor et al. 1990; Chen and Beversdorf 1991; Pomeroy et al. 1991), non-destructive selection methods using single cotyledons of a microspore-derived embryo for analysis of storage reserves are possible (Kott et al. 1996). The mutated lines having elevated oleic acids, those having reduced linolenic acid and those having reduced saturated fatty acid have been obtained in *B. napus* (Kott et al. 1996) and *B. rapa* (Ferrie and Keller 2002). Barro et al. (2001, 2002) reported the modification of erucic acid and glucosinolate contents through either UV or EMS treatment of *B. carinata*-cultured microspores.

Recently, in barley (*Hordeum vulgare*), 8.6–15.6% of morphological and physiological mutants such as dwarf, late heading and male sterility have been obtained by applying 10^{-5}–10^{-4} M sodium azide to the microspores for 1 h (Castillo et al. 2001).

3 Dry Artificial Seed

Microspore-derived embryos as well as somatic embryos can be utilized as artificial seeds. Although various types of artificial seeds have been proposed (Fujii et al. 1987), they are roughly divided into two types: hydrated and desiccated. However, many problems remain to be overcome, such as coating materials, long-term storage and plant conversion under various soil conditions. To date, hydrated artificial seeds using microspore-derived embryos have been reported in barley and wheat (Datta and Potrykus 1989; Datta and Schmid 1997). In comparison with the hydrated type, desiccated artificial seeds proposed by Kitto and Janick (1985) are considered to be a convenient form for practical utilization. If desiccated embryos had the same capacity as true seeds, naked desiccated embryos would be the simplest form of artificial seeds, and have the advantages of easier storage and a lower delivery cost.

3.1 Induction of Desiccation Tolerance

In vitro-derived embryos have similar morphological and physiological traits to zygotic embryos. However, an important difference between zygotic and in vitro-derived embryos is that the former lose water and acquire desiccation tolerance during their later developmental stage. In contrast, the latter lose their viability after desiccation. If embryo desiccation technology were developed, this not only would provide practical application such as dry artificial seeds and germplasm preservation, but also could be a model system to analyze the mechanism of desiccation tolerance of true seeds.

Successful induction of desiccation tolerance in microspore-derived embryos has been reported in *Brassica* spp. (*B. napus*, *B. oleracea* and *B. campestris*) (Anandarajah et al. 1991; Senaratna et al. 1991; Takahata et al. 1992, 1993; Brown et al. 1993; Wakui et al. 1994) and barley (Ryan et al. 1999), as well as in somatic embryos of several plants such as grape (Gray 1989), orchardgrass (Gray et al. 1987), celery (Kim and Janick 1991), alfalfa (Senaratna et al. 1989, 1990) and carrot (Iida et al. 1992). Desiccation tolerance of the embryos has been induced by various treatments such as exogenous application of abscisic acid (ABA), different kinds of carbohydrates, proline and temperature stress. In microspore-derived embryos of *Brassica*, ABA has a dramatic effect on the induction of desiccation tolerance (Brown et al. 1993; Takahata et al. 1993; Wakui et al. 1994). Induction of desiccation tolerance is not dependent on species or genotypes, but on ABA concentration, the length of exposure time to ABA and the developmental stage of embryos. The highest frequency of germination and plant conversion from desiccated embryos was obtained when embryos were treated with 100 μM ABA in *B. napus* and *B. oleracea* and 10 μM ABA in *B. campestris* (Brown et al. 1993; Takahata et al. 1993; Wakui et al. 1994). In contrast, in the absence of ABA, almost no embryos could survive after desiccation. Although 7 days' ABA exposure gave the highest induction of desiccation tolerance, 24-h exposure induced desiccation tolerance in *B. napus* and *B. oleacea* (Brown et al. 1993; Takahata et al. 1993). Wakui and Takahata (2002) observed that even though 7-day ABA exposure gave the highest frequency of germination (91.4% of *B. napus*, 79.7% of *B. campestris*) and plant regeneration (85.7% of *B. napus*, 70.2% of *B. campestris*), a 48-h exposure induced almost complete desiccation tolerance and a 12-h exposure induced some desiccation tolerance (Fig. 2). It has been reported that cotyledonary stage embryos are most responsive to ABA,

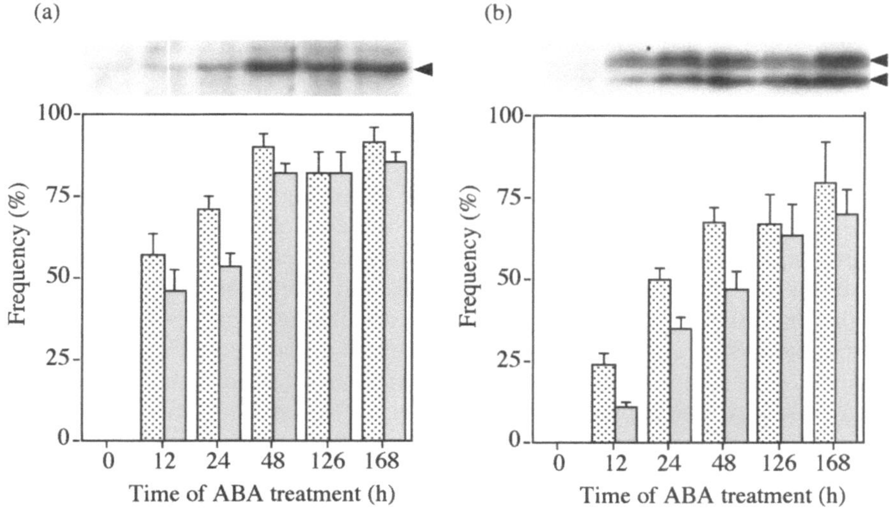

Fig. 2. Time course of the accumulation of ME-leaN4 and ME-leaC4 protein (*above*) and induction of desiccation tolerance (*below*) in microspore-derived embryos of *B. napus* (**a**) and *B. campestris* (**b**), which were treated with 100 and 10 μM ABA, respectively. Western blot analysis was carried out using antibody raised against ME-leaN4 polypeptides. *Arrowhead* indicates Lea protein. *Dotted shading* Germination; *solid shading* plant regeneration. (Wakui and Takahata 2002)

while earlier stage embryos, the globular and heart stages, are not responsive in microspore-derived embryos of *B. oleracea* (Takahata et al. 1993) and *B. napus* (Brown et al. 1993), or in somatic embryos of several crops (Kitto and Janick 1985; Senaratna et al. 1991) and zygotic embryos (Bartels et al. 1988).

Osmotic stress caused by high concentrations of sorbitol or thermal stress also induced desiccation tolerance of the embryos, but its effectiveness was much less than that of ABA (Anandarjah et al. 1991; Wakui et al. 1994). In barley microspore-derived embryos, 0.6 M trehalose was reported to have a similar effect to that of 10 μM ABA for induction of desiccation tolerance (Ryan et al. 1999).

3.2 Requirements for Artificial Seeds

It is thought that desiccated embryos have to possess many capacities such as conversion to seedling under various soil conditions, long-term storage and easy handling of delivery for practical utilization as dry artificial seeds. In order to better understand these characteristics, several studies have been carried out.

Desiccated microspore-derived embryos had germination ability when they were directly sown in soil. Rapeseed and Chinese cabbage desiccated embryos were able to recover in vermiculite fertilized with B5 salt, although the germination speed in soil was slower than that under in vitro conditions (Takahata et al. 1992; Wakui et al. 1994). The frequency of germination and plant conversion in desiccated embryos of Chinese cabbage was 50 and 39.3 %, respectively (Wakui et al. 1994). In order to achieve higher and more stable conversion rates in soil, it is necessary that the embryos accumulate more storage reserves and/or are coated with suitable coating materials.

The development of embryo desiccation technology would allow long-term storage and germplasm preservation. In hydrated artificial seeds of barley, Datta and Potrykus (1989) reported that non-desiccated microspore-derived embryos encapsulated in sodium alginate and stored at 4 °C for 6 months maintained their germination ability with about a 50 % decrease of germination rates. When desiccated microspore-derived embryos were stored at room temperature, those of *B. oleracea* and *B. napus* maintained their viability for 3 and 6 months, respectively (Takahata et al. 1992, 1993). Wakui et al. (1998) examined the longevity of desiccated embryos of *B. campestris* under various storage conditions. The survival rates of the embryos depended on the storage temperature. Lower temperature conditions prolonged the viability of the embryos. The embryos stored at room temperature and -20 °C maintained their ability for plant conversion for 6 and 26 months, though their viability decreased with length of the storage period. The embryos stored at -80 °C maintained their viability during a 36-month period without any decrease in the frequencies of germination and plant regeneration of embryos. Similar results were obtained in somatic embryos of carrot (Shiota et al. 1999).

Loss of water allows the embryos to cryopreserve due to prevention of intracellular ice crystallization and of injury to the plant cells themselves (Guy 1990). Desiccated embryos of *B. napus*, which were directly immersed in liquid nitrogen (LN), converted to plants without any decrease in their viability (Wakui et al. 1998). In contrast, no survival was obtained when non-desiccated embryos were directly immersed in LN. Similar results have been reported in somatic embryos of melon (Shimonishi et al. 1991). Successful cryopreservation of somatic embryos is reported to require treatments of cryoprotectant components and/or prefreezing (Florin et al. 1993). A desiccation method, however, enables omission of these requirements which are essential to the success of embryo cryopreservation. Although Wakui et al. (1998) carried out cryopreservation of the desiccated embryos for only 7 days, it is expected that storage for a semipermanent period would be possible. Naked desiccated embryos would have the advantages of easier handling of germplasm preservation and a lower delivery cost.

3.3 Mechanism of Induction of Desiccation Tolerance

The exact mechanism by which desiccation tolerance is induced by ABA or other treatments is unknown. If this mechanism were known, the basis for desiccation tolerance of seeds could be known as well as the expanded appli-

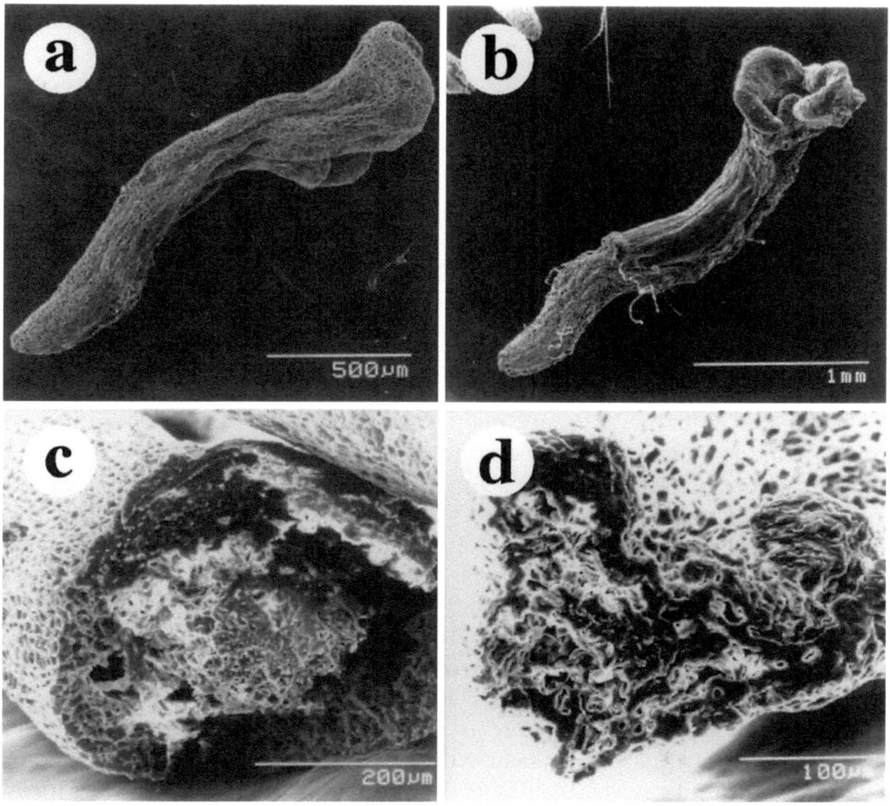

Fig. 3. SEM of desiccated microspore-derived embryos. **a, c** Desiccation-tolerant embryo. **b, d** Desiccation-non-tolerant embryo. **c, d** Freeze-fractured surface of the embryo. (Wakui et al. 1999)

cation of dry artificial seeds. Wakui et al. (1999) made a morphological comparison of desiccation-tolerant and -non-tolerant microspore-derived embryos of *B. napus*. Scanning electron microscopy observation revealed that though the external surface was shriveled due to severe dehydration, the external and internal tissue systems of desiccation-tolerant embryos were preserved (Fig. 3). In contrast, the tissue systems of desiccation-non-tolerant embryos were collapsed due to dehydration. Maintenance of tissue structures during dehydration may be attributed to the degree of cell damage. The stabilization of membranes is considered to be an important property in protecting cells against dehydration and can be increased by ABA (Putosvoitova 1987). Leakage of sugars was detected in desiccation-non-tolerant embryos after rehydration, but not in desiccation-tolerant ones (K. Wakui and Y. Takahata, unpubl.). Leakage of electrolytes and sugars from cells has been observed in dehydration-injured tissues and deteriorated seed, possibly due to damage of the membrane (Takayanagi 1980; Senaranta and McKersie

1983). This shows that the cell-membrane system of desiccation non-tolerant embryos is destroyed by severe dehydration.

Various genes are expressed during acquisition of desiccation tolerance of embryos. Of these, storage materials such as proteins and fatty acids of microspore-derived embryos were reported to be increased by application of ABA or osmotic stress (Ronald et al. 1990; Taylor et al. 1990; Holbrook et al. 1992; Pomeroy et al. 1994; Wakui et al. 1994). When the embryos supplied with either ABA or a high concentration of sorbitol (desiccation tolerance-inducing treatments) were treated with fluoridone, an inhibitor of ABA synthesis, the former embryos showed induction of desiccation tolerance and an increase in storage proteins, but the latter lost desiccation tolerance in spite of an increase in storage proteins (Sato et al. 1997). This indicated that storage proteins are not related to induction of desiccation tolerance.

In zygotic embryos of many plants, Lea (late embryogenesis abundant) mRNAs and proteins are considered to play an important role in the process of desiccation tolerance of seeds (Dure et al. 1989; Dure 1993; Xu et al. 1996). Wakui and Takahata (2002) reported isolation of *Lea* genes, *ME-leaN4* and *ME-leaC4* from desiccation-tolerant microspore-derived embryos of *B. napus* and *B. campestris*, respectively, and there is a close relationship between induction of desiccation tolerance of the embryos and Lea expression (Fig. 2). The two *Lea* genes have an open reading frame of 226 amino acids with a predicted molecular weight of 25 kDa and share a high homology of 92.9% between the two deduced proteins. These proteins had the same traits as the group 3 Lea family which is characterized by predominant hydrophilic amino acids and is principally consistent with a tandemly repeated 11-mer amino acid motif (Harada et al. 1989; Dure 1993). Acquisition of desiccation tolerance of the microspore-derived embryos and accumulation of mRNA and the translated products of the *Lea* genes began within 12 h of ABA treatment, and reached high levels after ABA treatment for 48–168 h (Fig. 2). A histological study using in situ hybridization and immunocytochemistry showed that the *ME-leaN4* mRNA and its protein were expressed in desiccation-tolerant embryos but not in desiccation-non-tolerant ones (Wakui et al. 2000). In desiccation tolerant embryos, the expression of *ME-leaN4* mRNA was observed in procambium and meristem tissues, and the protein was observed in the cytoplasm of almost all the cells, and was especially high in external tissues. Similar localizations of Lea proteins were reported in zygotic embryos of barley and cotton (Roberts et al. 1993; Marttila et al. 1996).

Although accumulation of Lea proteins is correlated with desiccation tolerance of embryos, the exact mechanism by which desiccation tolerance is induced by ABA and the function of Lea proteins are still unclear. Our preliminary study showed that *E. coli* expressing ME-leaN4 protein became osmotic and salt-tolerant (K. Wakui et al., unpubl.).

4 Transformation

Compared to methods using diploid cells or tissues as targets, transformation procedures using haploid targets have several advantages. The main advantage is that transformation of a haploid target followed by chromosome doubling provides a method for rapid genetic fixation of the introduced gene(s). In some plant species transformation of haploid targets is also advantageous where haploid culture is currently the most efficient in vitro culture system in these species. For example, in rapeseed microspore-derived embryo culture more than 10% of isolated uninucleate microspores undergo the first cell division within a day and develop into globular embryos within a week (Nitta et al. 1997). The length of time the microspores are in a dedifferentiated state appears to be very short, and a low spontaneous mutation rate is therefore expected. It is for these reasons that in the last 15 or more years many groups have attempted to transfer genes into isolated microspores and microspore-derived embryos. A number of these attempts have succeeded in generating fertile dihaploid transgenic plants (Table 1).

One of the most important factors for the successful utilization of isolated microspores as targets for genetic transformation is the efficiency of gene introduction. There have been many reports in which immature or mature pollen grains have been used as transformation targets. The various transformation methods to introduce exogenous DNA have been examined such as electroporation, particle bombardment and *Agrobacterium*-mediated DNA delivery (reviewed by Harwood et al. 1996). Fennell and Hauptmann (1992) reported using a CAT gene driven by a ubiquitin promoter as a reporter in which DNA could be delivered into maize (*Zea mays*) microspores using electroporation and PEG-mediated methods. The conditions used to deliver foreign DNA into protoplasts were also sufficient for maize microspores, despite the presence of a cell wall and developing exine in pollen grains. To date, however, particle bombardment is considered to be the most reliable and reproducible method (Stöger et al. 1992). Most of the attempts were in fact oriented toward obtaining transgenic plants through pollination by transgenic pollen grains (van der Leede-Plegt et al. 1992). Recently, establishment of a male germ line transformation procedure has been reported in which transgenic plants were produced using microspores transformed by particle bombardment followed by crossing to wild-type plants (Touraev et al. 1997; Aziz and Machray 2003).

For the gene delivery into microspores targeting production of genetic transformants through microspore embryogenesis, many studies have reported the use of various techniques. Since efficient microspore culture is currently only possible in a very limited number of plant species, most studies have been performed in highly responsive species such as *Nicotiana*, *Brassica* and barley. Numerous attempts have been made to obtain transient expression of marker genes in microspores by electroporation, microinjec-

Table 1. Transformation using microspore culture system

Target	Species	Method	Genes	Recovery	Reference
Microspore	*Brassica napus*	Agroinfection	nptII	Transgenic plant	Huang (1992)
		Agroinfection	nptII	Transgenic plant	Pechan (1989)
		Agroinfection	PAT, GUS	Transgenic plant	Dormann et al. (1995)
		Microinjection	GUS	Transient expression	Jones-Villeneuve et al. (1995)
		Electroporation	GUS	Transient expression	Jardinaud et al. (1993)
		Particle bombardment	Luciferase	Transgenic plant	Fukuoka et al. (1998)
		Particle bombardment	GUS, GFP	Transient expression	Nehlin et al. (2000)
	Brassica rapa	Agroinfection	PAT, GUS	Transgenic plant	Dormann et al. (1995)
	Hordeum vulgare	Particle bombardment	GUS	Transgenic plant	Jähne et al. (1994)
		Particle bombardment	GFP	Transgenic plant	Carlson et al. (2001)
	Nicotiana tabacum	Particle bombardment	GUS	Transgenic plant	Stöger et al. (1995)
	Triticum aestivum	Particle bombardment	C1, B-peru	Transient expression	Mentewab et al. (1999)
		Particle bombardment	GUS	Transgenic plant	Folling and Olesen (2001)
	Zea mays	Electroporation	CAT	Transient expression	Fennell and Hauptman (1992)
		PEG-mediated transfer	CAT	Transient expression	Fennell and Hauptman (1992)
Microspore-derived embryo	*Brassica napus*	Microinjection	nptII	Transgenic plant	Neuhaus et al. (1987)
		Agroinfection	nptII	Transgenic plant	Pechan (1989)
		Agroinfection	nptII	Transgenic plant	Swanson and Erickson (1989)
		Agroinfection	nptII	Transgenic plant	Huang (1992)
		Agroinfection	PAT	Transgenic plant	Oelck et al. (1991)
		Agroinfection	nptII, TGG	Transgenic plant	Troczynska et al. (2003)
		Particle bombardment with/without DNA Imbibition	nptII	Transgenic plant	Chen and Beversdorf (1994)
	Datura innoxia	Agroinfection	GUS, nptII	Transgenic plant	Sangwan et al. (1993)
	Nicotiana tabacum	Agroinfection	GUS, nptII	Transgenic plant	Sangwan et al. (1993)
	Triticum aestivum	Particle bombardment	GUS	Transient expression	Loeb and Reynolds (1994)
		Particle bombardment	GUS	Transient expression	Ingram et al. (1999)
		Particle bombardment	C1, B-peru	Transgenic chimera	Mentewab et al. (1999)

tion and polyethylene glycol-mediated gene transfer; however, stable transformants have never been obtained (Fennell and Hauptmann 1992; Jardinaud et al. 1993; Jones-Villeneuve et al. 1995). *Agrobacterium*-mediated transformation is in general a much more efficient method, but, again, no reproducible method for *Agrobacterium*-mediated gene transfer into isolated microspores has been reported. Huang (1992) reported that only one transformant was obtained from more than 50 co-cultivation experiments. The transformant developed from a culture in which bacteria were visible several days after co-cultivation, and thus the precise timing of *Agrobacterium* infection was not clear. Although detailed procedure or condition was not described, freshly isolated microspores were reported to become competent to *Agrobacterium*-mediated gene transfer with cellulolytic enzyme treatment, and transgenic plants were obtained at the frequency of two transgenics per 1×10^6 microspores in *B. napus* (Dormann et al. 1995).

To date, most of the successful transformation procedures have utilized particle bombardment for gene introduction. In *Nicotiana tabacum*, 5 out of 10^4 microspores were reporter gene (GUS) positive, and 1–5% of these developed into transgenic multicellular structures (Stöger et al. 1995). In total, two transgenic embryos were manually selected out of 0.4 million embryos that had been obtained from 80 million microspores and 800 flowers. Jähne et al. (1994) reported development of a microspore transformation system in barley in which on average one transgenic plant could be expected from 2.8×10^6 microspores. Carlson et al. (2001) described screening of transgenic barley microspore embryos using the green fluorescent protein gene as a visual marker. In rapeseed, four transgenic embryos were identified and selected from 7.7×10^4 embryos obtained from 5.7 million microspores using a non-destructive luciferase marker gene (Fukuoka et al. 1998). Recently, in wheat, Folling and Olesen (2001) have reported that a maximum of 3.5 transgenic embryos were obtained from 1×10^6 microspores. Nehlin et al. (2000) assessed the efficiency of particle bombardment-mediated microspore transformation using a transient assay and concluded that 1.1×10^6 microspores would be necessary to obtain one stable transgenic embryo, which was largely consistent with the experimental results reported by other laboratories.

In the case of tobacco and rapeseed, the use of antibiotics as selective agents for transgenic events failed. Transgenic embryos had to be identified among a large population of non-transgenic embryos by either exhausting manual selection under a microscope using a non-destructive GUS substrate, ImaGeneGreen, or a luciferase-based method, which required an expensive photon-counting image analyzer. Stöger et al. (1995) suggested that transformants do not develop efficiently under antibiotic selection conditions because the growth of the majority of the non-transformed embryos was suppressed, resulting in a drastic decrease of the density of vital cells. An additional possibility is that the commonly used CaMV 35S promoter does not drive expression of the selectable marker gene during the early stages of embryogenesis. Custers et al. (1999) reported that the CaMV 35S promoter was silent during early microspore embryogenesis in rapeseed. The utilization of promoters active from the initiation of embryogenesis may be a solution to overcome the difficulty of drug-selection of transgenic embryos. Establishment of an efficient selection method should be essential for practical use of microspore transformation.

In contrast, attempts to take advantage of secondary embryogenesis from microspore-derived haploid embryos as an alternative target have led to a more efficient haploid transformation system. Microspore culture in vitro provides sufficient numbers of immature embryos for genetic transformation compared to zygotic embryos collected from immature seeds. Neuhaus et al. (1987) subjected rapeseed microspore-derived pro-embryos to direct gene delivery by microinjection and obtained transgenic plants by selection of secondary embryos that arose from chimeric primary embryos. Using a secondary embryogenesis system regenerated from hypocotyls of microspore-derived embryos in rapeseed, Chen and Beversdorf (1994) reported that a combination of particle bombardment and DNA imbibition enhanced the transformation frequency four-fold compared to the efficiency obtained by the bombardment alone. While *Agrobacterium*-mediated transformation was reported to be non-reproducible for direct microspore transformation, successful production of transgenic plants has been reported in various species using microspore-derived embryos as targets for gene introduction. Sangwan et al. (1993) reported efficient production of fertile transgenic dihaploid plants in *Datura innoxia* and *Nicotiana tabacum* by *Agrobacterium*-mediated gene introduction to immature microspore-derived embryos. When cotyledon stage embryos were inoculated, more than 70 % of the embryos gave rise to one or more transgenic calli/shoots, and most of these structures regenerated transgenic plants. Although a much lower transformation efficiency and reproducibility were obtained in these experiments, perhaps due to the low regeneration rate of the culture system and/or low transgenic potential of the *Agrobacterium* strain that was used, transgenic plants were obtained by *Agrobacterium* infection of microspore-derived embryos in rapeseed (Pechan 1989; Swanson and Erickson 1989; Oelck et al. 1991; Huang 1992; Troczynska et al. 2003; J. Custers, pers. comm.). In wheat (*Triticum aestivum*), particle bombardment of immature microspore-derived embryos was examined, and both transient expression on embryos (Loeb and Reynolds 1994; Ingram et al. 1999) and chimeric stable transformation (Mentewab et al. 1999) have been reported.

5 Conclusions

There is an enormous potential for the use of microspores and microspore-derived embryos for genetic engineering such as mutation, selection, gene transfer and artificial seeds as well as developmental research of plant embryogenesis, because of the single-cell system of microspore culture. In comparison with genetic manipulation using somatic diploid cells and tissues as targets, the major advantage of using haploid targets is the easy selection of recessive mutants and rapid genetic fixation after mutagenesis and transformation. The production of more uniform embryos in microspore culture also offers the potential for application of artificial seed technology. Development of effective diplodization technology of embryos during microspore culture and knowledge of seed development such as accumulation of seed storage products and the mechanism of desiccation tolerance will allow practical application of artificial seeds. In some species, in which the microspore culture system is the most effective in vitro culture, gene transfer using this system is attractive. The microspores and derived embryos are considered to be readily subjected to various methods of gene transfer. However, the efficiency of transformation is still low and unstable. Development of more effective and reproducible techniques is needed for plant breeding and a study of functional genomics. The increase in efficiency of microspore embryogenesis is indispensable to the utilization of microspore-derived embryos for genetic manipulation and the basic science of development. With further improvements to microspore culture methods it is possible that in the future they will provide more efficient targets for genetic engineering in a range of crop species.

References

Ahmad I, Day JP, MacDonald MV, Ingram DS (1991) Haploid culture and UV mutagenesis in rapid-cycling *Brassica napus* for the generation of resistance to chlorsulfuron and *Alternaria brassicicola*. Ann Bot 67:521–525

Anandarajah KKL, Kotto L, Beversdorf WD, McKersie BD (1991) Induction of desiccation tolerance in microspore-derived embryos of *Brassica napus* L. by thermal stress. Plant Sci 77:119–123

Aziz N, Machray GC (2003) Efficient male germ line transformation for transgenic tobacco production without selection. Plant Mol Biol 51:203–211

Bajaj YPS (1990) In vitro production of haploids and their use in cell genetics and plant breeding. In: Bajaj YPS (ed) Biotechnology in agriculture and forestry, vol 12. Haploids in crop improvement I. Springer, Berlin Heidelberg New York, pp 3–44

Barro F, Fernadez-Escobar J, de la Vega M, Martin A (2001) Doubled haploid lines of *Brassica carinata* with modified erucic acid content through mutagenesis by EMS treatment of isolated microspores. Plant Breed 120:262–264

Barro F, Fernadez-Escobar J, de la Vega M, Martin A (2002) Modification of glucosinolate and erucic acid contents in doubled haploid lines of *Brassica carinata* by UV treatment of isolated microspores. Euphytica 129:1–6

Bartels D, Singh M, Salamini F (1988) Onset of desiccation tolerance during development of the barley embryo. Planta 175:485–492

Brown DCW, Watson EM, Pechan PM (1993) Induction of desiccation tolerance in microspore-derived embryos of *Brassica napus*. In Vitro Cell Dev Biol 29P:113–118

Carlson AR, Letarte J, Chen J, Kasha KJ (2001) Visual screening of microspore-derived transgenic barley (*Hordeum vulgare* L.) with green-fluorescent protein. Plant Cell Rep 20:331–337

Castillo AM, Cistue L, Valles MP, Sanz JM, Romagosa L, Molina-Cano JL (2001) Efficient production of androgenic doubled-haploid mutants in barley by the application of sodium azide to anther and microspore cultures. Plant Cell Rep 20:105–111

Chen JL, Beversdorf WD (1991) Evaluation of microspore-derived embryos as models for studying lipid biosynthesis in seed of rapeseed (*Brassica napus* L.). Euphytica 58:145–155

Chen JL, Beversdorf WD (1994) A combined use of microprojectile bombardment and DNA imbibition enhances transformation frequency of canola (*Brassica napus* L.). Theor Appl Genet 88:187–192

Custers JBM, Snepvangers SCHJ, Jansen HJ, Zhang L, van Lookeren-Campagne MM (1999) The 35S-CaMV promoter is silent during early embryogenesis but activated during nonembryogenic sporophytic development in microspore culture. Protoplasma 208:257–264

Datta SK, Potrykus I (1989) Artificial seeds in barley: encapsulation of microspore derived embryos. Theor Appl Genet 77:820–824

Datta SK, Schmid J (1997) Prospects of artificial seeds from microspore-derived embryos of cereals. In: Mohan Jain S, Sopory SK, Veilleux RE (eds) In vitro haploid production in higher plants, vol 2. Fundamental aspects and methods. Kluwer, Dordrecht, pp 353–365

Dormann M, Wang HM, Dalta N, Ferrie AMR, Keller WA, Oelck MM (1995) Transformation of freshly isolated *Brassica* microspores and regeneration to fertile homozygous plants. Proc 9th Int Rapeseed Congr 3:816–818

Dure L III (1993) A repeating 11-mer amino acid motif and plant desiccation. Plant J 3:363–369

Dure L III, Crouch M, Harada J, Ho THD, Mundy J, Quatrano R, Thomas T, Sung ZR (1989) Common amino acid sequence domains among the LEA proteins of higher plants. Plant Mol Biol 12:475–486

Fennell A, Hauptmann R (1992) Electroporation and PEG delivery of DNA into maize microspores. Plant Cell Rep 11:567–570

Ferrie AMR, Keller WA (2002) Application pf doubled haploidy and mutagenesis in *Brassica*. In: Proc 13th Cruciferae Genetic Worksh, Abstracts, http://vric.ucadavis.edu/crucifer/abstracts/vegetable/ferrie.html

Finkelstein RR, Somerville C (1989) Abscisic acid or high osmoticum promote accumulation of long-chain fatty acids in developing embryos of *Brassica napus*. Plant Sci 61:213–217

Florin B, Tessereau H, Leocouteux C, Didier C, Petiard V (1993) Long-term preservation of somatic embryos. In: Redenbaugh K (ed) Synseeds. CRC Press, Boca Raton, pp 133–162

Folling L, Olesen A (2001) Transformation of wheat (*Triticum aestivum* L.) microspore-derived callus and microspores by particle bombardment. Plant Cell Rep 20:629–636

Fujii JA, Slade DT, Redenbaugh K, Walker KA (1987) Artificial seeds for plant propagation. TIBTECH 5:335–339

Fukuoka H, Ogawa T, Matsuoka M, Ohkawa Y, Yano H (1998) Direct gene delivery into isolated microspores of rapeseed (*Brassica napus* L.) and the production of fertile transgenic plants. Plant Cell Rep 17:323–328

Gray DJ (1989) Effects of dehydration and exogenous growth regulators on dormancy, quiescence and germination of grape somatic embryos. In Vitro Cell Dev Biol 25:1173–1178

Gray DJ, Conger BV, Songstad DD (1987) Desiccation quiescent somatic embryos of orchardgrass for use as synthetic seeds. In Vitro Cell Dev Biol 23:29–33

Guy CL (1990) Cold acclimation and freezing stress tolerance: role of protein metabolism. Annu Rev Plant Physiol Plant Mol Biol 41:187–223

Harada J, Delisle A, Baden C, Crouch M (1989) Unusual sequence of an abscisic acid-inducible mRNA which accumulates late in *Brassica napus* seed development. Plant Mol Biol 12:395–401

Harwood WA, Chen DF, Creissen GP (1996) Transformation of pollen and microspores. In: Mohan Jain S, Sonory SK, Veilleux RE (eds) In vitro haploid production in higher plants, vol 2. Kluwer, Dordrecht, pp 53–71

Holbrook LA, Magus JR, Taylor DC (1992) Abscisic acid induction of elongase activity, biosynthesis and accumulation of very long chain monounsaturated fatty acids and oil body proteins in microspore-derived embryos of *Brassica napus* L. Reston. Plant Sci 84:99–115

Huang B (1992) Genetic manipulation of microspores and microspore-derived embryos. In Vitro Cell Dev Biol 28P:53–58

Iida Y, Watabe K, Kamada H, Harada H (1992) Effect of abscisic acid on the induction of desiccation tolerance in carrot somatic embryos. J Plant Physiol 140:356–360

Ingram HM, Power JB, Lowe KC, Davey MR (1999) Optimisation of procedures for microprojectile bombardment of microspore-derived embryos in wheat. Plant Cell Tissue Organ Cult 57:207–210

Jähne A, Becker D, Brettschneider R, Lörz H (1994) Regeneration of transgenic, microspore-derived, fertile barley. Theor Appl Genet 89:525–533

Jardinaud MF, Souvré A, Alibert G (1993) Transient GUS gene expression in *Brassica napus* electroporated microspores. Plant Sci 93:177–184

Jones-Villeneuve E, Huang B, Prudhomme I, Bird S, Kemble R, Hattori J, Miki B (1995) Assessment of microinjection for introducing DNA into uninuclear microspores of rapeseed. Plant Cell Tiss Org Cult 40:97–100

Keller WA, Arnison PG, Cardy BJ (1987) Haploids from gametophytic cells – recent development and future prospects. In: Green CE, Somers DA, Hackett WP, Biesboer DD (eds) Plant tissue and cell culture. Liss, New York, pp 223–241

Khush GS, Virmani SS (1997) Haploids in plant breeding. In: Mohan Jain S, Sopory SK, Veilleux RE (eds) In vitro haploid production in higher plants, vol 1. Fundamental aspects and methods. Kluwer, Dordrecht, pp 11–33

Kim YH, Janick J (1991) Abscisic acid and proline improve desiccation tolerance and increase fatty acid content of celery somatic embryos. Plant Cell Tissue Organ Cult 24:83–89

Kitto SL, Janick J (1985) Production of synthetic seeds by encapsulation asexual embryos of carrot. J Am Soc Hort Sci 110:277–282

Kott L, Wong R, Swanson E, Chen J (1996) Mutation and selection for improved oil and meal quality in *Brassica napus* utilizing microspore culture. In: Mohan Jain S, Sopory SK, Veilleux RE (eds) In vitro haploid production in higher plants, vol 2. Fundamental aspects and methods. Kluwer, Dordrecht, pp 151–167

Loeb TA, Reynolds TL (1994) Transient expression of the uidA gene in pollen embryoids of wheat following microprojectile bombardment. Plant Sci 104:81–91

MacDonald MV, Ahmad I, Menten JOM, Ingram DS (1991) Haploid culture and in vitro mutagenesis (UV light, X-rays and gamma rays) of rapid cycling *Brassica napus* for improved resistance to disease. In: Proc Int Symp on the Contribution of Plant Mutation Breeding to Crop Improvement, Plant Mutation Breeding for Crop Improvement, vol 2, International Atomic Energy Agency and Food and Agriculture Organization, Vienna, pp 129–138

Maluszynski M, Szarejko I, Sigurbjornsson B (1996) Haploidy and mutation techniques. In: Mohan Jain S, Sopory SK, Veilleux RE (eds) In vitro haploid production in higher plants, vol 1. Fundamental aspects and methods. Kluwer, Dordrecht, pp 67–93

Marttila S, Tenhola T, Mikkonen A (1996) A barley (*Hordeum vulgare* L.) LEA3 protein, HVA1, is abundant in protein storage vacuoles. Planta 199:602–611

Mentewab A, Letellier V, Marque C, Sarrafi A (1999) Use of anthocyanin biosynthesis stimulatory genes as markers for the genetic transformation of haploid embryos and isolated microspores in wheat. Cereal Res Comm 27:17–24

Mikami T, Takahata Y, Kaizuma N (1993) Induction of artificial mutation and in vitro selection of NaCl tolerance using microspore culture of *Brassica napus*. Jpn J Breed 43(Suppl 2):81

Nehlin L, Möllers C, Bergman P, Glimelius K (2000) Transient *β-gus* and *gfp* gene expression and viability analysis of microprojectile bombarded microspores of *Brassica napus* L. J Plant Physiol 156:175–183

Neuhaus G, Spangengerg G, Mittelsten Scheid O, Schwieger HG (1987) Transgenic rapeseed plants obtained by the microinjection of DNA into microspore-derived embryos. Theor Appl Genet 75:30–36

Nitta T, Takahata Y, Kaizuma N (1997) Scanning electron microscopy of microspore embryogenesis in *Brassica* spp. Plant Cell Rep 16:406–410

Oelck MM, Phan CV, Eckers P, Donn G, Rakow G, Keller WA (1991) Field resistance resistance of canola transformants (*Brassica napus* L.) to Ignite® (Phosphinotricin). In: Proc 8th Int Rapeseed Congr, Saskatoon, Saskatchewan, pp 292–297

Pechan PM (1989) Successful cocultivation of *Brassica napus* microspores and proembryos with *Agrobacterium*. Plant Cell Rep 8:387–390

Pomeroy MK, Kramer JKG, Hunt DJ, Keller WA (1991) Fatty acid changes during development of zygotic and microspore-derived embryos of *Brassica napus*. Physiol Plant 81:447–454

Pomeroy K, Brown DCW, Takahata Y (1994) Response of *Brassica napus* L. microspore-derived embryos to exogenous abscisic acid and desiccation. In Vitro Cell Dev Biol 30P:196–203

Pustovoitova T (1987) Effect of abscisic acid on batacyanin leakage from plant tissues. Biol Plant 29:338–341

Roberts JK, DeSimone NA, Lingle WL, Dure L III (1993) Cellular concentrations and uniformity of cell-type accumulation of two lea protein in cotton embryos. Plant Cell 5:769–780

Ronald WW, Mandel RM, Pharis RP, Holbrook LA, Moloney MM (1990) Effect of abscisic acid and high osmoticum on storage protein gene expression in microspore embryos of *Brassica napus*. Plant Physiol 94:875–881

Ryan AB, Castillo AM, Valles MP, Sanz JM, Cistue L (1999) Desiccated doubled-haploid embryos obtained from microspore culture of barley cv. Igri. Plant Cell Rep 18:924–928

Sangwan RS, Ducrocq C, Sangwan-Norreel B (1993) *Agrobacterium*-mediated transformation of pollen embryos in *Datura innoxia* and *Nicotiana tabacum*: production of transgenic haploid and fertile homozygous dihaploid plants. Plant Sci 95:99–115

Sato C, Takahata Y, Kaizuma N (1997) Effect of high osmoticum and fluridone on desiccation tolerance in microspore-derived embryos of *Brassica napus*. Breed Sci 47(Suppl 2):362

Senaranta T, McKersie BD (1983) Characterization of solute efflux from dehydration injured soybean (*Glycine max* L. Merr) seeds. Plant Physiol 72:911–914

Senaratna T, Mckersie BD, Bowley SR (1989) Desiccation tolerance of alfalfa (*Medicago sativa* L.) somatic embryos. Influence of abscisic acid, stress pretreatments and drying rates. Plant Sci 65:253–259

Senaratna T, Mckersie BD, Bowley SR (1990) Artificial seed of alfalfa (*Medicago sativa* L.). Induction of desiccation tolerance in somatic embryos. In Vitro Cell Dev Biol 26:85-90

Senaratna T, Kott L, Beversdorf WD, McKersie BD (1991) Desiccation of microspore derived embryos of oilseed rape. Plant Cell Rep 10:342–344

Shimonishi K, Ishikawa M, Suzuki S, Oosawa K (1991) Cryopreservation of melon somatic embryos by desiccation method. Jpn J Breed 41:347–351

Shiota H, Tachibana K, Watabe K, Kamada H (1999) Successful long-term preservation of abscisic-acid-treated and desiccated carrot somatic embryos. Plant Cell Rep 18:749–753

Stöger E, Moreno RMB, Ylstra B, Vincente O, Heberle-Bors E (1992) Comparison of different techniques for gene transfer into mature and immature tobacco pollen. Transgenic Res 1:71–78

Stöger E, Fink C, Pfosser M, Heberle-Bors E (1995) Plant transformation by particle bombardment of embryogenic pollen. Plant Cell Rep 14:273–278

Swanson EB, Erickson LR (1989) Haploid transformation in *Brassica napus* using an octopine-producing strain of *Agrobacterium tumefaciens*. Theor Appl Genet 78:831–835

Swanson EB, Coumans MP, Brown GL, Patel JD, Beversdorf WD (1988) The characterization of herbicide tolerant plants in *Brassica napus* L. after in vitro selection of microspores and protoplasts. Plant Cell Rep 7:83–87

Swanson EB, Herrgesell MJ, Arnoldo M, Sippell DW, Wong RSC (1989) Microspore mutagenesis and selection: canola plants with field tolerance to the imidazolinones. Theor Appl Genet 78:525–530

Takahata Y (1997) Microspore culture. In: Kalia HR, Gupta SK (eds) Recent advances in oilseed brassicas. Kalyani, Ludhiana, pp 162–181

Takahata Y, Wakui K, Kaizuma N, Brown DCW (1992) Dry artificial seed system for *Brassica* crops. Acta Hortic 319:317–322

Takahata Y, Brown DCW, Keller WA, Kaizuma N (1993) Dry artificial seeds and desiccation tolerance induction in microspore-derived embryos of broccoli. Plant Cell Tissue Organ Cult 35:121–129

Takayanagi K (1980) Seed storage and viability tests. In: Tsunoda S, Hinata K, Gómez-Campo C (eds) *Brassica* crops and wild allies. Japan Scientific Societies Press, Tokyo, pp 303–321

Taylor DC, Weber N, Underhill EW, Pomeroy MK, Keller WA, Scowcroft WR, Wilen RW, Moloney MM, Holbrook LA (1990) Storage-protein regulation and lipid accumulation in microspore embryos of *Brassica napus* L. Planta 181:18–26

Touraev A, Stöger E, Voronin V, Heberle-Bors E (1997) Plant male germ line transformation. Plant J 12:949–956

Troczynska J, Drozdowska L, Cegielska-Taras T (2003) Transformation of microspore-derived embryos to study the myrosinase-glucosinolate system in *Brassica napus* L. Proc 11th Int Rapeseed Congr 1:175–177

Van der Leede-Plegt LM, van de Ven BCE, Bino RJ, van der Salm TPM, van Tunen AJ (1992) Introduction and differential use of various promoters in pollen grains of *Nicotiana glutinosa* and *Lilium longiflorum*. Plant Cell Rep 11:20–24

Wakui K, Takahata Y (2002) Isolation and expression of *Lea* gene in desiccation-tolerante microspore-derived embryos in *Brassica* spp. Physiol Plant 116:223–230

Wakui K, Takahata Y, Kaizuma N (1994) Effect of abscisic acid and high osmoticum concentration on the induction of desiccation tolerance in microspore-derived embryos of Chinese cabbage (*Brassica campestris* L.). Breed Sci 44:29–34

Wakui K, Sato C, Takahata Y, Kaizuma N (1998) Long-term storage and cryopreservation of desiccated microspore-derived embryos in *Brassica* spp. Plant Biotech 15:123–126

Wakui K, Takahata Y, Kaizuma N (1999) Scanning electron microscopy of desiccation-tolerant and -sensitive microspore-derived embryos of *Brassica napus* L. Plant Cell Rep 18:595–600

Wakui C, Tsuchiya T, Wakui K, Watanabe M, Takahata Y (2000) Expression of ME-leaN4 in desiccation tolerant microspore-derived embryo of *Brassica napus* and *B. campestris*. In: Proc 12th Crucifer Genetics Worksh, 5–9 Sept, Wellesbourne, UK, Abstracts. Horticulture Research International, Warwick

Xu D, Duan X, Wang B, Hong B, Ho THD, Wu R (1996) Expression of a late embryogenesis abundant protein gene, *HVA1*, from barley confers tolerance to water deficit and salt stress in transgenic rice. Plant Physiol 110:249–257

Zhang F, Takahata Y (1999) Microspore mutagenesis and in vitro selection for resistance to soft rot disease in Chinese cabbage (*Brassica campestris* L. ssp. *pekinensis*). Breed Sci 49:161–166

Section II Utilization of Haploids in the Improvement of Specific Crop Species

II.1 Haploids in the Improvement of Solanaceous Species

George C.C. Tai[1]

1 Introduction

The Solanaceae family includes five major cultivated crop plants (Simmonds 1974): pepper (*Capsicum*) in which *Capsicum annuum* L. is the economically important species, tomato (*Lycopersicum esculentum* Mill.), tobacco (*Nicotiana tabacum* L.), eggplant (*Solanum melongena* L.) and potato (*Solanum tuberosum* L.). The cultivated species of pepper, eggplant and tomato are diploid ($2n=2x=24$). Tobacco and potato are, respectively, allotetraploid ($2n=4x=48$) and autotetraploid ($2n=4x=48$). The basic chromosome number of all the above species is $x=12$. Except for potatoes, all the species are seed propagated, tolerant of inbreeding and bred as pure lines. Potato is an outbreeding species, suffers inbreeding depression and is mainly vegetatively propagated (see Simmonds 1979). Pepper, tomato and eggplant have a number of related species that are all diploids. The related species of tobacco are mainly diploids and tetraploids, whereas those of potatoes range from diploids to hexaploids.

Haploids are plants with the gametophytic chromosome number of the parents. Haploids from diploid or high ploidy parents should, according to the definition, all be called 'haploids'. To avoid confusion in practice, however, those derived from tetraploids are often referred to as 'dihaploids'. Haploids out of diploids are 'monohaploids' but often simply called 'monoploids'. Those obtained by doubling the chromosome number of the monoploids are referred to as 'double haploids'. Research work in haploids represents a very fruitful field in the Solanaceae crop species. Haploidization is achieved by several means, and haploids are now applied in a number of research areas including gene mapping, identification of major and quantitative trait loci (QTL), genetic transformation, somatic fusion and marker-assisted selection in breeding for new cultivars. Haploids are specifically used in ploidy manipulation in a unique breeding scheme for potatoes. The various applications of haploids in genetic and breeding research in tobacco are reviewed by Wernsman (1993) and Peloquin and colleagues on potatoes (Peloquin et al. 1989, 1990; Peloquin and Ortiz 1993).

[1] Potato Research Centre, Agriculture and Agri-Food Canada, P.O. Box 20280, Fredericton, New Brunswick, E3B 4Z7, Canada

Biotechnology in Agriculture and Forestry, Vol. 56
Haploids in Crop Improvement II (ed. by C.E. Palmer, W.A. Keller, and K.J. Kasha)

2 Induction of Haploids

There are extensive reports on induction of haploids in the Solanaceae species. Nitzsche and Wenzel (1977) provided a summary of the techniques for producing haploids of solanaceous crops before 1980. Haploids are produced by both androgenesis (paternal haploids) and gynogenesis (maternal haploids). Only the latest developments in tobacco, tomato, potatoes and pepper are briefly reviewed here.

2.1 Paternal Haploids by Androgenesis

Androgenesis is the process whereby an embryo is developed from a microspore. Microspores are haploid cells formed by meiosis. In the normal process, microspores undergo an unequal mitotic division leading to the formation of a generative and a vegetative nucleus. The generative one divides one more time to form two sperm nuclei, either before or after pollen grain maturation. Androgenesis as induced by anther culture alters the development process of microspores such that the first mitosis is often symmetrical in mode. This leads to the formation of two identical nuclei in the cells that develop into embryos or callus. Plantlets are then generated either directly from the embryos or indirectly by inducing roots and shoots from the callus. The various techniques for anther culture and induction of androgenesis are summarized by Chlyah et al. (1990) and Summers (1997) for tomato and Vagera (1990) for pepper respectively. A number of factors affect the success of anther culture.

2.1.1 Media

Anther culture media have been developed for Solanaceae crops over the past three decades. Collins and Genovesi (1982) presented a brief summary on the basal medium, stage of flower buds and response to culture media for crop species reported in various references. Bajaj (1990) presented the composition of four of the basal media commonly used for anther culture. Some more recent reports used media developed by Nakata and Tanaka (1968) and Nitsch and Nitsch (1969) for tobacco (Schnell and Wernsman 1986; Wernsman et al. 1989; Walker and Aycock 1994); a medium from Dumas de Vaulx et al. (1981), a modified Nitsch and Nitsch's H medium (Mityko et al. 1995; Dolcet-Sanjuan et al. 1997) and a Murashige and Skoog (MS) medium (Luz et al. 1998) for pepper; the Dumas de Vaulx et al. (1981) medium for eggplant (Rotino et al. 1987); and an MS medium (1962) for tomato (Shtereva et al. 1998; Zagorska et al. 1998). Johansson (1986) and Tiainen (1992) induced embryogenesis and the production of dihaploid plants from anther cultures of tetraploid *S. tuberosum* clones based on MS medium. Anther culture is used to obtain monoploids from diploid *S. tuberosum*, wild species and interspecific hybrids (Jacobsen and Ramanna 1994). Shen and Veilleux (1995) and Rokka et al. (1995) used a liquid medium and a semi-solid medium, respectively, to produce monoploids from diploid/dihaploid potatoes ($2x=24$). Rokka et al. (1997) reviewed the production of 'somatohaploid' by anther culture from interspecific somatic hybrids of potato. Both tobacco and potato have well-established procedures for haploid production and are now used as model plants for laboratory exercises (Reed 1996; Veilleux 1999).

2.1.2 Genotypes

Induction of callus from anthers and the subsequent regeneration of plants are genotype dependent in solanaceous crops (see, for example, Cappadocia et al. 1984; Powell and Uhrig 1987; Mityko et al. 1995; Zagorska et al. 1998; Aziz et al. 1999). Genetic factors are involved in determining the response to anther culture as frequency of monohaploids could be promoted by selection in potato crosses between responding and also non-responding parents (Jacobsen and Ramanna 1994). F_1 hybrids between poor and fairly responsive pepper parents showed fair responses (Mityko et al. 1995). Tomato cultivars containing the male sterility gene *ms* 10^{35} showed strong response to anther culture for both callus induction and regeneration of plants (Zagorska et al. 1998). More studies are needed to reveal the genetic mechanisms for haploid induction by androgenesis.

2.1.3 Donor Plant Conditions

Pepper plants grown in greenhouse in winter produced higher yields of haploids due to high moisture and short days (Shtereva et al. 1998). The optimum temperature in the greenhouse for obtaining embryos from young pepper plants (less than 9 weeks old) was reported to be 26 °C (Kristiansen and Andersen 1993). The diploid species *Solanum phureja* requires greenhouse conditions with 25 °C day/15 °C night, a 16-h photoperiod and high light intensity provided by high-pressure sodium vapor lamps (Veilleux 1999).

2.1.4 Developmental Stage of Microspores

Some experimental reports on the optimal development stage of microspores for anther culture indicate the period from prophase I (anther length <1.6 mm) (Summers 1992) to telophase II (anther length 1.7–2.5 mm) (Shtereva et al. 1998) during meiosis for tomato, the uninucleate stage for potatoes (anther length 2.5–3.5 mm) (Veilleux 1999) and pepper (Dolcet-Sanjuan et al.

1997) and uninucleate, mitosis and binucleate stages for tobacco (see Collins and Genovesi 1982).

2.1.5 Treatments of Anthers and Buds

Pretreatment of the flower buds of potatoes at low temperature (4–6 °C) for 2–4 days in the dark stimulated embryogenesis (Johansson 1986; Powell and Uhrig 1987; Tiainen 1992). Jaramillo and Summers (1991) studied the effect of dark–light regimes on anther callus initiation and growth of tomato. Calli grown for 10 weeks in darkness were 3.4 times larger than calli exposed to 5 weeks of darkness followed by 5 weeks of light. Shen and Veilleux (1995) used a treatment combining a high temperature shock (35 °C for 12 h) with elevated incubation temperature (30 °C 16 h/20 °C 8 h) which yielded 11 times more embryos than the control (20 °C). Dolcet-Sanjuan et al. (1997) reported the effects of culturing pepper anthers by using a two-phase system with activated charcoal in the solid phase and maltose as the sugar source, and by flushing the culture environment with CO_2. Shtereva et al. (1998) found that treating tomato anthers at 4 °C (2 days) and 10 °C (9 days) stimulated callus induction and regeneration of plants. Combined treatments of the anthers with 4 Gy gamma rays and 10 °C (9 days) proved to be most effective. Luz et al. (1998) found that the influence of thidizuron (TDZ) addition to the induction medium increased the embryoid frequency of sweet pepper.

2.2 Maternal Haploids by Gynogenesis

Maternal haploids are generated by the process of gynogenesis. Gynogenetic haploids of tobacco were first induced by using *N. africana* Mermuller and Buttler as pollen donor in interspecific hybridization with *N. tabacum* (Burk et al. 1979). Gynogenetic haploids are now produced by both in vitro and in vivo methods. *N. tabacum* plants were pollinated by irradiated pollen from *N. alata* Link & Otto and then followed by in vitro cultures of ovules (Kumashiro and Oinuma 1985). The unfertilized eggs underwent parthenogenetic development into haploids. An *N. tabacum* line homozygous for the dominant lethal gene Rac^- was developed (Muller et al. 1985). This mutant genotype was not able to develop a normal root system and had to be maintained by grafting onto a normal variety for flowering and seed harvesting (Pelletier et al. 1987). It was used in an in vivo system to generate haploids by crossing the mutant line with a normal variety. Only haploid plantlets in the progenies without fertilization survived under reduced watering due to their normal root system. Androgenetic haploids containing the nucleus of the mutant line and the cytoplasm of the normal parent or gynogenetic haploids were produced when the mutant line was used as the female or male parent respectively.

A conditional lethal marker was used for producing gynogenetic haploids of tomato in vivo (Hamza et al. 1993). Transgenic plants with the *aux2* gene from *Agrobacterium rhizogenes* were not able to develop roots in the presence of naphthalene acetamide (NAM). Progenies obtained by using the transformed line as male to cross with wild type were grown in the nutrient solution with NAM added as a toxic substrate. Only haploid plants survived as heterozygous diploid plants carrying *aux2* were killed by NAM.

Both dominant and conditional lethal markers require the screening of a large population of progenies grown in the greenhouse. This represents a serious limitation because the spontaneous haploid plants occurred at a very low frequency. A proficient parthenogenetic system for producing in vivo gynogenetic haploids was developed in potatoes (see Peloquin et al. 1990). Special genotypes from the diploid species *S. phureja* ($2n=2x=24$) are used as the 'pollinator' to hybridize with the tetraploid *S. tuberosum* cultivars (Hougas et al. 1964; Peloquin et al. 1996). Maternal (di)haploids ($2n=2x=24$) are obtained through parthenogenesis by the union of two chromosome sets of *S. phureja* with the polar nuclei but lack of fertilization of the egg. The use of a genetic morphological marker for an embryo spot (haploids lack the spot on the seed; see Caligari et al. 1988) or electrophoretic isozyme pattern (absence of isozyme marker; see Liu and Douches 1993) enables the easy and early detection of the haploids after $4x$–$2x$ crosses. Both the tuberosum genotype and the pollen source affect the frequency of haploid production (Hougas et al. 1964). Haploids have also been generated from a number of *Solanum* species with ploidy and endosperm balance number (EBN) ranging from $2x$ (2EBN) to $8x$ (4EBN). The pollinator effect, EBN and maternal influence are factors affecting the occurrence of pseudogamous parthenogenetic haploid production from these species (Singsit and Hanneman 1991). The parthenogenesis method was also used to produce monoploids ($2n=x=12$) from diploid *S. tuberosum* and diploid *Solanum* species (Uijtewaal et al. 1987a,b).

3 Development and Use of Doubled Haploids

Haploids derived from tobacco, tomato, pepper and eggplant are monoploids. Doubled haploids (DH) are generated after doubling of the chromosome set of the monoploids (see Kasha on chromosome doubling, Chap. I.7, this Vol.). Chromosome doubling is usually achieved by means of colchicine (see Ross et al. 1967; Burk et al. 1972). DH lines are expected to be homozygous. This represents an instant way to produce inbred lines which lead to the production of superior F_1 hybrids.

The DH lines of pepper are generally homogeneous and stable (Dumas de Vaulx 1990). DH lines obtained from F_1 hybrids between two parents allow for detection of useful genes by genetic analysis based on homozygous genotypes. An additional advantage is that the genotypes can be multiplied for repeated tests. A DH line resistant to pepper veinal mottle virus (PVMV) was recovered from the F_1 hybrid of two susceptible parents (Caranta and Palloix 1996). The resistance was from two recessive complementary genes, each one coming from one of the parents. DH progenies from F_1 hybrids of two parents (Caranta and Palloix 1996) and DH lines, F_1, F_2 and backcross progenies to a resistant parent (Dogimont et al. 1996) were used to examine specific and polygenic factors involved in resistance to potyviruses. Jiang and Li (1984) compared androgenic DH lines from F_1 of sweet × hot pepper parents. The later generation DH were uniform within lines with yields not higher than that of the F_1. Chen (1985) obtained hybrids between androgenic DH lines with high yields.

DH obtained from androgenetic and gynogenetic haploids of tobacco showed significant differences for a number of agronomic traits. Gynogenetic DHs were agronomically superior and had yielding ability close to the cultivars (Kumashiro and Oinuma 1985; Wernsman et al. 1989). DH lines generated from androgenetic haploids of near homozygous cultivars exhibited among-line genetic variability. Their performances were inferior to the parental cultivars (Burk and Matzinger 1976; Schnell and Wernsman 1986). Aberrant cytological behavior in androgenetic DH × parental cultivar hybrids indicated amplification of DNA in DH lines without changes in chromosome numbers (Reed et al. 1991). Two or more cycles of selection in full-sib families obtained from DH lines of a cultivar achieved a population equal in yielding ability to the parent (Schnell and Wernsman 1986). Comparing genotypes for several quantitative traits revealed close association of performance between haploids and doubled haploids of the same genotypes, suggesting the feasibility of mass selection for quantitative traits among haploids followed by chromosome doubling of selected individuals (Witherspoon and Wernsman 1989). Walker and Aycock (1994) worked on combined disease resistance in androgenetic DH lines. They suggested that screening for desired traits can be more efficiently conducted at the haploid level before chromosome doubling.

DH lines were generated from 1*x* to 2*x* and from 2*x* to 4*x* from diploid potato parents (Uijtewaal et al. 1987b). Increase in vigor was observed at the first increase of ploidy but less clearly at the second increase of ploidy. M'Ribibu and Veilleux (1991) observed significant high correlations between the monoploids and their DHs of *S. phureja*, with the latter having higher values for a majority of observed traits. Paz and Veilleux (1999) investigated the use of in vitro shoot regeneration to generate DH from a diploid potato species *S. phureja*.

4 Development and Use of Dihaploids

Dihaploids from the 4*x* *S. tuberosum* are genetically heterozygous. They are used as parents to cross with other *Solanum* species to introduce new genes through diploid, tetraploid and even higher ploidy levels (Peloquin and Ortiz 1993; Carputo et al. 2000). Tetraploid hybrids obtained by unilateral (4*x* × 2*x* matings) and bilateral (2*x* × 2*x* matings) polyploidization process are now a major breeding method for new cultivars.

Several methods were developed to obtain hybrids between dihaploids and other *Solanum* species.

4.1 Diploid Hybrids

Dihaploids of *S. tuberosum* were used as parents in crosses with 2*x* (2 EBN) wild species for germplasm enhancement of agronomic and tuber quality traits (Hermundstad and Peloquin 1985; Yerk and Peloquin 1990a; Serquen and Peloquin 1996; Santini et al. 2000). A hybrid of 2*x* *S. tuberosum* and *S. chacoense* ($2n=2x=24$) was used to cross with a *S. phureja* clone ($2n=2x=24$). QTL in the diploid progenies were mapped and used for marker-assisted selection of specific gravity (Freyre and Douches 1994) and tuber dormancy (Freyre et al. 1994). Hybrids from DH × wild species were evaluated for cold-chipping ability (Oltmans and Novy 2002).

4.2 Sexual Polyploidization

Dihaploids extracted from 4*x* *S. tuberosum* are crossed with other diploid species ($2n=2x=24$) to produce elite diploid hybrids for use in the breeding program (see Yerk and Peloquin 1990b; Darmo and Peloquin 1991). Tetraploid progenies are obtained through sexual polyploidization of the diploid parents that form unreduced 2*n* gametes. The two major modes of 2*n* gamete formation are the first division restitution (FDR) and second division restitution (SDR) (Veilleux 1985; Peloquin et al. 1989). The 2*n* pollen and 2*n* egg are caused by different abnormal cytological events during meiosis (see Tai 1994). Formation of 'parallel/fused spindles' during anaphase II of meiosis prevents cell division and consequently two 2*n* microspores are formed (Mok and Peloquin 1975; Ramanna 1979; Veilleux et al. 1982). This induces FDR gametes. A 'premature cytokinesis' following the first division prevents the occurrence of second division during meiosis. Consequently a dyad of two 2*n* microspores is formed (Mok and Peloquin 1975). This induces SDR gametes. Synaptic mutants are found which cause poor pairing and/or reduced chiasma frequencies in microsporogenesis (Jongedijk and Ramanna 1988; Peloquin et al. 1989). When combined with 'parallel spindles', FDR 2*n* pollen

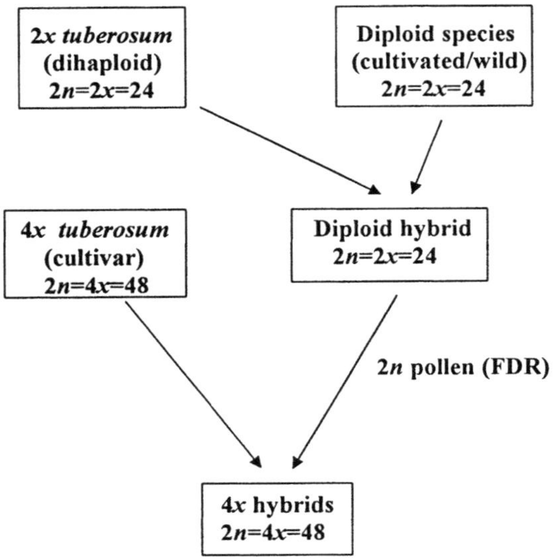

Fig. 1. Breeding scheme for generating 4*x* hybrid progenies from a 4*x*–2*x* cross

without crossing over, referred to as FDR-NCO gametes, are formed (Okwuagwu and Peloquin 1981; Hermundstad and Peloquin 1987). Parallel spindles, premature cytokinesis and synaptic mutants have all been identified in microsporogenesis and thus are all mechanisms for the formation of 2*n* pollen. Omission of the second meiotic division is the predominant mechanism of 2*n*-egg formation (Jongedijk 1985; Stelly and Peloquin 1986). There are extensive works on the merit of breeding for new cultivars using unilateral (4*x*–2*x* or 2*x*–4*x*) and bilateral (2*x*–2*x*) mating schemes (see Yerk and Peloquin 1990b; de Jong and Tai 1991; Werner and Peloquin 1991; Ortiz 1998; Buso et al. 1999, 2000). Figure 1 shows a scheme used in breeding for new cultivars through the unilateral (4*x*–2*x*) mating system.

4.3 Asexual Polyploidization – Chromosome Doubling

Subaxillary meristems of young plants (25–30 cm tall) are treated with colchicine (0.25–0.5 %) to obtain shoots with a doubled chromosome set. The procedure was described by Ross et al. (1967). Chromosome doubling of haploids is effective in increasing the homozygosity of the tetraploid parent (de Maine and Jervis 1989). *S. commersonii* is used as a bridge species to overcome crossability barriers after its 1EBN number is increased to 2EBN by chromosome doubling (Bamberg et al. 1994). Tetraploid progenies produced by diploid hybrid parents and their chromosome doubled counterpart parents have been compared in 2*x*–2*x* 4*x*–2*x* and 4*x*–4*x* crosses (de Maine 1994; Tai and de Jong 1997).

4.4 Asexual Polyploidization – Somatic Fusion

Somatic fusion uses asexual means to combine two nuclear genomes. Haploidization followed by somatic hybridization is especially useful for obtaining interspecific hybrids between incompatible species. Dihaploid *S. tuberosum* was used for protoplast fusion with sexually incompatible wild species. Protoplast fusion can be induced chemically or via electrofusion (Wenzel 1994). The latter appears to be the preferred method. Leaflets are digested by enzymes and protoplasts isolated. Parental protoplasts are mixed and subjected to electrofusion (see Thieme et al. 1997). The protoplasts are cultured in regeneration media (see Wenzel 1994). Shoots are grown from protoplast-derived calli. Verification of successful somatic hybrids is carried out by cytological analysis, isozyme analysis and species-specific molecular markers (see Thach et al. 1993; Novy and Helgeson 1994; Thieme et al. 1997; Barone et al. 2002). A set of chromosome-specific cytogenetic DNA markers has been developed for chromosome identification in potato (Dong et al. 2000). The somatic hybrids were investigated for the possibilities to incorporate late blight from wild species into potato breeding lines (Thieme et al. 1997). There is extensive literature on production and inheritance in tetraploid somatic hybrids in *Solanum* species.

5 Research Topics with Application of Haploidization

The following are examples of research topics in which the application of haploidization is a key component.

5.1 Endosperm Balance Number

Crosses between interspecific and interploidy parents may or may not be successful. The endosperm balance number (EBN) was proposed as a mechanism responsible for the success or failure of hybridization (Johnson and Hanneman 1982). EBN must be in a 2 maternal:1 paternal ratio for normal endosperm development. Dihaploids of *S. tuberosum* have 2EBN and thus are able to tap the genetic diversity from related wild species from hybrids of dihaploid × species matings (Yerk and Peloquin 1988, 1990a). More complex manipulation of ploidy levels to change EBN numbers has made it possible for direct in vivo germplasm introgression to *S. tuberosum* from sexually isolated species with EBN=1 (Carputo et al. 1997).

5.2 Chromosome Engineering

Wernsman (1993) described a procedure to generate amphiploid progenies by crossing an *N. tabacum* cultivar (McNair 944) with *N. africana* ($2n$=46), which possesses resistance to potato virus Y, followed by chromosome doubling. The amphiploid was backcrossed three times to the cultivar. Haploid and aneuhaploid plants were obtained by in vitro anther culture. Chromosome doubled lines were obtained by in vitro culture of leaf midvein explants. Two lines had $2n$=50 with 24 chromosome pairs from tobacco and one pair from *N. africana*. A scheme was presented by Ortiz (1998) to develop chromosome addition or substitution lines of potato from non-tuberous and tuberous species through manipulating EBN to overcome crossing barriers.

5.3 Tetrasomic Inheritance

The genetic architecture of the heterozygous autotetraploid potatoes is complex. There are four alleles at a locus that give tetrasomic segregation patterns. The distance of a locus to the centromere of the chromosome further complicates the segregation ratios. The segregation pattern of dihaploid genotypes, however, is much simpler. Table 1 compares the segregation ratios in $2n$ gametes generated by autotetraploid and dihaploid parental genotypes with all possible allelic combinations at a locus. The α and β parameters represent, respectively, the coefficients of double reduction ($0 \leq \alpha \leq 1/6$) and rate of single exchange tetrads ($0 \leq \beta \leq 1$) in autotetraploids and dihaploids. A locus located close to the centromere has $\alpha \sim 0$ and $\beta \sim 0$. Extensive work has been undertaken on the genetic consequences of sexual polyploidization through the $2n$ gametes at both theoretical and experimental levels (see Mendiburu et al. 1974; Hermsen 1984; Peloquin et al. 1989; Ortiz and Peloquin 1994; Tai 1994; David et al. 1995 as examples). This includes estimation of the coefficient of double reduction (Haynes and Douches 1993) and mating designs for the estimation of first- (Tai 1982a,b, 1986) and second-degree genetic parameters in the quantitative inheritance of traits (Boudec et al. 1989; Haynes 1990, 1992; David et al. 1995). A combination of the use of haploidization, chromosomal doubling and somatic fusion, as shown in Fig. 2, enables the creation of a group of autotetraploid genotypes with all possible tetrasomic combinations of alleles from four monoploid parents. This allows in-depth investigation of various genic effects of quantitative traits in terms of tetrasomic inheritance.

5.4 Gene Mapping - Monoploids and DH

Monoploids and their DH lines from a single genotype are expected to show the (1:1) ratio of segregation over all heterozygous loci without the interference of the dominance relationship between alleles in a locus. This simplifies

Table 1. Frequencies of *2n* gametes of various genotypes in a locus produced by a tetraploid parent with five types of genotypes, and an FDR or SDR diploid parent with two types of genotypes. *DR* Produced by double reduction event; *Non-DR* produced by non-double reduction event; α coefficient of double reduction, $0 \leq \alpha \leq 1/6$; β rate of single exchange tetrads, $0 \leq \beta \leq 1$

Allelic pattern	Genotype of parent	Genotype	2x gametes frequency DR	Non-DR	Total
Monoallelic	$A_1A_1A_1A_1$	A_1A_1	–	–	1.0
Unbalanced diallelic	$A_1A_1A_1A_2$	A_1A_1	$3\alpha/4$	$(1-\alpha)/2$	$(2+\alpha)/4$
		A_2A_2	$\alpha/4$	0	$\alpha/4$
		A_1A_2	0	$(1-\alpha)/2$	$(1-\alpha)/2$
Balanced diallelic	$A_1A_1A_2A_2$	A_1A_1	$\alpha/2$	$(1-\alpha)/6$	$(1+2\alpha)/6$
		A_2A_2	$\alpha/2$	$(1-\alpha)/6$	$(1+2\alpha)/6$
		A_1A_2	0	$(2-2\alpha)/3$	$(2-2\alpha)/3$
Triallelic	$A_1A_1A_2A_3$	A_1A_1	$\alpha/2$	$(1-\alpha)/6$	$(1+2\alpha)/6$
		A_2A_2	$\alpha/4$	0	$\alpha/4$
		A_3A_3	$\alpha/4$	0	$\alpha/4$
		A_1A_2	0	$(1-\alpha)/3$	$(1-\alpha)/3$
		A_1A_3	0	$(1-\alpha)/3$	$(1-\alpha)/3$
		A_2A_3	0	$(1-\alpha)/6$	$(1-\alpha)/6$
Tetra-allelic	$A_1A_2A_3A_4$	A_1A_1	$\alpha/4$	0	$\alpha/4$
		A_2A_2	$\alpha/4$	0	$\alpha/4$
		A_3A_3	$\alpha/4$	0	$\alpha/4$
		A_4A_4	$\alpha/4$	0	$\alpha/4$
		A_1A_2	0	$(1-\alpha)/6$	$(1-\alpha)/6$
		A_1A_3	0	$(1-\alpha)/6$	$(1-\alpha)/6$
		A_1A_4	0	$(1-\alpha)/6$	$(1-\alpha)/6$
		A_2A_3	0	$(1-\alpha)/6$	$(1-\alpha)/6$
		A_2A_4	0	$(1-\alpha)/6$	$(1-\alpha)/6$
		A_3A_4	0	$(1-\alpha)/6$	$(1-\alpha)/6$
Monoallelic (FDR/SDR)	A_1A_1	A_1A_1			1
Diallelic (FDR)	A_1A_2	A_1A_1			$\beta/4$
		A_1A_2			$(2-\beta)/2$
		A_2A_2			$\beta/4$
Diallelic (SDR)	A_1A_2	A_1A_1			$(1-\beta)/2$
		A_1A_2			β
		A_2A_2			$(1-\beta)/2$

a great deal of gene mapping. DH progenies were used to construct an intraspecific integrated molecular markers (RFLP and RAPD) linkage map of pepper (Lefebvre et al. 1995). Several genomic regions showed aberrant segregation ratios. Distorted segregation of genetic markers was also reported in monoploids from diploid hybrids of potatoes (Rivard et al. 1996; Chani et al. 2002). Tai et al. (2000) investigated the possible underlying genetic mechanisms for the distortion. A method of mapping was developed to accommodate the phenomenon of segregation-distorted loci in the genome of monoploids.

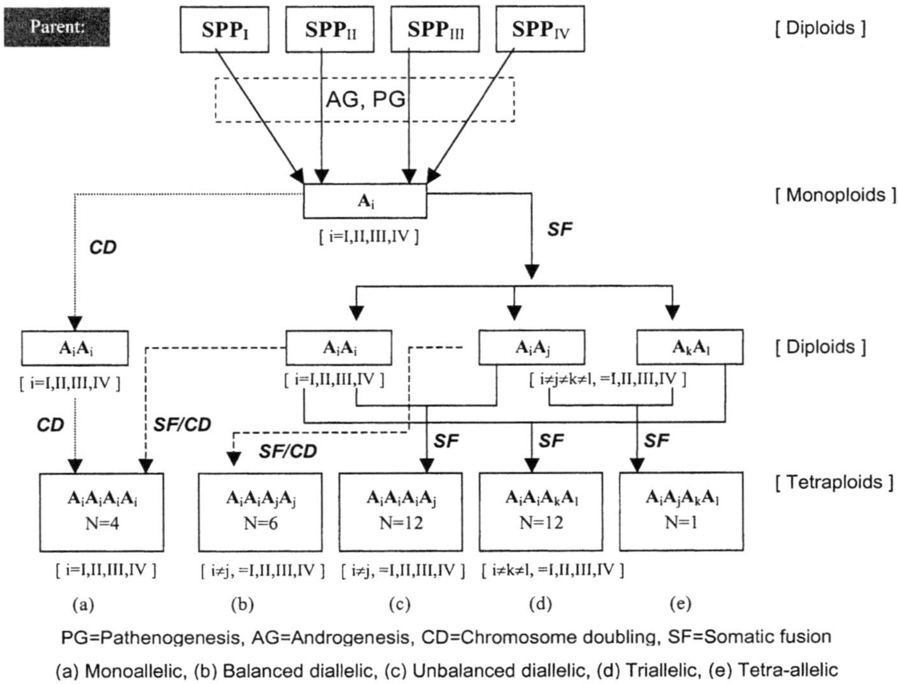

Fig. 2. Asexual production of tetraploid progenies of all possible genotypes from four diploid species parents

5.5 Gene Mapping: 4*x*–2*x* Hybrids

A gene-centromere mapping method based on progenies from 4*x* *S. tuberosum* × diploid hybrid was developed and used by Mendiburu and Peloquin (1979), and further developed for the location of specific gene loci with respect to the centromere of the chromosome (Douches and Quiros 1987; Wagenvoort and Zimnoch-Guzowska 1991; Ortiz and Peloquin 1993; Tai 1994; Bastiaanssen et al. 1996). The method, it appears, has the potential to be extended for mapping over the whole length of chromosomes.

6 Conclusions

Research work on haploids in the Solanaceae species over the past three decades has been extensive and fruitful. A large quantity of haploids are generated for the Solanaceae species due to successful methods of androgenesis and gynogenesis. This promotes the various aspects of research activities on haploids in the improvement of solanaceous species. Reduction of genome size through haploidization represents the first step toward ploidy manipulation for crop improvement. Haploids are used to produce homozygous DH lines for the tetraploid (tobacco) and diploid (tomato, pepper and eggplant) species. Dihaploids produced from the tetraploid species *S. tuberosum* enable the introduction of genes from the related diploid species. Both are valuable for genetic studies and breeding purposes. More studies on the genetic mechanisms, developmental stage and pretreatment strategies of donor plants and organs, culture media, and other factors are needed in order to have an effective process using haploidization process over any chosen genotypes in the Solanaceae species. A combination of haploidization and polyploidization in a special mating scheme like the one shown in Fig. 2 would lead to further research works on chromosomal engineering, gene mapping, transfer of genes between species, QTL identification, tetrasomic inheritance and development of new breeding procedures.

References

Aziz AN, Seabrook JEA, Tai GCC, de Jong H (1999) Screening diploid *Solanum* genotypes responsive to different anther culture conditions and ploidy assessment of anther-derived roots and plantlets. Am J Potato Res 76:9–16

Bajaj YPS (1990) In vitro production of haploids. In: Bajaj YPS (ed) Biotechnology in agriculture and forestry, vol 12. Haploids in crop improvement I. Springer, Berlin Heidelberg New York, pp 3–30

Bamberg JB, Hanneman RE Jr, Palta JP, Harbage JF (1994) Using disomic 4*x* (2EBN) potato species' germplasm via bridge species *Solanum commersonii*. Genome 37:866–870

Barone A, Li J, Sebastiano A, Cardi T, Frusciante L (2002) Evidence for tetrasomic inheritance in a tetraploid *Solanum commersonii* (+) *S. tuberosum* somatic hybrid through the use of molecular markers. Theor Appl Genet 104:539–546

Bastiaanssen HJM, Ramanna MS, Sawor S, Mincione A, Steen A, Jacobsen E (1996) Pollen markers for gene-centromere mapping in diploid potato. Theor Appl Genet 93:1040–1047

Boudec P, Masson M, Dattee Y (1989) A quantitative genetics model for the estimation of variances, covariances between relatives in crosses using 2n gametes in potato. In: Louwes KM, Toussaint HAJM, Dellaert LMW (eds) Parental line breeding and selection in potato breeding. PUDOC, Wageningen, pp 43–48

Burk LG, Gwynn GR, Chaplin JE (1972) Diploidized haploids from aseptically cultured anthers of *Nicotiana tabacum* L. A colchicine method applicable to plant breeding. J Hered 63:355–360

Burk LG, Matzinger DF (1976) Variation among anther-derived doubled haploids from an inbred line of tobacco. J Hered 67:381–384

Burk LG, Gerstel DU, Wernsman EA (1979) Maternal haploids of *Nicotiana tabacum* L. from seed. Science 206–585

Buso JA, Boiteux LS, Peloquin SJ (1999) Comparison of haploid *Tuberosum–Solanum chacoense* versus *Solanum phureja*-haploid *Tuberosum* hybrids as staminate parents of 4*x*–2*x* progenies evaluated under distinct crop management systems. Euphytica 109:191–199

Buso JA, Boiteux LS, Tai GCC, Peloquin SJ (2000) Direct estimation of the effects of meiotic recombination on potato traits via analysis of 4*x*–2*x* progenies from synaptic mutants with 2*n*-pollen formation by FDR without crossing over. Theor Appl Genet 101:139–145

Caligari PDS, Powell W, Liddell W, De Maine MJ, Swan EEL (1988) Methods and strategies for detecting *Solanum tuberosum* dihaploids in interspecific crosses with *S. phureja*. Ann Appl Biol 115:323–328

Cappadocia M, Cheng DSK, Ludlum-Simonette R (1984) Plant regeneration from in vitro culture of anthers of *Solanum chacoense* Bitt. and interspecific diploid hybrids *S. tuberosum* L. × *S. chacoense* Bitt. Theor Appl Genet 69:139–143

Caranta C, Palloix A (1996) Both common and specific genetic factors are involved in polygenic resistance of pepper to several potyviruses. Theor Appl Genet 92:15–20

Carputo D, Barone A, Cardi T, Sebastiano A, Frusciante L, Peloquin SJ (1997) Endosperm balance number manipulation for direct in vivo germplasm introgression to potato from a sexually isolated relative (*Solanum commersonii* Dun.). Proc Natl Acad Sci USA 94:12013–12017

Carputo D, Barone A, Frusciante L (2000) 2n gametes in the potato: essential ingredients for breeding and germplasm transfer. Theor Appl Genet 101:805–813

Chani E, Ashkenzai V, Hillel J, Veilleux RE (2002) Microsatellite marker analysis of an anther-derived potato family: skewed segregation and gene-centromere mapping. Genome 45:236–242

Chen XS (1985) Determination of combining ability and analysis of heterosis in pollen lines of *Capsicum annuum* var. *grosssum* Sendt. Acta Hortic Sin 12:267–272

Chlyah A, Taarji H, Chlyah H (1990) Tomato (*Lycopersicon esculentum* L.): anther culture and induction of androgenesis. In: Bajaj YPS (ed) Biotechnology in agriculture and forestry, vol 12. Haploids in crop improvement I. Springer, Berlin Heidelberg New York, pp 442–456

Collins GB, Genovesi AD (1982) Anther culture and its application to crop improvement. In: Tomas DT, Ellis BE, Harmey PM, Kasha KJ, Peterson RL (eds) Application of plant cell and tissue culture to agriculture and industry. University of Guelph, Ontario, pp 1–24

Darmo E, Peloquin SJ (1991) Use of 2*x* *tuberosum* haploid-wild species hybrids to improve yield and quality in 4*x* cultivated potato. Euphytica 53:1–9

David JL, Boudecand P, Gallais A (1995) Quantitative genetics of 4*x*–2*x* hybrid population with first-division restitution and second-division restitution 2*n* gametes produced by diploid parents. Genetics 139:1797–1803

De Jong H, Tai GCC (1991) Evaluation of potato hybrids obtained from tetraploid–diploid crosses I. Parent–offspring relationships. Plant Breed 107:177–182

De Maine MJ (1994) Comparison of tetraploid progenies of potato dihaploids, their chromosome-doubling derivatives and second generation dihaploids. Potato Res 37:173–181

De Maine MJ, Jervis L (1989) The use of dihaploids in increasing homozygosity of tetraploid potatoes. Euphytica 44:37–42

Dogimont C, Palloix A, Daubze AM, Marchoux G, Gebre-Selassie K, Pochard E (1996) Genetic analysis of broad spectrum resistance to potyviruses using doubled haploid lines of pepper (*Capsicum annuum* L.). Euphytica 88:231–239

Dolcet-Sanjuan R, Claveria E, Huerta A (1997) Androgenesis in *Capsicum annuum* L. – effects of carbohydrate and carbon dioxide enrichment. J Am Soc Hortic Sci 122:468–475

Dong F, Song J, Naess SK, Helgeson JP, Gebhardt C, Jiang J (2000) Development and applications of a set of chromosome-specific cytogenetic DNA markers in potato. Theor Appl Genet 101:1001–1007

Douches DS, Quiros CF (1987) Transmission of heterozygosity through meiotic mechanisms – comparison of theoretical and experimental data. Am Potato J 64:434

Dumas de Vaulx R (1990) Haploidy and pepper breeding: a review. Capsicum Newsl 8/9:13–17

Dumas de Vaulx R, Cambonnet RD, Pochard E (1981) Culture in vitro d'antheres de pimet (*Capsicum annuum* L): amelioration des taux d'obtention de plantes chez differents genotypes par des traitements <+35 °C. Agronomie 1:859–864

Freyre R, Douches DS (1994) Development of a model for marker-assisted selection of specific gravity in diploid potato across environments. Crop Sci 34:1361–1368

Freyre R, Warnke S, Sosinski B, Douches DS (1994) Quantitative trait locus analysis of tuber dormancy in diploid potato (*Solanum* spp.). Theor Appl Genet 89:474–480

Hamza S, Camilleri JM, Pollien JM, Vaucheret H, Bourgin JP, Chupeau Y (1993) Selection for spontaneous tomato haploids using a conditional lethal marker. Theor Appl Genet 86:657–664

Haynes KG (1990) Covariances between diploid parent and tetraploid offspring in tetraploid × diploid crosses of *Solanum tuberosum* L. J Hered 81:208–210

Haynes KG (1992) Covariance between haploid-species hybrid and *tuberosum* × haploid-species hybrid in 4*x*–2*x* crosses of *Solanum tuberosum* L. J Hered 83:119–122

Haynes KG, Douches DS (1993) Estimation of the coefficient of double reduction in the cultivated tetraploid potato. Theor Appl Genet 85:857–862

Hermsen JGT (1984) The potential of meiotic polyploidization in breeding allogamous crops. Iowa State J Res 58(4):435–448

Hermundstad SA, Peloquin SJ (1985) Germplasm enhancement with potato haploids. J Hered 76:463–467

Hermundstad SA, Peloquin SJ (1987) Breeding at the 2*x* level and sexual polyploidization. In: Jellis GJ, Richardson DR (eds) The production of new potato varieties: technological advances. Cambridge University Press, Cambridge, pp 197–210

Hougas RW, Peloquin SJ, Gabert AC (1964) Effect of seed parent and pollinator on frequency of haploids in *Solanum tuberosum*. Crop Sci 4:593–595

Jacobsen E, Ramanna MS (1994) Production of monohaploids of *Solanum tuberosum* L. and their use in genetics, molecular biology and breeding. In: Bradshaw JE, Mackay GR (eds) Potato genetics. CAB International, Wallingford, pp 155–170

Jaramillo J, Summers WL (1991) Dark-light treatments influence induction of tomato anther callus. HortScience 26:915–916

Jiang ZR, Li CL (1984) Observations and experiments on later generations of sweet × hot pepper derived by anther culture. Acta Hortic Sin 11:191–194

Johansson L (1986) Improved methods for induction of embryogenesis in anther cultures of *Solanum tuberosum*. Potato Res 29:179–190

Johnson SA, Hanneman RE Jr (1982) Manipulations of endosperm balance number overcome crossing barriers between diploid *Solanum* species. Science 217:446–448

Jongedjik E (1985) The pattern of megasporogenesis and megagametogenesis in diploid *Solanum* species hybrids: its relevance to the origin of *n* eggs and the induction of apomixis. Euphytica 34:599–611

Jongedjik E, Ramanna MS (1988) Synaptic mutants in potato, *Solanum tuberosum* L. 1. Expression and identity of genes for desynapsis. Genome 30:664–670

Kristiansen K, Andersen SB (1993) Effects of donor plant temperature, photoperiod, and age on anther culture response of *Capsicum annuum* L. Euphytica 67:105–109

Kumashiro T, Oinuma T (1985) Comparison of genetic variability among anther-derived and ovule-derived doubled haploid lines of tobacco. Jpn J Breed 35:301–310

Lefebvre V, Palloix A, Caranto C, Pochard E (1995) Construction of an intraspecific integrated linkage map of pepper using molecular markers and doubled-haploid progenies. Genome 38:112–121

Liu CA, Douches DS (1993) Production of haploids of potato (*Solanum tuberosum* subsp. *tuberosum*) and their identification with electrophoresis analysis. Euphytica 70:113–126

Luz JM, Pinto JE, Ehlert P, Cerqueira ES (1998) In vitro embryo induction in sweet pepper anthers. Hort Bresileira 16:56–60

Mendiburu AO, Peloquin SJ (1979) Gene-centromere mapping by 4*x*–2*x* matings in potatoes. Theor Appl Genet 54:177–180

Mendiburu AO, Peloquin SJ, Mok DWS (1974) Potato breeding with haploids and 2*n* gametes. In: Kasha KJ (ed) Haploids in higher plants. Guelph University Press, Guelph, Ontario, pp 249–258

Mityko J, Andrasfalvy A, Csillery G, Fari M (1995) Anther-culture response in different genotypes and F_1 hybrids of pepper (*Capsicum annuum* L.). Plant Breed 114:78–80

Mok DWS, Peloquin SJ (1975) Three mechanisms of 2*n* pollen formation in diploid potatoes. Can J Gene Cytol 17:217–225

M'Ribibu HK, Veilleux RE (1991) Phenotypic variation and correlation between monoploids and doubled monoploids of *Solanum phureja*. Euphytica 54:279–284

Muller DT, Goujaud J, Caboche M (1985) Isolation in vitro of naphthaleneacetic acid-tolerant mutants of *Nicotiana tabacum*, which are impaired root morphogenesis. Mol Gen Genet 199:194–200

Murashige T, Skoog F (1962) A revised medium for rapid growth and bioassays with tobacco tissue culture. Physiol Plant 15:473–497

Nakata K, Tanaka M (1968) Differentiation of embryoids from developing germ cells in anther culture of tobacco. Jpn J Genet 43:65–71

Nitsch JP, Nitsch C (1969) Haploid plants from pollens grains. Science 163:85–87

Nitzsche W, Wenzel G (1977) Haploids in plant breeding. Parey, Berlin

Novy RG, Helgeson JP (1994) Somatic hybrids between *Solanum tuberosum* and diploid, tuber bearing *Solanum* clones. Theor Appl Genet 89:775–782

Okwuagwu CO, Peloquin SJ (1981) A method of transferring the intact parental genotype to the offspring via meiotic mutants. Am Potato J 58:512–513

Oltmans SM, Novy RG (2002) Identification of potato (*Solanum tuberosum* L.) haploid × wild species hybrids with the capacity to cold-chip. Am J Potato Res 79:263–268

Ortiz R (1998) Potato breeding via ploidy manipulations. Plant Breed Rev 16:14–87

Ortiz R, Peloquin SJ (1993) Mapping the flower pigmentation locus in potato. J Genet Breed 47:171–173

Ortiz R, Peloquin SJ (1994) Use of 24-chromosome potatoes (diploids and dihaploids) for genetical analysis. In: Bradshaw JE, Mackay GR (eds) Potato genetics. CAB International, Wallingford, pp 133–154

Paz MM, Veilleux RE (1999) Influence of culture medium and in vitro conditions on shoot regeneration in *Solanum phureja* monoploids and fertility of regenerated doubled monoploids. Plant Breed 118:53–57

Pelletier G, Ferault M, Goujaud G, Vedel F, Caboche M (1987) The use of rootless mutants for the screening of spontaneous androgenetic and gynogenetic haploids in *Nicotiana tabacum*: evidence for the direct transfer of cytoplasm. Theor Appl Genet 75:13–15

Peloquin SJ, Ortiz R (1993) Techniques for introgressing unadapted germplasm to breeding populations. In: Stalker HT, Murphy JP (eds) Plant breeding in the 1990s. CAB International, Wallingford, pp 485–507

Peloquin SJ, Yerk GL, Werner JE (1989) Ploidy manipulation in the potato. In: Adolph KW (ed) Chromosomes: eukaryotic, prokaryotic, and viral, vol 2. CRC Press, Boca Raton, pp 167–178

Peloquin SJ, Werner JE, Yerk GL (1990) The use of potato haploids in genetics and breeding. In: Gupta PK, Tsuchiya T (eds) Chromosome engineering in plants, part B. Elsevier, Essex, pp 79–92

Peloquin SJ, Gabert AC, Ortiz R (1996) Nature of „pollinator" effect in potato haploid production. Ann Bot 77:539–542

Powell W, Uhrig H (1987) Anther culture of *Solanum* genotypes. Plant Cell Tissue Organ Cult 11:13–24

Ramanna MS (1979) A re-examination of the mechanisms of 2*n* gamete formation in potato and its implications for breeding. Euphytica 28:537–561

Reed SM (1996) Production of haploid tobacco plants using anther culture. In: Trigano R, Gray D (eds) Plant tissue culture concepts and laboratory exercises. CRC Press, Boca Raton, pp 241–246

Reed SM, Wernsman EA, Burns JA (1991) Aberrant cytological behavior in tobacco androgenic doubled haploid × parental cultivar hybrids. Crop Sci 31:97–101

Rivard SR, Cappadocia M, Landry BS (1996) A comparison of RFLP maps based on anther culture derived, selfed, and hybrid progenies of *Solanum chacoense*. Genome 39:611–621

Rokka VM, Valkonen JPT, Pehu E (1995) Production and characterization of haploids derived from somatic hybrids between *Solanum brevidens* and *S. tuberosum* through anther culture. Plant Sci 112:85–95

Rokka VM, Valkonen JPT, Pehu E (1997) Somatohaploid production by anther culture of interspecific somatic hybrids and their prospects in potato breeding. In: Jain SM, Sopory SK, Veilleux RE (eds) In vitro haploid production in higher plants. Kluwer, Dordrecht, pp 23–243

Ross RW, Dionne LA, Hougas RW (1967) Doubling the chromosome number of selected *Solanum* genotypes. Eur Potato J 10:37–52

Rotino GL, Falavigna A, Restaino F (1987) Production of anther-derived plantlets of eggplant. Capsicum Newsl 6:89–90

Santini M, Camadro EL, Marcellán ON, Erazzú LE (2000) Agronomic characterization of diploid hybrid families derived from crosses between haploids of the common potato and three wild Argentinian tuber-bearing species. Am J Potato Res 77:211–218

Schnell RJ, Wernsman EA (1986) Androgenic somaclonal variation in tobacco and estimation of its value as a source of novel genetic variability. Crop Sci 26:84–88

Serquen FC, Peloquin SJ (1996) Variation for agronomic and processing traits in *Solanum tuberosum* haploids × wild species hybrids. Euphytica 89:185–191

Shen LY, Veilleux RE (1995) Effect of temperature shock and elevated incubation temperature on androgenic embryo yield of diploid potato. Plant Cell Tissue Organ Cult 43:29–35

Shtereva LA, Zagorska NA, Dimitrov BD, Kruleva MM, Oanh HK (1998) Induced androgenesis in tomato (*Lycopersicon esculentum* Mill.) II. Factors affecting induction of androgenesis. Plant Cell Rep 18:312–317

Simmonds NW (1974) Evolution of crop plants. Longman, London

Simmonds NW (1979) Principles of crop improvement. Longman, London

Singsit C, Hanneman RE Jr (1991) Haploid induction in Mexican polyploid species and colchicine-doubled derivatives. Am Potato J 68:551–555

Stelly DM, Peloquin SJ (1986) Formation of $2n$ megagametophytes in diploid tuber-bearing *Solanums*. Am J Bot 73:1351–1363

Summers WL (1992) Microspore developmental stage and anther length influence the induction of tomato anther callus. HortScience 27:838-840

Summers WL (1997) Haploid plantlet production in tomato. In: Jain SM, Sopory SK, Veilleux RE (eds) In vitro haploid production in higher plants. Kluwer, Dordrecht, pp 219–231

Tai GCC (1982a) Estimation of double reduction and genetic parameters in autotetraploids. Heredity 49:63–70

Tai GCC (1982b) Estimation of double reduction and genetic parameters in autotetraploids based on $4x$–$2x$ and $4x$–$4x$ matings. Heredity 49:331–335

Tai GCC (1986) Biometrical genetical analysis of tetrasomic inheritance based on matings of diploid parents which produce $2n$ gametes. Heredity 57:315–317

Tai GCC (1994) Use of $2n$ gametes. In: Bradshaw JE, Mackay GR (eds) Potato genetics. CAB International, Wallingford, Oxon, pp 109–132

Tai GCC, de Jong H (1997) A comparison of performance of tetraploid progenies produced by diploid and their vegetatively doubled (tetraploid) counterpart parents. Theor Appl Genet 94:303–308

Tai GCC, Seabrook JEA, Aziz A N (2000) Linkage analysis of anther-derived monoploids showing distorted segregation of molecular markers. Theor Appl Genet 101:126–130

Thach NQ, Frei U, Wenzel G (1993) Somatic fusion for combining virus resistances in *Solanum tuberosum* L. Theor Appl Genet 85:863–867

Thieme R, Darsow U, Gavrilenko T, Dorokhov D, Tieman H (1997) Production of somatic hybrids between *S. tuberosum* L. and late blight resistant Mexican wild potato species. Euphytica 97:189–200

Tiainen T (1992) The influence of culture conditions on anther culture response of commercial varieties of *Solanum tuberosum* L. Plant Cell Tissue Organ Cult 30:211–219

Uijtewaal BA, Huigen DJ, Hermsen JGT (1987a) Production of potato monohaploids ($2n$=x=12) through prickle pollination. Theor Appl Genet 73:751–758

Uijtewaal BA, Jacobsen E, Hermsen JGT (1987b) Morphology and vigour of monoploid potato clones, their corresponding homozygous diploids and tetraploids and their heterozygous diploid parent. Euphytica 36:745–753

Vagera J (1990) Pepper (*Capsicum* spp.): in vitro induction of haploids. In: Bajaj YPS (ed) Biotechnology in agriculture and forestry, vol 12. Haploids in crop improvement I. Springer, Berlin Heidelberg New York, pp 374–391

Veilleux RE (1985) Diploid and polyploid gametes in crop plants: mechanisms of formation and utilization in plant breeding. Plant Breed Rev 3:253–288

Veilleux RE (1999) Anther culture of potato and molecular analysis of anther-derived plants as laboratory exercises for plant breeding courses. Hortic Technol 9:585–588

Veilleux RE, McHale NA, Lauer FL (1982) $2n$ gametes in diploid *Solanum*: frequency and types of spindle abnormalities. Can J Genet Cytol 24:301–314

Wagenvoort M, Zimnoch-Guzowska E (1991) Gene-centromere mapping in potato by half-tetrad analysis: map distances of H_1, *Rx* and *Ry* and their possible use for ascertaining the mode of $2n$-pollen formation. Genome 35:1–7

Walker DR, Aycock MK Jr (1994) Development of anther-derived dihaploids to combine disease resistance in Maryland tobacco. Crop Sci 34:335–338

Wenzel G (1994) Tissue culture. In: Bradshaw JE, Mackay GR (eds) Potato genetics. CAB International, Wallingford, Oxon, pp 173–195

Werner JE, Peloquin SJ (1991) Yield and tuber characteristics of $4x$ progeny from $2x \times 2x$ crosses. Potato Res 34:261–267

Wernsman EA (1993) Varied roles for the haploid sporophyte in plant improvement. In: Stalker HT, Murphy JP (eds) Plant breeding in the 1990s. CAB International, Wallingford, pp 461–484

Wernsman EA, Matzinger DF, Rufty RC (1989) Androgenetic vs. gynogenetic double haploids of tobacco. Crop Sci 29:1151–1155

Witherspoon WD Jr, Wernsman EA (1989) Feasibility of selection for quantitative traits among haploid tobacco sporophytes. Crop Sci 29:125–129

Yerk GL, Peloquin SJ (1988) $2n$ pollen in eleven $2x$, 2EBN wild species and their haploid × wild species hybrids. Potato Res 31:581–589

Yerk GL, Peloquin SJ (1990a) Selection of potato haploid parents for use in crosses with $2x$ (2 endosperm balance number) wild species. Crop Sci 30:943–946

Yerk GL, Peloquin SJ (1990b) Performance of haploid × wild species, $2x$ hybrids (involving five newly evaluated species) in $4x \times 2x$ families. Am Potato J 67:405–417

Zagorska NA, Shtereva A, Dimitrov BD, Kruleva MM (1998) Induced androgenesis in tomato (*Lycopersicon esculentum* Mill.) I. Influence of genotype on androgenetic ability. Plant Cell Rep 17:968–973

II.2 Haploids in the Improvement of Crucifers

WOLFGANG FRIEDT and M. KARIM ZARHLOUL[1]

1 Introduction

The production of doubled haploids (DH) in plants is a biotechnological approach that has been used to produce homozygous breeding lines and varieties. The production of haploids can be achieved either by gynogenesis or by androgenesis. While the formation of haploids in the first case proceeds from the embryo sack (megagametophyte), in the second case microspores are used as target tissue.

The use of haploids in producing new cultivars of crucifers (Brassicaceae) is widespread now, particularly in *Brassica napus* (oilseed rape). For the *Brassica* species the first attempt at producing DH lines was made using the anther culture system (Thomas and Wenzel 1975; Keller and Armstrong 1977). However, later research showed that in *Brassica* spp., particularly *B. napus*, higher frequencies of regenerated plants can be achieved by isolated microspore culture (Chuong and Beversdorf 1985; Pechan and Keller 1988; Telmer et al. 1992; Barro and Martin 1999). Biotechnological DH line production offers various advantages for plant breeders, including the possibility to obtain homozygous lines rapidly, as well as easy selection due to the absence of heterozygosity. It also facilitates genetic studies, particularly regarding quantitative traits. Furthermore, the use of DH progeny as mapping population(s) for the development of molecular markers is very advantageous since it enhances the efficiency of detecting markers, particularly for quantitative traits.

Within the genus *Brassica*, most work on the development of DH lines has been devoted to *B. napus*. This is not surprising since it is one of the most important oilseed crops worldwide. Furthermore, *B. napus* is much easier to handle in tissue culture than other Brassicaceae spp. (Weber et al. 2000). Anther culture as a method of producing DHs is still widely used, e.g. in cereals such as barley (e.g. Friedt and Rasmussen 2002), but it is accompanied by the problem that regeneration from somatic cells of the anther tissue can take place leading to the formation of diploid (heterozygous) progenies. Contrary to anther culture, in the case of microspore culture the target material is defi-

[1] Institute of Crop Science and Plant Breeding I, Justus-Liebig-University of Giessen, Heinrich-Buff-Ring 26–32, 35392 Giessen, Germany

Biotechnology in Agriculture and Forestry, Vol. 56
Haploids in Crop Improvement II (ed. by C.E. Palmer, W.A. Keller, and K.J. Kasha)

nitely derived from haploid gametic cells and is therefore homozygous. Consequently, the efficiency of obtaining haploids and DH lines via microspore culture is usually higher than with anther culture. Furthermore, microspore culture can be easily combined with other biotechnologies and related methods, such as gene transfer or mutagenesis, in order to create novel genetic variation in the starting material.

In this chapter, we will first describe the steps of producing DH lines using the microspore culture system. Furthermore, the achievements made and the impediments observed regarding the creation of DHs in crucifers will be highlighted. In addition, we will consider the role of haploids, and more precisely of DHs, in breeding programs of crucifers.

2 Progress and Status of Haploid Production

2.1 Protocol Description

First success in embryogenesis from a sperm cell precursor was reported by Guha and Maheshwari (1964) when culturing anthers of *Datura innoxia*. Since then, haploid plants have been produced from a large number of species belonging to various plant families using this technique (cf. Kott et al. 1990; Kott 1998). In crucifers, the most frequently studied plant species with regard to haploid production belong to the genus *Brassica*. The microspore culture technique was first successfully used by Lichter (1982) for the induction of embryogenesis from isolated microspores. Today, it is the method of choice for producing DH lines since further research has resulted in the development of highly efficient haploid embryogenic systems (Chuong and Beversdorf 1985; Pechan and Keller 1988; Telmer et al. 1992). In general, the procedure can be subdivided into the following steps:

1. Preparation of buds
2. Isolation of microspores
3. Embryo induction and regeneration of embryos
4. Chromosome doubling.

The various steps of the haploid production procedure in *Brassica* will be briefly described below. It can be easily reproduced in most laboratories, although modifications and optimization may be necessary when establishing the method depending on species, genotype, greenhouse or growth cabinet facilities and laboratory conditions.

1. *Preparation of buds*: the buds are harvested when they are at a developmental stage corresponding to a bud length of 2.5–5.5 mm, surface sterilized for 10 s in 96% ethanol, followed by sterilization for 15 min in 5.6% sodium hypochlorite (NaOCl). The buds are then rinsed with sterile water.

2. *Isolation of microspores*: buds are transferred to plastic bags containing 50–100 ml B5 medium (Gamborg et al. 1968), and the microspores are then released by maceration. The resulting cloudy solution now contains the microspores and is sieved through a colander with a pore size of 50 μm in order to separate the microspores from the debris of the bud cells. The filtrate is centrifuged for 10 min at 1,000 rotations per minute (rpm). The pellet, composed of accumulated microspores, is resuspended in 50 ml B5 medium. This whole procedure is repeated three times.
3. *Embryo induction and regeneration of embryos*: microspores are resuspended in 10 ml hormone-free NLN medium (Lichter 1981), and the concentration of microspores is adjusted to 90,000 microspores ml^{-1}. After heat shock treatment at 30 °C, the Petri dishes are cultured on a rotary shaker (60 rpm).
4. *Chromosome doubling*: the chromosome number of regenerated haploid plants can be efficiently doubled with an antimitotic agent such as colchicine (0.1 % water solution).

2.2 Factors Influencing Embryo Yield

Using the above-described steps as a basis, the method has been continuously optimized. In this context parameters such as developmental stage of the donor plants and buds (Pechan and Keller 1988; Baillie et al. 1992), microspore density (Huang et al. 1990; Barro and Martin 1999; Ferrie et al. 1999), sucrose or polyethylene glycol as an osmoticum (Baillie et al. 1992; Ilic-Grubor et al. 1998; Ferrie et al. 1999; Lionneton et al. 2001) and temperature optimization during the culture period (Lo and Pauls 1992; Cegielska-Taras et al. 2002) have been the object of various studies (cf. Table 1).

Microspores are generally embryogenic at the uninucleate stage (Kott et al. 1990; Telmer et al. 1992; Kott 1998), but those at an early binucleate stage are still able to change their developmental program from pollen maturation to

Table 1. Factors influencing the efficiency of haploid production in *B. oleracea*

Factor	Reference
Genotype (subspecies, variety, cultivar, line)	Ockendon (1984, 1985); Orton and Browers (1985); Ockendon and Sutherland (1987); Duijs et al. (1992); Ferrie et al. (1999); Wang et al. (1999); Dias (2001); Penhuizic et al. (2001)
Plants within genotype	Keller and Armstrong (1983); Ockendon (1984, 1988)
Anther and microspore development stage	Orton and Browers (1985); Lichter (1989); Dias (2001)
Sucrose concentration	Yang et al. (1992); Ferrie et al. (1999)
Temperature regime	Keller and Armstrong (1983); Duijs et al. (1992); Dias (2001); Dias and Correia (2002)

the embryogenic pathway (Pechan and Keller 1988; Huang et al. 1990). In this case a different cytological development is observed (Kott et al. 1988; Zaki and Dickinson 1990; Lo and Pauls 1992; Telmer et al. 1993), for example, an abundance of starch grains has been described (Telmer et al. 1993). However, morphological differences between embryogenic and non-embryogenic microspores have also been observed (Zaki and Dickinson 1991; Nitta et al. 1997). The importance of the cytoskeleton in the embryogenic process has been shown by Simmonds and Keller (1999). They studied microtubule organization during the first 24 h of heat induction in the embryogenic *B. napus* cv. Topas and the non-embryogenic *B. napus* breeding line #0025. Their results showed that preprophase bands of microtubules appeared in Topas microspores in late uninucleate microspores and in prophase after 4–8 h of heat treatment, but not in non-induced microspores. Furthermore, the continuous pre-prophase bands are required for embryogenesis.

Another observed phenomenon in microspores subjected to stress is the formation of heat-shock proteins (Smykal and Pechan 2000; Pechan and Smykal 2001). The former authors focused their research on the possible connection between stress-induced heat-shock proteins and the induction of embryogenesis in microspores. They found that heat-shock proteins may have a role to play in androgenesis, but only in combination with other factors, such as the appropriate phase of cell cycle, the stage of the differentiation process and inherent level of stress tolerance (Smykal and Pechan 2000). In the sections below we will describe the achievements made in this area.

2.2.1 Embryo Induction

The first step after preparing the target tissue, i.e. preparation of buds and isolation of microspores or whole anthers, is the treatment of the isolated microspores or the microspores within the anthers in order to induce embryogenesis. This has been achieved using different methods: In most cases heat shock was applied (Keller and Armstrong 1981; Georg and Rao 1982; Lichter 1982; Keller and Armstrong 1983; Ockendon 1984, 1985; Pechan and Keller 1988; Custers et al. 1996; Schulze and Pauls 1998; Simmonds and Keller 1999), but in some cases the induction of embryogenesis was achieved using colchicine as a stress agent (Zaki and Dickinson 1995; Zhao et al. 1996a,b; Zhou et al. 2002), or gamma irradiation in combination with ethanol (Pechan and Keller 1989). The temperature regime that is used to induce embryogenesis has been studied for the different *Brassica* species and for different genotypes within species. Georg and Rao (1982) found that only anthers of *B. juncea* which were subjected to a combination of cold treatment and elevated temperature shock responded by embryogenesis. Pechan et al. (1991) demonstrated that stress treatment such as high temperature is necessary to initiate the redirection process of microspores. During the application of high temperature stress, the first 8 h (prior to the first embryogenic nuclear division) are critical for the induction of microspore embryogenesis (Pechan et al. 1991). The activation of the process regulating the induction and maintenance of microspore embryogenesis takes place within this time frame. Simmonds and Keller (1999) observed that the first division in heat-treated microspores is a symmetric division. This is in contrast to the normal asymmetric division that occurs after the first pollen mitosis in plants or in microspores cultured continuously at 25 °C. However, a split temperature regime was also found to be beneficial in some cases; for example, microspores of *B. napus* were embryogenic if cultured at 32.5 °C for 18–24 h and then at 25 °C for the remaining culture period (Simmonds and Keller 1999). In *B. campestris*, anther culture at elevated temperature stimulated embryogenesis (Keller and Armstrong 1979). More recent work in this species evaluated various time and temperature combinations, and cultivation of isolated microspores at 32 °C for 48 h yielded the most embryos, although a genotypic variation of embryo yield was also observed (Baillie et al. 1992). Similarly, elevated temperature before incubation at 25 °C was observed to be essential in radish (*Raphanus sativus*) for the induction of microspore embryogenesis (Takahata et al. 1996). In a previous study aimed at finding a procedure to induce embryogenesis in more recalcitrant crucifer species, i.e. *B. nigra*, *B. oleracea* and *R. sativus* (Lichter 1989), experiments were carried out which revealed that efficient embryoid yields are obtained when aeration of the medium by agitating the Petri dishes on a shaker was started about 3 days after incubation (Lichter 1989). Regarding the species mentioned above (*B. nigra*, *B. oleracea*, *R. sativus*) most work has concentrated on optimizing haploid creation in *B. oleracea*. In this species the application of high temperature to induce embryogenesis has also been shown to be a prerequisite for embryogenesis (Keller and Armstrong 1981, 1983; Ockendon 1984). Keller and Armstrong (1981) found that in the autotetraploid marrowstem (*B. oleracea* var. *acephala*) maximum embryo yields were also obtained by an initial high temperature treatment at 35 °C for 1 day followed by continuous culture at 25 °C. In broccoli (*B. oleracea* var. *italica*), however, the optimum heat treatment was found to take place at 35 °C for 2 days prior to maintenance at 25 °C (Keller and Armstrong 1983), while in Brussels sprouts (*B. oleracea* var. *gemmifera*) the best results were achieved when anthers were heat treated at 35 °C for 16 h (Ockendon 1984). This indicates that there is a different response of various morphological forms of *B. oleracea* to heat treatment intensity and exposure. Furthermore, in a recent study on isolated microspores, the effect of incubation regimes at 27.5 and 30 °C for 2 days or 32.5 and 35 °C for 1 and 2 days, respectively, on embryo production in tronchuda cabbages (*B. oleracea* L. var. *tronchuda*) was studied. Genotypic variation was found between the varieties 'Couve Grelo', 'Penca da Póvoa' and 'Penca de Mirandela' used in this study (Dias and Correia 2002).

As mentioned above, microspore embryogenesis was also achieved in Brassicaceae spp. by employing colchicine. The results of a recent study in which the effect of colchicine on embryogenesis and chromosome doubling were evaluated show that the combination of embryogenic induction and chromosome doubling by the use of colchicine is very suitable at least in *B. napus* (Zhou et al. 2002). In this study a chromosome doubling efficiency of 91% was achieved by applying 500 mg l^{-1} colchicine for 15 h. It is to be hoped that more research will be carried out in other Brassicaceae species in order to optimize this combined treatment for inducing embryogenesis and parallel duplication of the chromosome set. Advantages of this combination comprise a further shortening of the DH production time and reduction in the number of polyploid and chimeric plants formed. Moreover, the latter authors claim that plant treatment for chromosome doubling significantly delays plant growth and development, and often results in the formation of chimeric plants with relatively small diploid sectors if compared to the immediate colchicine treatment of microspores (Zhou et al. 2002).

2.2.2 Development Stage of Microspores

Attempts were made to find the optimum development stage of buds in which the growing microspores are at a beneficial stage for embryogenesis. Baillie et al. (1992) evaluated the utility of bud size in *B. campestris* for this purpose. They examined bud sizes of 2, 2–2.9 and 3–3.9 mm, respectively, where embryos were obtained only from buds of 2–2.9 mm length. The optimal bud size of *B. napus* is obviously different, since it has been demonstrated that the best embryogenic microspores came from buds having a size of 4–4.2 mm (Pechan and Keller 1988). Takahata et al. (1996) observed that bud size also influenced microspore embryogenesis in radish. Though optimum bud size was different between genotypes, the microspore populations represented in these buds contained uninucleate and binucleate microspores. It was found in a comparative study in which different *Brassica* species were compared that the microspore size varied at the stage where they were most suitable for induction: *B. nigra* had the smallest microspore size followed by *B. napus*, with *B. oleracea* having the largest of the three species (Lichter 1989). In broccoli (*B. oleracea* var. *italica*), Orton and Browers (1985) observed that one of the factors that impacted embryo recovery was the interaction between genotype and anther development stage of the donor plant.

2.2.3 Culture Density

Another factor influencing the efficiency of obtaining embryos from microspores is culture density. It has been demonstrated for *B. napus* that a microspore density of 30,000–40,000 ml^{-1} in the first 2–4 days of culture followed by a dilution to 10,000 microspores ml^{-1} is crucial for embryogenesis and that a culture density higher than 100,000 microspores ml^{-1} is inhibitory to embryogenesis (Huang et al. 1990). Among *B. oleracea* accessions, Ferrie et al. (1999)

found that the lower density of 50,000 microspores ml^{-1} produced a higher embryogenic frequency than a higher density of 100,000 microspores ml^{-1}. However, this was not the case for two accessions, Bo-1 and Bo-4, where the latter density was more beneficial for embryogenesis. This demonstrates that a general statement on optimal microspore density cannot be made for all brassicas, but there is a tendency for a density of 10,000–40,000 microspores ml^{-1} to be more favourable, depending on the species and genotype. In *B. carinata* it was found that for buds of 2.5–3 mm versus buds of 3–3.5 mm a plating density of 100,000 or 150,000, respectively, was necessary to achieve the highest embryo frequency (Barro and Martin 1999).

2.2.4 Osmoticum

Sucrose has been used as a carbohydrate supplement and as an osmoticum in both anther and microspore culture of *Brassica* ssp. (Palmer et al. 1996a). Baillie et al. (1992) working with *B. campestris* and Ferrie et al. (1999) studying *B. oleracea* both demonstrated that culture with 17% sucrose for 48 h followed by a media change to 10% sucrose increased the frequency of microspore embryogenesis. Corresponding effects of sucrose level on microspore embryogenesis of *B. juncea* genotypes have recently been described by Lionneton et al. (2001). Here, with the exception of two genotypes (BJ-99 and BJ-7), the decrease from 17% sucrose during the first 48 h to 10% thereafter favoured an increase in embryo production. Sucrose in high concentrations can act as an osmotic stress for embryo induction (Palmer et al. 1996a). The latter authors suggested that it would be better to discriminate between the osmotic and nutritional roles of sucrose in order to better exploit the carbohydrate role in the embryo formation. Evidence of the sucrose double effect on embryo formation is provided by the study of Ilic-Grubor et al. (1998). In this work the possible effects of osmoticum used on the morphological development of microspore-derived embryos have demonstrated that at the heart, torpedo and early cotyledonary stages, microspore-derived embryos on polyethylene glycol (PEG) closely resembled their zygotic counterparts. In contrast, the external morphology of embryos induced on high sucrose medium differs from that of PEG and zygotic embryos, indicating that a high concentration of sucrose in culture has a morphological effect on microspore-derived embryos of *B. napus* (Ilic-Grubor et al. 1998). In the course of their evaluation of the microspore culture response of several *B. oleracea* accessions, Ferrie et al. (1999) found that sucrose concentration is the most important medium component influencing embryogenesis.

2.2.5 Other Factors

A number of other factors, besides the ones described above, can be optimized in order to obtain a sufficient number of embryos and consequently of haploid plants for the production of DHs. We will consider only some of these factors in this chapter. For more details readers are invited to consult the review of Palmer et al. (1996a).

Tuning of temperature, i.e. temperature optimization, applied during microspore culture is not only used to induce embryogenesis but has also been applied successfully to initiate conversion of microspore embryos to plantlets. Cegielska-Taras et al. (2002) reported that in *B. napus* a germination rate of 74.5–86.5% was achieved when microspore-derived embryos were grown at 1 °C for 14 days, and then at 24 °C for the next 21 days. Moreover, the development of these embryos was also better as cotyledons developed separately, in contrast to microspore-derived embryos from the other temperature variants where cotyledons were fused. The donor plant also influences microspore embryogenesis: In rapeseed (*B. napus*) this influence has been studied by Lo and Pauls (1992). They demonstrated that significantly higher embryo yields were obtained in microspore cultures initiated from donor plants grown at 15/12 °C instead of 23/18 °C in a light/dark cycle of 16/8 h.

In many Brassicaceae, such as *B. napus* and *B. oleracea*, plant growth regulators have been used in microspore or anther culture (Lichter 1982; Charne and Beversdorf 1988; Gland et al. 1988; Lichter 1989; Arnison et al. 1990). In some cases plant growth regulators had an effect on the development of embryos. For example, in *B. oleracea* haploid embryos produced through anther culture showed a high tendency towards direct production of somatic embryos in response to 2 mg l^{-1} benzyladenine (BA) and 0.1 mg l^{-1} naphthaleneacetic acid (NAA) (Prabhudesai and Bhaskaran 1993). The plant genotype used as a donor for the production of haploid plants must be taken into consideration and must not be underestimated. Such a genotype-dependent response has been seen, for example, in Brussels sprouts (Ockendon 1984, 1985; Ockendon and Sutherland 1987), cauliflower (Ockendon 1988) and broccoli (Orton and Browers 1985; Wang et al. 1999; Dias 2001; Penhuizic et al. 2001).

3 Use of Haploids in Breeding of Crucifers

3.1 Haploids in Breeding and Variety Development

The haploid technique has become a common tool for plant breeders (Palmer et al. 1996b). In breeding and production of cultivars the use of haploids – particularly via microspore culture – has many advantages. Firstly, by artificial doubling of the haploid chromosome set completely homozygous lines can be obtained within less than 12 months from isolating the microspores to regeneration of DH plants (Zarhloul, unpubl. data). This leads to a reduction of the breeding period by about 4 years in comparison to classical breeding procedures, for example the pedigree method, as well as to a rapid fixation of desired breeding traits, for example quality characters. Furthermore, through 100% homozygosity the genome can be liberated from gene alleles that have a lethal effect on plant development, enabling increased plant survival rates of the DH lines produced. In self-incompatible *Brassica* species like *B. oleracea* and *B. campestris* the creation of pure lines by producing DHs is an effective approach (Palmer et al. 1996b).

Moreover, the technique facilitates selection for recessive characteristics because genes are not overshadowed by the dominant allele and the number of plants with recessive genes in populations derived from F1 microspores is higher than in inbred populations. This means that in DH populations genetic segregations are less complex than in sexual progenies. For example, erucic acid content of high erucic acid genotypes is encoded by two additive dominant genes (Harvey and Downey 1964). Due to this additive digenic inheritance a segregation ratio of 1:4:6:4:1 for erucic acid content is expected in the F2 generation of a high by low-erucic cross, and only 1/16 of the segregating plants are homozygous recessive. In comparison, the corresponding segregation ratio for the same trait in a DH population would be 1:2:1. Therefore, the number of individuals homozygous for all loci necessary for the desired trait is four times higher (1/16 vs. 1/4) in a DH population than in a conventional F2. Moreover, recessive traits can be immediately identified amongst DH progeny and the additive effect of alleles is generally more apparent in DH lines.

With regard to hybrid seed production in cross-pollinating plants where inbred lines are difficult to achieve, this method constitutes a very powerful tool for creating new parent plants for hybrid seed production.

In the creation of DH lines two strategies are followed: (1) either haploid production is done using F1 microspores followed by a stringent selection; in this case the selection aims at trait fixation and is more efficient for traits that are more dependent on environmental factors and where selection in the F2 generation is difficult; (2) in the alternative case haploidization is carried out after one or two initial selection steps: this method is more efficient for highly heritable traits and enables the breeder to quickly eliminate undesired genotypes (Picard et al. 1994).

However, the use of the microspore culture technique is also faced with some problems. The most important limitation of this technique is seen in the fact that the formation of microspore-derived plants from certain genotypes can be very weak. Evidence of genotypic influence has been provided by Zhang and Takahata (2001). They examined the mode of inheritance of microspore response in vitro in *B. napus* and *B. campestris* by diallelic crosses between four cultivars of each species, where genotypic differential responses with regard to embryo formation were observed. The results of this study suggest that embryogenesis in oilseed rape (*B. napus*) is mainly controlled by two multiple gene loci with additive effects. Furthermore, because of the high heritability estimated for microspore embryogenesis the authors suggest that the embryogenic ability can be easily transferred from high- to low-responsive genotypes. Consequently, incorporating androgenesis in breeding programs with such genotypes or species can lead to the loss of gene alleles and therefore genetic variation may be reduced. Such a DH population would not be valuable for plant breeding (Choo et al. 1985). For example, Thiagarajah and Stringam (1993) examined the unintentional selection of gametes and compared the genetic segregation in traditional populations with DH populations of *B. juncea*. They found that the segregation of the various genes controlling seed colour and leaf hairiness studied in DHs did not differ from that expected in traditional populations.

For *B. oleracea* it has been shown by Rudolf et al. (1999) that haploid induction in cabbage (*B. oleracea* var. *capitata*) was less efficient in comparison to other morphological forms such as broccoli, Brussels sprouts and cauliflower. The same authors demonstrated the possible improvement of non-responsive genotypes with regard to embryogenic efficiency by crossing with a responsive genotype. This can be very useful, as the absence of genotype-independent protocols for haploid production limits the use of the DH procedure in cabbage breeding (Rudolf et al. 1999). Microspore or anther culture is now routinely used for the production of homozygous lines in broccoli breeding, and regenerated plant populations from anther culture represent a mixture of ploidy (Farnham et al. 1998; Wang et al. 1999). In addition to characterization of regenerants derived from microspore culture, Wang et al. (1999) compared the results with those obtained via anther culture. They found that the percentage of broccoli DHs that produce seed after selfing did not differ between both culture methods. An expected advantage of microspore over anther culture was the lower frequency of ploidy variation within clonal groups derived from individual embryos.

3.2 Use of Haploids in the Development of Novel Traits

The most important agronomic or horticultural trait is yield and yield stability. In oil-producing plants the oil content in combination with a high seed yield is a very important selection criterion during cultivar development. High yield and yield stability are complex traits that are influenced not only by different yield components but also by the whole plant development and its capability to resist biotic and abiotic stress. In the paragraphs below we will highlight some of the achievements made with regard to quality traits and disease resistances as a prerequisite to augmenting crop performance.

The quality of oils and fats is determined by the composition of fatty acids, i.e. their chain length, degree of desaturation, kind and number of functional groups, etc. For industrial purposes either oils or fats with high amounts of a single or unique fatty acid, or vegetable oils containing unusual fatty acids or novel compositions are required (Friedt and Lühs 1998). Work regarding the development of distinct fatty acid compositions in rapeseed was one of the important aims of molecular breeding in recent years; examples are high oleic acid (Hitz et al. 1995) or elevated erucic acid contents (Lühs and Friedt 1994; Weier et al. 1997; Han et al. 2001). Microspores can be used in order to select for a desired fatty acid composition if a segregating population is available. The microspore-derived plants of a specific cross represent in the ideal case all possible gametes, i.e. allelic combinations. Consequently, the derived DH plants segregate with regard to the desired trait. Many studies of seed oil formation have been conducted with microspore-derived embryos, as it was found that such embryos and zygotic embryos accumulate the triglycerides (seed oil) in a similar way (Taylor and Weber 1994). Consequently, Albrecht et al. (1995) and Möllers et al. (2000) have shown that it is possible to select superior genotypes in vitro at an early stage during microspore culture by using one of the two cotyledons for fatty acid analysis while retaining the rest of the embryo. In *B. carinata*, Barro et al. (2002) developed DH lines with modified glucosinolate and erucic acid contents in the seeds by combining induced mutagenesis via UV light irradiation and microspore culture in order to obtain lines with high erucic acid and low glucosinolate content. Three different groups with modified glucosinolate and erucic acid content have been created in the course of that study. For a better exploitation of their material in breeding the authors suggested that further genetic analyses were required to establish the basis of the phenotypic changes. In *B. juncea* some studies have also been conducted in order to determine the number of genes involved in the inheritance of glucosinolates using DHs. For example, Stringam and Thiagarajah (1995) predicted that five to nine genes were involved in the inheritance of glucosinolate content. Similar results have been described by Sodhi et al. (2002) who found that six or seven genes control the trait in a population of 752 DH plants; by using a BC1DH population of 1,263 individuals they unambiguously showed a seven-gene control.

Furthermore, Hawkins et al. (2002) have shown by using DH populations derived from crosses between the *B. napus* cultivars 'Cascade' and 'Rebel', with different freezing tolerance and vernalization requirements (Cascade has an obligate vernalization requirement and is highly freezing tolerant, while Rebel has a weak vernalization requirement and little freezing tolerance) that both traits can be separated and the positive trait expressions combined. The authors suggest that one of the developed DH lines is of particular importance because it lacks a vernalization requirement (Vern-) and possesses a high degree of freezing tolerance. This line has canola quality and good agronomic performance; furthermore, it shows potential as a spring cultivar. Moreover, its winter lineage makes it a desirable resource to extend the genetic diversity into spring rape seed material.

Disease resistance is also very important for crop cultivars. Resistance to light leaf spot disease (*Cylindrosporium concentricum*) was introduced into oilseed rape cultivars through the development of a synthetic amphidiploid (genome AACC) (Bradburne et al. 1999). Resistance to this disease was found within wild accessions of both *B. oleracea* and *B. rapa*. In the final step of the creation of resistant lines, microspore culture was successfully used and several agronomically acceptable DH lines were developed which expressed very high levels of adult plant resistance (Bradburne et al. 1999). For *B. oleracea*, black rot caused by *Xanthomonas campestris* is the most important disease (Vicente et al. 2002), where six different *Xanthomonas* races are recognized. Using *B. oleracea* DHs, Vicente et al. (2002) demonstrated that resistance to race 3 in the DH line BOH 85C was controlled by a single dominant locus. In the *italica* group of *B. oleracea* downy mildew caused by *Peronospora parasitica* is one of the economically most important diseases (Wang et al. 2000, 2001). The characterization of the resistance reaction of DH lines derived from the resistant hybrid 'Everest' lead to the identification of three different reaction phenotypes to infection: the first group showed susceptibility at both the cotyledon stage and the true-leaf stage, the second group was resistant at both the cotyledon and true-leaf stage, and the third group expressed susceptibility at the cotyledon stage but resistance at the true-leaf stage (Wang et al. 2000). Further work with the same material showed that resistance is controlled by two complementary dominant genes which can be easily incorporated into F1 hybrids and used commercially to prevent downy mildew (Wang et al. 2001). In broccoli, 20 anther culture-derived DH lines (Keller et al. 1975; Keller and Armstrong 1977; Jensen et al. 1999) were evaluated for resistance to *P. parasitica*. The latter study showed that it is possible to identify partial resistance to downy mildew in breeding material derived from modern broccoli cultivars with well-adapted horticultural traits.

4 *Brassica* Haploids as a Tool in Breeding Research

Not only are haploids very useful in applied breeding for crop improvement but also they can be efficiently used in basic studies, for example in breeding research based on molecular markers. Microspore-derived plants have been used, for example in genetic analysis and mapping, mutagenesis and genetic transformation (Jardinaud et al. 1993; Kuginuki et al. 1997; NiemirowiczSzczytt 1997; Voorrips and Kanne 1997; Fukuoka et al. 1998). Subsequently, the achievements in basic genetic studies using microspore-derived plants and focusing on molecular markers are highlighted.

For the development of molecular markers, useful populations segregating for the trait of interest are required. For example, using a *B. rapa* DH population derived from a cross between a susceptible (Homei P09) and a resistant parent (Siloga S2), Kuginuki et al. (1997) developed markers linked to club root (*Plasmodiophora brassicae*) resistance. A major advantage of using DH lines in comparison to a segregating F2 population is the fact that DH lines can be easily reproduced and maintained by self-pollination, so that different but genetically identical seedlings (groups) can be used for testing the reaction to different *P. brassicae* pathotypes as well as for other tests. In contrast, testing a segregating F2 population is not very reliable since the resistance test would have to be conducted with single plants (Kuginuki et al. 1997). Regarding white rust (*Albugo candida*), an important disease of *B. juncea*, two markers linked to resistance have been developed in a DH population by Prabhu et al. (1998). More recently, a high-density genetic linkage map of *B. juncea* has been constructed using an F1-derived DH population of 123 individuals. This map could be used for the dissection and transfer of agronomically important traits and favourable quantitative trait loci (QTL) from exotic germplasm to cultivated Indian varieties by recurrent backcrossing through marker-assisted breeding (Pradhan et al. 2003).

Furthermore, other studies have been conducted using DH lines in order to develop markers for many traits. For example, a DH population of *B. napus* derived from the F1 of the cross 'Apollo' (black-seeded) × YN90–1016 (yellow-seeded) was analysed via bulked segregant analysis (BSA) to identify molecular markers associated with the yellow-seed trait for future implementation in marker-assisted breeding of oilseed rape (Somers et al. 2001) or to identify random amplified polymorphic DNA (RAPD) markers associated with low linolenic acid loci (Somers et al. 1998). Ecke et al. (1995) mapped two erucic acid genes using an F1 microspore-derived DH population from the cross 'Mansholt's Hamburger Raps' × 'Samourai'. Furthermore, three QTL for seed oil content have been mapped in the latter study and additive effects of these QTL were shown to explain approx. 51% of the phenotypic variation observed for seed oil content in this population. Here, two of the QTL for oil content showed a close linkage to the two erucic acid genes, leading the authors to suggest that a direct effect of the erucic acid genes on oil content is probable (Ecke et al. 1995). With regard to seed glucosinolate content as another important quality trait, Uzunova et al. (1995) mapped four QTL that explain approx. 61% of the phenotypic variance in the same DH mapping population. Also, BSA was employed for the development of molecular markers for low linolenic acid, which are useful to facilitate the breeding of low linolenic rapeseed. Two RAPD markers – RM350 and RM574 – associated with linolenic acid were identified in a DH population segregating for linolenic and erucic acid levels (Rajcan et al. 1999). The authors suggest that these RAPD markers should be useful tools in the early detection of low linolenic or low/high erucic acid genotypes in rapeseed breeding programs based on DHs.

B. juncea can be used both as an oil plant and for the production of condiment. Segregating DH populations have been used for the determination of markers linked to certain traits (QTL) and for genetic map construction (Cheung et al. 1997, 1998; Prabhu et al. 1998; Lionneton et al. 2002). For the production of condiment, high seed oil content induces manufacturing problems (Lionneton et al. 2002). The combined use of DH plants and molecular markers (amplified fragment length polymorphism, AFLP) proved to be very suitable since AFLP produces a large number of polymorphic loci and dominant and co-dominant markers have equal information in DHs. The authors expect that a combination of QTL mapping and candidate gene approach will lead to an improved marker-assisted selection procedure in order to accelerate breeding for low oil content in brown mustard (Lionneton et al. 2002).

B. oleracea vegetable forms represent the most diverse morphological group in the genus *Brassica*. Here, investigations on the inheritance of morphological traits are very important (Sebastian et al. 2002). Research on morphological characters was carried out using both F2 populations (Kennard et al. 1994; Lan and Paterson 2000) and DH lines (Massie et al. 1996; Sebastian et al. 2002). In this context, a segregating population of F1-derived DH lines of *B. oleracea* was used to detect and locate QTL controlling 27 morphological and developmental traits, including leaf morphology, flowering, axillary bud formation and stem characters (Sebastian et al. 2002). Further research in *B. oleracea* focused on the identification of markers and genes or QTL for other traits using DHs (Voorrips et al. 1997; Vicente and King 2001). For example, in the study of Voorrips et al. (1997) a genetic map covering 615 cM in 12 linkage groups was developed on the basis of a DH population by using 92 RFLP and AFLP markers; one DH line was found to be resistant to clubroot (*P. brassicae*) and two loci for clubroot resistance were identified (Voorrips et al. 1997). Such examples show that DH populations are a powerful tool for the identification of molecular markers and for molecular breeding.

We ourselves have been using microspore-derived embryos in several research areas. Exploitation of genetic potential with regard to seed and oil yield is a primary breeding objective in order to further increase the economic value of winter oilseed rape (*B. napus*). With the advent of hybrid rapeseed breeding, seed yield has undergone considerable enhancement in recent years. In addition to individual yield performance the availability of useful genetic diversity between the potential crossing partners as a prerequisite for superior combining ability is necessary. We aim at the development of a new divergent gene pool based on high-erucic acid and high-glucosinolate rapeseed (HEAR) which can be utilized in hybrid breeding of double-low winter oilseed rape (canola). The yield performance of inter-pool hybrids (double-low quality × HEAR) in a three-location field trail (data not shown) was used to estimate the general combining ability (GCA) of 20 selected HEAR lines. In order to identify gene loci contributing to 'combining ability' with regard to relevant quantitative traits, such as seed and oil yield as well as oil content, a segregating mapping population of about 200 DH lines was developed by microspore culture. The population derived from a cross between a 'good combiner' (cv. Express, double-low quality) and a 'poor combiner' (V8, HEAR quality) is now used for subsequent genetic mapping by AFLP and SSR markers (Spiller et al. 2003 see also, Snowdon and Friedt 2004).

Another primary aim of our research is elucidating the genetic control of yellow seed colour. In order to identify gene loci contributing to this trait, a *B. napus* genome map was produced for a DH population segregating for seed colour expression using AFLP and SSR markers. In the course of this study one large QTL accounting for more than 50% of the phenotypic variation and two small QTLs for seed colour were detected (Badani et al. 2003).

5 Haploids in Combination with Other Biotechnological Methods

Microspore culture can be easily combined with other breeding techniques such as mutagenesis or gene transfer in order to create new genetic variation in the starting material. The latter is a prerequisite for producing new breeding lines and crop cultivars. Therefore, microspore culture has been used in conjunction with other biotechnological approaches for the creation of novel genotypes and cultivars.

5.1 Mutagenesis

Several mutagens, both chemical and physical agents, can be used to induce mutations and to create valuable mutants. The most widely used chemical mutagens are sodium azide (NaN_3) and ethylmethanesulfonate (EMS), while gamma or UV radiation constitute promising physical means for microspore mutagenesis. The combination of microspore culture and mutagenesis treatment has the following advantages: even recessive mutation events become apparent during selection and due to the high number of cells treated with the mutagen (millions of microspores are feasible with few isolation steps), the probability of identifying promising (recessive) mutants is high (Kott 1998).

In this context, it was possible in Chinese cabbage (*B. rapa*) to produce plants resistant to soft rot (*Erwinia carotovora*) disease by in vitro-mutagenesis and in vitro-selection using a microspore culture system in combination with UV irradiation (Zhang and Takahata 1999). From this study the importance of high numbers of potential mutants for obtaining the desired trait becomes clear: to be able to apply stringent selection on a medium containing a soft rot culture filtrate it was necessary to produce 6,657 embryos derived from UV-irradiated microspores in order to finally obtain 250 plantlets. When their progenies were assessed for soft rot reaction only three lines showed a markedly increased resistance to the disease (Zhang and Takahata 1999). Similarly, plants from microspores of *B. napus* derived after mutagenic treatment were shown to be able to survive under salt (NaCl) stress (Rahman et al. 1995). A total of 10 salt-surviving embryos were obtained out of a potential pool of 834,000 embryos that were mutagenized. In the area of quality improvement, chemical mutagenesis was used successfully, for example, to alter erucic acid content (Barro et al. 2001). Physical mutagenesis was also applied to microspores with the aim of modifying the erucic acid and glucosinolate contents (Barro et al. 2002). For further details on the improvement of oil and meal quality in *B. napus* using microspore culture, readers are referred to the review of Kott et al. (1996).

5.2 Genetic Transformation

Many factors influence the efficiency of genetic transformation of plants. This is the case for both *Agrobacterium*-mediated transformation and other methods of direct gene delivery (e.g. biolistic approach). Some factors influence the gene transfer as such and others affect the regeneration of transgenic plants from the transformed cells (Weber et al. 2000). Using *Agrobacterium tumefaciens* the following prerequisites must be fulfilled: successful bacterial colonization, induction of bacterial virulence, generation of T-DNA transfer complex and T-DNA transfer and integration of T-DNA into the plant genome (de la Riva et al. 1998). The latter represents a tissue culture aspect and depends on a whole range of factors including selectable markers, antibiotics for Agrobacteria elimination and culture conditions, e.g. light intensity, subculture interval, concentration and combinations of applied growth phytohormones, etc.

The use of genetic transformation to efficiently transfer specific foreign genes into the nuclear genome of higher plants offers plant breeders alternative routes to transfer relevant genes from unrelated species into crop plants, like oilseed rape, and opens new possibilities for the expansion of genetic variation in breeding material (cf. Friedt and Lühs 1998). Just a few studies have been reported dealing with gene transfer into microspores of *Brassica* spp. (Jardinaud et al. 1993; Dormann et al. 1995; Fukuoka et al. 1998; Nehlin et al. 2000), indicating that it still remains difficult to genetically modify microspores and that there is a need to establish an efficient transformation system for this kind of cell. Furthermore, in two of the above-mentioned studies stable transformation of microspores was confirmed (Dormann et al. 1995; Fukuoka et al. 1998). In the latter, direct gene delivery was achieved by particle bombardment, whereas Dormann et al. (1995) successfully applied *A. tumefaciens*-mediated transformation.

In principle, the utilization of haploid cells like microspores would substantially facilitate breeding with regenerated transgenic plants after doubling the chromosome number. This is due to the fact that after gene transfer to the microspore, whether by *A. tumefaciens* or via the biolistic approach, the newly introduced trait can simply be fixed in a homozygous state just by colchicine treatment. Evidence of this is provided by Fukuoka et al. (1998) who achieved direct gene transfer into isolated microspores of rapeseed

(*B. napus*), using the firefly luciferase (Luc) gene as a non-destructive marker and the biolistic approach, to finally develop fertile transgenic plants. Diploidized plants obtained from the haploid embryos were self-pollinated and all of the offspring tested were Luc-positive, indicating rapid fixation of the transgene as mentioned above.

Cogan et al. (2001) used DH lines from commercial F1 cultivars by anther culture to identify genotypes with both improved and reduced transformation capacities, irrespective of the responsiveness of the initial F1hybrid. This shows the possibility of identifying DH lines with improved explant transformation efficiency, indicating that the production of haploids is an effective means of manipulating the plant genotype for transformation. Furthermore, the latter authors developed DH lines of cv. Hawke segregating in a quantitative manner regarding explant transformation, which can be used for identifying and characterizing the genes involved in plant transformation. More recently, Cogan et al. (2002) identified QTL for transgenic and adventitious root production using a *B. oleracea* DH mapping population.

6 Conclusions, Future Trends and Perspectives

Man has been modifying plants for thousands of years simply by adaptation to the requirements of a sustained agriculture. Conventional methods of plant breeding and, in recent years, modern biotechnology have evolved into powerful tools for developing improved crop species and novel superior varieties. The production of DHs via biotechnological approaches such as anther and microspore culture offers the possibility of accelerating the breeding process, as well as facilitating basic scientific research work. Major advantages of these methods in comparison to the conventional breeding method via repeated self-pollination include reductions of time and space for breeding and ultimately reduction of the costs for variety development. From a methodical point of view, the production of haploids in crucifers can nowadays be considered as a routine practice. Nevertheless, there are still factors to be evaluated and optimized for different species and genotypes within a species like oilseed rape. Innovations that have resulted in improvements with regard to genetic engineering of the cultivated microspores may be employed to produce transgenic plants with newly introduced and fixed traits. This area of genetic engineering using isolated microspores merits further research in order to be able to improve this technique and apply it on a larger scale in research and practical breeding.

References

Albrecht S, Möllers C, Röbbelen G (1995) Selection in vitro for erucic acid content in segregating populations of microspore derived embryoids of *Brassica napus*. Plant Breed 114:210–214

Arnison PG, Donaldson P, Jackson A, Semple C, Keller WA (1990) Genotype specific response of cultured broccoli (*Brassica oleracea* var. *italica*) anthers to cytokinins. Plant Cell Tissue Organ Cult 20:217–222

Badani AG, Snowdon R, Baetzel R, Lühs W, Horn R, Friedt W (2003) QTL mapping for yellow seed colour in oilseed rape (*Brassica napus*). In: Proc 11th Int Rapeseed Congr, Copenhagen, Denmark, Publ 1, pp 85–87

Baillie AMR, Epp DJ, Hutcheson D, Keller WA (1992) In vitro culture of isolated microspores and regeneration of plants in *Brassica campestris*. Plant Cell Rep 11:234–237

Barro F, Martin A (1999) Response of different genotypes of *Brassica carinata* to microspore culture. Plant Breed 118:79–81

Barro F, Fernandez-Escobar J, de la Vega M, Martin A (2001) Doubled haploid lines of *Brassica carinata* with modified erucic acid content through mutagenesis by EMS treatment of isolated microspores. Plant Breed 120:262–264

Barro F, Fernandez-Escobar J, de la Vega M, Martin A (2002) Modification of glucosinolate and erucic acid contents in doubled haploid lines of *Brassica carinata* by UV treatment of isolated microspores. Euphytica 129:1–6

Bradburne R, Majer D, Magrath R, Werner CP, Lewis B, Mithen R (1999) Winter oilseed rape with high levels of resistance to *Pyrenopeziza brassicae* derived from wild *Brassica* sp. Plant Pathol 48:550–558

Cegielska-Taras T, Tykarska T, Szala T, Kuras M, Krzymanski J (2002) Direct plant development from microspore-derived embryos of winter oilseed rape *Brassica napus* L. ssp. *oleifera* (DC.) Metzger. Euphytica 124:341–347

Charne DG, Beversdorf WD (1988) Improving microspore culture as a rapeseed breeding tool: the use of auxins and cytokinins in an induction medium. Can J Bot 66:1671–1675

Cheung WY, Friesen L, Rakow, GFW, Seguin-Swartz G, Landry BS (1997) A RFLP-based linkage map of mustard (*Brassica juncea* (L.) Czern. and Coss.). Theor Appl Genet 94:841–851

Cheung WY, Gugel RK, Landry BS (1998) Identification of RFLP markers linked to the white rust resistance gene (Acr) in mustard (*Brassica juncea* (L.) Czern, and Coss.). Genome 41:626–628

Choo TM, Reinbergs E, Kasha KJ (1985) Use of haploids in breeding barley. Plant Breed Rev 3:219–252

Chuong PV, Beversdorf WD (1985) High frequency embryogenesis through isolated microspore culture in *Brassica napus* L. and *B. carinata* Braun. Plant Sci 39:219–226

Cogan N, Harvey E, Robinson H, Lynn J, Pink D, Newbury HJ, Puddephat I (2001) The effects of anther culture and plant genetic background on *Agrobacterium rhizogenes*-mediated transformation of commercial cultivars and derived doubled-haploid *Brassica oleracea*. Plant Cell Rep 20:755–762

Cogan NOI, Lynn JR, King GJ, Kearsey MJ, Newbury HJ, Puddephat IJ (2002) Identification of genetic factors controlling the efficiency of *Agrobacterium rhizogenes*-mediated transformation in *Brassica oleracea* by QTL analysis. Theor Appl Genet 105:568–576

Custers JBM, Cordewener JHG, Dons HJM, Lookeren MM-V (1996) Regulation of the inductive phase of microspore embryogenesis in *Brassica napus*. Acta Hort 407:209–217

De la Riva GA, González-Cabrera J, Vázquez-Padrón R, Ayra-Pardo C (1998) *Agrobacterium tumefaciens*: a natural tool for plant transformation. Electr J Biotechnol 1, http://www.ejb.org

Dias JS (2001) Effect of incubation temperature regimes and culture medium on broccoli microspore culture embryogenesis. Euphytica 119:389–394

Dias JS, Correia MC (2002) Effect of medium renovation and incubation temperature regimes on tronchuda cabbage microspore culture embryogenesis. Sci Hortic 93:205–214

Dormann M, Wang HM, Dalta N, Ferrie AMR, Keller WA, Oelck MM (1995) Transformation of freshly isolated *Brassica* microspores and regeneration to fertile homozygous plants. In: Proc 9th Int Rapeseed Congr, 4–7 July, Cambridge, UK, pp 816–818

Duijs JC, Voorrips RE, Visser DL, Custers JBM (1992) Microspore culture is successful in most crop types of *Brassica oleracea* L. Euphytica 60:45–55

Ecke W, Uzunova M, Weissleder K (1995) Mapping the genome of rapeseed (*Brassica napus*). 2. Localisation of genes controlling erucic acid synthesis and seed oil content. Theor Appl Genet 91:972–977

Farnham MW, Caniglia EJ, Thomas CE (1998) Efficient ploidy determination of anther-derived broccoli. HortScience 33:323–327

Ferrie AMR, Taylor DC, MacKenzie SL, Keller WA (1999) Microspore embryogenesis of high sn-2 erucic acid *Brassica oleracea* germplasm. Plant Cell Tissue Organ Cult 57:79–84

Friedt W, Lühs W (1998) Recent developments and perspectives of industrial rapeseed breeding. Fett/Lipid 100:219–226

Friedt W, Rasmussen M (2002) Modern European barley cultivars: genetic progress in resistance, quality and yield. In: Marè C, Faccioli P, Stanca AM (eds) Proc EUCARPIA Cereal Section Meeting, From Biodiversity to Genomics: Breeding Strategies for Small Grain Cereals in the Third Millennium. Exp Inst for Cereal Res, Fiorenzuola d'Arda, Italy, pp 73–78

Fukuoka H, Ogawa T, Matsuoka M, Ohkawa Y, Yano H (1998) Direct gene delivery into isolated microspores of rapeseed (*Brassica napus* L.) and the production of fertile transgenic plants. Plant Cell Rep 17:323–328

Gamborg OL, Miller RA, Ojima J (1968) Nutrient requirements of suspension cultures of soybean root cells. Exp Cell Res 50:151–158

Georg L, Rao PS (1982) In vitro induction of pollen embryos and plantlets in *Brassica juncea* through anther culture. Plant Sci Lett 26:111–116

Gland A, Lichter R, Schweiger HG (1988) Genetic and exogenous factors affecting embryogenesis in isolated microspore culture of *Brassica napus* L. J Plant Physiol 132:613–617

Guha S, Maheshwari SC (1964) In vitro production of embryoids from anthers of *Datura*. Nature 204:497

Han J, Lühs W, Sonntag K, Zähringer U, Borchardt DS, Wolter FP, Heinz E, Frenzen M (2001) Functional characterisation and promoter analysis of β-ketoacyl-CoA synthase genes from *B. napus* L. Plant Mol Biol 46:229–239

Harvey BL, Downey RK (1964) The inheritance of erucic acid content in rapeseed (*Brassica napus*). Can J Plant Sci 44:104–111

Hawkins GP, Deng Z, Kubik TJ, Johnson-Flanagan AM (2002) Characterization of freezing tolerance and vernalization in Vern-, a spring-type *Brassica napus* line derived from a winter cross. Planta 216:220–226

Hitz WD, Yadav NS, Reiter RS, Mauvais CJ, Kinney AJ (1995) Reducing polyunsaturation in oils of transgenic canola and soybean. In: Kader JC, Mazliak P (eds) Plant lipid metabolism. Kluwer, Dordrecht, pp 506–508

Huang B, Bird S, Kemble R, Simmonds D, Keller W, Miki B (1990) Effects of culture density, conditioned medium feeder cultures on microspore embryogenesis in *Brassica napus* L. cv. Topas. Plant Cell Rep 8:594–597

Ilic-Grubor K, Attree SM, Fowke LC (1998) Comparative morphological study of zygotic and microspore-derived embryos of *Brassica napus* L. as revealed by scanning electron microscopy. Ann Bot 82:157–165

Jardinaud MF, Souvre A, Alibert G (1993) Transient gus gene expression in *Brassica napus* electroporated microspores. Plant Sci 93:177–184

Jensen BD, Vaerbak S, Munk L, Andersen SB (1999) Characterization and inheritance of partial resistance to downy mildew, *Peronospora parasitica*, in breeding material of broccoli, *Brassica oleracea* convar. *botrytis* var. *italica*. Plant Breed 118:549–554

Keller WA, Armstrong KC (1977) Embryogenesis and plant regeneration in *Brassica napus* anther cultures. Can J Bot 55:1383–1388

Keller WA, Armstrong KC (1979) Stimulation of embryogenesis and haploid production in *Brassica campestris* anther cultures by elevated temperature treatments. Theor Appl Genet 55:65–67

Keller WA, Armstrong KC (1981) Production of anther-derived dihaploid plants in autotetraploid marrowstem kale (*Brassica oleracea* var. *acephala*). Can J Genet Cytol 23:259–265

Keller WA, Armstrong KC (1983) Production of haploids via anther culture in *Brassica oleracea* var. *italica* Broccoli. Plant Breed 32:151–159

Keller WA, Rajhathy T, Lacapra J (1975) In vitro production of plants from pollen in *Brassica campestris*. Can J Genet Cytol 17:655–666

Kennard WC, Slocum MK, Figdore SS, Osborn TC (1994) Genetic analysis of morphological variation in *Brassica oleracea* using molecular markers. Theor Appl Genet 87:721–732

Kott LS (1998) Application of double haploid technology in breeding of oilseed *Brassica napus*. AgBiotech News Inf 10(3):69N–74N

Kott LS, Polsoni L, Beversdorf WD (1988) Cytological aspects of isolated microspore culture in *Brassica napus*. Can J Bot 66:1658–1664

Kott LS, Erickson LR, Beversdorf WD (1990) The role of biotechnology in canola/rapeseed research. In: Shahidi F (ed) Canola and rapeseed: production, chemistry, nutrition and processing technology. Van Nostrand Reinhold, New York, pp 47–78

Kott LS, Wong R, Swanson E, Chen J (1996) Mutation and selection for improved oil and meal quality in *Brassica napus* utilizing microspore culture. In: Jain SM, Sopory SK, Veilleux RE (eds) In vitro haploid production in higher plants, vol 2. Applications. Kluwer, Dordrecht, pp 151–167

Kuginuki Y, Ajisaka H, Yui M, Yoshikawa H, Hida K, Hirai M (1997) RAPD markers linked to a clubroot-resistance locus in *Brassica rapa* L. Euphytica 98:149–154

Lan TH, Paterson AH (2000) Comparative mapping of quantitative trait loci sculpting the curd of *Brassica oleracea*. Genetics 155:1927–1954

Lichter R (1981) Anther culture of *Brassica napus* in a liquid medium. Z Pflanzenphysiol 103:229–237

Lichter R (1982) Induction of haploid plants from isolated pollen of *Brassica napus*. Z Pflanzenphysiol 105:427–434

Lichter R (1989) Efficient yield of embryoids by culture of isolated microspores of different Brassicacea species. Plant Breed 103:119–123

Lionneton E, Beuret W, Delaitre C, Rancillac M (2001) Improved microspore culture and doubled-haploid plant regeneration in the brown condiment mustard (*Brassica juncea*). Plant Cell Rep 20:126–130

Lionneton E, Ravera S, Sanchez L, Aubert G, Delourme R, Ochatt S (2002) Development of an AFLP-based linkage map and localization of QTLs for seed fatty acid content in condiment mustard (*Brassica juncea*). Genome 45:1203–1215

Lo KH, Pauls KP (1992) Plant growth environment effects on rapeseed microscope development and culture. Plant Physiol 99:468–472

Lühs W, Friedt W (1994) Stand und Perspektiven der Züchtung von Raps (*Brassica napus* L.) mit hohem Erucasäure-Gehalt im Öl für industrielle Nutzungszwecke. Fat Sci Technol 96:137–146

Massie IH, Astley D, King GJ (1996) Patterns of genetic diversity and relationships between regional groups and populations of Italian landrace cauliflower and broccoli (*Brassica oleracea* L. var. *botrytis* L. and var. *italica* Plenck). Acta Hort 407:45–53

Möllers C, Rücker B, Schierholt A (2000) In vitro selection for oleic and linoleic acid content in segregating populations of microspore derived embryos of *Brassica napus*. Euphytica 112:195–201

Nehlin L, Mollers C, Bergman P, Glimelius K (2000) Transient beta-gus and gfp gene expression and viability analysis of microprojectile bombarded microspores of *Brassica napus* L. J Plant Physiol 156:175–183

NiemirowiczSzczytt K (1997) Excessive homozygosity in doubled haploids – advantages and disadvantages for plant breeding and fundamental research. Acta Physiol Plant 19:155–167

Nitta T, Takahata Y, Kaizuma N (1997) Scanning electron microscopy of microspore embryogenesis in *Brassica* spp. Plant Cell Rep 16:404–410

Ockendon DJ (1984) Anther culture in Brussels sprouts (*Brassica oleracea* var. *gemmifera*). I. Embryo yields and plant regeneration. Ann Appl Biol 105:285–291

Ockendon DJ (1985) Anther culture in Brussels sprouts (*Brassica oleracea* var. *gemmifera*). II. Effect of genotype on embryo yields. Ann Appl Biol 107:101–104

Ockendon DJ (1988) The ploidy of plants obtained from anther culture of cauliflowers (*Brassica oleracea* var. *botrytis*). Ann Appl Biol 113:319–325

Ockendon DJ, Sutherland RA (1987) Genetic and non-genetic factors affecting anther culture of Brussels sprouts (*Brassica oleracea* var. *gemmifera*). Theor Appl Genet 74:566–570

Orton TJ, Browers MA (1985) Segregation of genetic markers among plants regenerated from cultured anthers of broccoli (*Brassica oleracea* var. *italica*). Theor Appl Genet 69:637–643

Palmer CE, Keller WA, Arnison PG (1996a) Experimental haploidy in *Brassica* species. In: Jain SM, Sopory SK, Veilleux RE (eds) In vitro haploid production in higher plants, vol 3. Important selected plants. Kluwer, Dordrecht, pp 143–171

Palmer CE, Keller WA, Arnison PG (1996b) Utilization of *Brassica* haploids. In: Jain SM, Sopory SK, Veilleux RE (eds) In vitro haploid production in higher plants, vol 3. Important selected plants. Kluwer, Dordrecht, pp 143–171

Pechan PM, Keller WA (1988) Identification of potentially embryogenic microspores in *Brassica napus*. Physiol Plant 74:377–384

Pechan PM, Keller WA (1989) Induction of microspore embryogenesis in *Brassica napus* L. by gamma irradiation and ethanol stress. In Vitro Cell Dev Biol J 25:1073–1074

Pechan PM, Smykal P (2001) Androgenesis: affecting the fate of the male gametophyte. Physiol Plant 111:1–8

Pechan PM, Bartels D, Brown DCW, Schell J (1991) Messenger-RNA and protein changes associated with induction of *Brassica* microspore embryogenesis. Planta 184:161–165

Penhuizic CLL, Chatelet C, Kloareg B, Potin P (2001) Carrageenan oligosaccharides enhance stress-induced microspore embryogenesis in *Brassica oleracea* var. *italica*. Plant Sci 160:1211–1220

Picard E, Crambes E, Mihamou-Ziyyat A (1994) L'haplodisation: un outil multi-usage pour la genetique et l'amelioration des cereales. Quel avenir pour l'amélioration des plantes? Ed. AUPELF-UREF. John Libbey Eurotext, Paris, pp 355–369

Prabhu KV, Somers DJ, Rakow G, Gugel RK (1998) Molecular markers linked to white rust resistance in mustard *Brassica juncea*. Theor Appl Genet 97:865–870

Prabhudesai V, Bhaskaran S (1993) A continuous culture system of direct somatic embryogenesis in microspore-derived embryos of *Brassica juncea*. Plant Cell Rep 12:289–292

Pradhan AK, Gupta V, Mukhopadhyay A, Arumugam N, Sodhi YS, Pental D (2003) A high-density linkage map in *Brassica juncea* (Indian mustard) using AFLP and RFLP markers. Theor Appl Genet 106:607–674

Rahman MH, Krishnaraj S, Thorpe TA (1995) Selection for salt tolerance in-vitro using microspore derived embryos of *Brassica napus* cv. Topas, and the characterisation of putative tolerant plants. In Vitro Cell Dev Biol Plant 31:116–121

Rajcan I, Kasha KJ, Kott LS, Beversdorf WD (1999) Detection of molecular markers associated with linolenic and erucic acid levels in spring rapeseed (*Brassica napus* L.). Euphytica 105:173–181

Rudolf K, Bohanec B, Hansen M (1999) Microspore culture of white cabbage, *Brassica oleracea* var *capitata* L: genetic improvement of non-responsive cultivars and effect of genome doubling agents. Plant Breed 118:237–241

Schulze D, Pauls KP (1998) Flow cytometric characterisation of embryogenic and gametophytic development in *Brassica napus* microspore cultures. Plant Cell Physiol 39:226–234

Sebastian RL, Kearsey MJ, King GJ (2002) Identification of quantitative trait loci controlling developmental characteristics of *Brassica oleracea* L. Theor Appl Genet 104:601–609

Simmonds DH, Keller WA (1999) Significance of preprophase bands of microtubules in the induction of microspore embryogenesis of *Brassica napus*. Planta 208:383–391

Smykal P, Pechan PM (2000) Stress, as assessed by the appearance of sHsp transcripts, is required but not sufficient to initiate androgenesis. Physiol Plant 110:135–143

Snowdon RJ, Friedt W (2004) Molecular markers in *Brassica* oilseed breeding: current status and future possibilities. Plant Breeding 123:1–8

Sodhi YS, Mukhopadhyay A, Arumugam N, Verma JK, Gupta V, Pental D, Pradhan AK (2002) Genetic analysis of total glucosinolate in crosses involving a high glucosinolate Indian variety and a low glucosinolate line of *Brassica juncea*. Plant Breed 121:508–511

Somers DJ, Friesen KRD, Rakow G (1998) Identification of molecular markers associated with linoleic acid desaturation in *Brassica napus*. Theor Appl Genet 96:897–903

Somers DJ, Rakow G, Prabhu VK, Friesen KRD (2001) Identification of a major gene and RAPD markers for yellow seed coat colour in *Brassica napus*. Genome 44:1077–1082

Spiller T, Lühs W, Baetzel R, Snowdon R, Friedt W (2003) Development of improved rapeseed populations for quality breeding and QTL mapping of agronomic characters. In: Proc 11th Int Rapeseed Congr, Copenhagen, Denmark, Publ 1, pp 94–96

Stringam GR, Thiagarajah MR (1995) Inheritance of alkenyl glucosinolates in traditional and microspore-derived doubled haploid populations of *Brassica juncea* L. Czern and Coss. In: Proc 9th Int Rapeseed Congr, Cambridge, UK, Publ 3, pp 804–806

Takahata Y, Komatsu H, Kaizuma N (1996) Microspore culture of radish (*Raphanus sativus* L.): influence of genotype and culture conditions on embryogenesis. Plant Cell Rep 16:163–166

Taylor DC, Weber N (1994) Microspore-derived embryos of Brassicaceae – model system for studies of storage lipid bioassembly and its regulation. Fat Sci Technol 96:228–235

Telmer CA, Simmonds DH, Newcomb W (1992) Determination of developmental stage to obtain high frequencies of embryogenic microspores in *Brassica napus*. Physiol Plant 84:417–424

Telmer CA, Newcomb W, Simmonds DH (1993) Microspore development in *Brassica napus* and the effect of high temperature on division in vivo and in vitro. Protoplasma 172:154–165

Thiagarajah MR, Stringam GR (1993) A comparison of genetic segregation in traditional and microspore-derived populations of *Brassica juncea* L. Czern and Coss. Plant Breed 111:330–334

Thomas E, Wenzel G. (1975) Embryogenesis from microspores of *Brassica napus*. Z Pfanzenzüchtung 74:77–81

Uzunova M, Ecke W, Weissleder K, Robbelen G (1995) Mapping the genome of rapeseed (*Brassica napus*). 1. Construction of an RFLP linkage map and localisation of QTLs for seed glucosinolate content. Theor Appl Genet 90:194–204

Vicente JG, King GJ (2001) Characterisation of disease resistance gene-like sequences in *Brassica oleracea* L. Theor Appl Genet 102:555–563

Vicente JG, Taylor JD Sharpe AG, Parkin IAP, Lydiate DJ, King GJ (2002) Inheritance of race specific resistance to *Xanthomonas campestris* pv. *campestris* in *Brassica* genomes. Phytopathology 92:1134–1141

Voorrips RE, Kanne HJ (1997) Genetic analysis of resistance to clubroot (*Plasmodiophora brassicae*) in *Brassica oleracea*. II. Quantitative analysis of root symptom measurements. Euphytica 93:41–48

Voorrips RE, Jongerius MC, Kanne HJ (1997) Mapping of two genes for resistance to clubroot (*Plasmodiophora brassicae*) in a population of doubled haploid lines of *Brassica oleracea* by means of RFLP and AFLP markers. Theor Appl Genet 94:75–82

Wang M, Farnham MW, Nannes JSP (1999) Ploidy of broccoli regenerated from microspore culture versus anther culture. Plant Breed 118:249–252

Wang M, Farnham MW, Thomas CE (2000) Phenotypic variation for downy mildew resistance among inbred broccoli. HortScience 35:925–929

Wang M, Farnham MW, Thomas CE (2001) Inheritance of true leaf stage downy mildew resistance in broccoli. J Am Soc Hortic Sci 126:727–729

Weber S, Zarhloul MK, Friedt W (2000) Modification of oilseed quality by genetic transformation. Progress in botany, vol 62. Springer, Berlin Heidelberg New York, pp 140–174

Weier D, Hanke C, Eickelkamp A, Lühs W, Dettendorfer J, Schaffert E, Möllers C, Friedt W, Wolter FP, Frentzen M (1997) Trierucoylglycerol biosynthesis in transgenic plants of rapeseed (*Brassica napus* L.). Fett/Lipid 99:160–165

Yang Q, Chauvin JE, Herve Y (1992) A study of factors affecting anther culture of cauliflower (*Brassica oleracea* var. *botrytis*). Plant Cell Tissue Organ Cult 28:289–296

Zaki M, Dickinson HG (1990) Structural changes during the first division of embryos resulting from anther and microspore culture in *Brassica napus*. Protoplasma 156:149–162

Zaki MAM, Dickinson HG (1991) Microspore-derived embryos in *Brassica*: the significance of division symmetry in pollen mitosis I to embryogenic development. Sex Plant Reprod 4:48–55

Zaki M, Dickinson H (1995) Modification of cell development in vitro – the effects of colchicine on anther and isolated microspore culture in *Brassica napus*. Plant Cell Tissue Organ Cult 40:255–270

Zhang FI, Takahata Y (1999) Microspore mutagenesis and in vitro selection for resistance to soft rot disease in Chinese cabbage (*Brassica campestris* L. ssp *pekinensis*). Breed Sci 49:161–166

Zhang FL, Takahata Y (2001) Inheritance of microspore embryogenesis ability in *Brassica* crops. Theor Appl Genet 103:254–258

Zhao JP, Simmonds DH, Newcomb W (1996a) Induction of embryogenesis with colchicine instead of heat in microspores of *Brassica napus* L. cv. Topas. Planta 198:433–439

Zhao JP, Simmonds DH, Newcomb W (1996b) High frequency production of doubled haploid plants of *Brassica napus* cv. Topas from colchicine-induced microspore embryogenesis without heat shock. Plant Cell Rep 15:668–671

Zhou WJ, Hagberg P, Tang GX (2002) Increasing embryogenesis and doubling efficiency by immediate colchicine treatment of isolated microspores in spring *Brassica napus*. Euphytica 128:27–34

II.3 Haploids in the Improvement of Poaceae

PIERRE DEVAUX[1] and RICHARD PICKERING[2]

1 Introduction

In 2001, cereals covered 45% of the world's cultivated land, representing more than 0.67 billion ha producing 2.1 billion tonnes (from FAOSTAT, ‚http://apps.fao.org/‚). Although yield differed between developed and developing countries, 3.7 and 2.8 tonnes ha^{-1}, respectively, huge variations exist among countries, ranging from 0.12 to 7.6 tonnes ha^{-1}, which partly reflects the different environmental conditions over the globe. Despite the huge range, both yield and quality have greatly increased through the release of well-adapted new cultivars over the last few decades. In most of the cultivated Poaceae, the development of superior cultivars has been a continuing breeding process for many different characters. These include yield, quality of the harvested products, factors related to consistent yield and quality such as resistance to pests and diseases and tolerance to abiotic stress, which all contribute to more sustainable agriculture.

Among the Poaceae, there are both self- and cross-pollinated species, and landraces, natural populations and modern cultivars have been cultivated as synthetic, F_1 hybrids or homozygous selections, which require different breeding schemes depending on the species. Homozygosity is often a requirement for registering new self-pollinated cultivars and superior lines prior to crossing but is also useful as a tool to decrease the frequency of undesirable alleles in populations of cross-pollinated species by phenotypic selection. Homozygosity can be achieved by means of several inbreeding systems including selfing, full- and half-sib mating and backcrossing, all of which differ in the time needed to attain homozygosity and opportunities for selection (Hallauer and Miranda 1981). However, the fastest route to homozygosity is doubled haploidy and, therefore, doubled-haploid plants (DHs) have been very attractive to many scientists and breeders. DHs can be produced by several methods, which vary in their efficiency and are species-dependent. In some species, such as barley for which thousands of DHs can be produced

[1] Florimond Desprez, Biotechnology Laboratory, 3 rue Florimond Desprez, P.O. Box 41, 59242 Cappelle en Pévèle, France

[2] New Zealand Institute for Crop and Food Research Limited, Private Bag 4704, Christchurch, New Zealand

Biotechnology in Agriculture and Forestry, Vol. 56
Haploids in Crop Improvement II (ed. by C.E. Palmer, W.A. Keller, and K.J. Kasha)

routinely at moderate cost, doubled haploidy has been used widely in breeding programs and has contributed to the release of many cultivars. In other species, for example maize, low efficiency has limited its use. Homozygous DH populations have also enabled the mapping of many molecular markers as well as qualitative and quantitative trait loci (QTL) for subsequent marker-assisted selection (MAS). These have contributed to gene isolation. To a lesser extent haploidy has been used in mutation breeding and in genetic transformation.

Since we published a review on barley DHs 10 years ago (Pickering and Devaux 1992), several excellent books related to doubled haploidy have been published (Jain et al. 1996/1997; Chupeau et al. 1998). In this chapter, we focus on the latest developments in doubled haploidy of the Poaceae for optimum production and usage. In this respect, most of the cited references are more recent than 1992 since many important ones have already been noted in our 1992 review.

2 Doubled Haploid Production

2.1 Anther Culture

2.1.1 Donor Plant Growth Conditions

It has been widely recognized that the conditions under which the donor plants have been raised are critical to the success of tissue culture and more specifically of anther and microspore culture. To ensure better reproducibility or to eliminate some of the environmental parameters that adversely influence the technique, controlled environment growth chambers have been preferred (Afele and Kannenberg 1990; Hoekstra et al. 1992; Barceló et al. 1994; Orshinsky and Sadasivaiah 1997), although conventional greenhouses have been the most commonly used environment (Devaux et al. 1993b; Alemanno and Guiderdoni 1994; Saisingtong et al. 1996; González et al. 1997; Puolimatka and Pauk 2000). However, a few authors have collected their material from field-grown plants (Karsai et al. 1994; Lentini et al. 1995; Machii et al. 1998; Tuvesson et al. 2000) and perhaps, therefore, their investigation has been limited by availability of growth rooms and the success rates subject to more variation. Three optimal temperatures in growth rooms and greenhouses have been used depending upon the species. The lowest temperature range is for barley, usually between 12 and 15 °C (Ziauddin et al. 1992; Hou et al. 1994) with a possible reduction during the night (Kintzios and Fischbeck 1994). A higher range of temperatures, 15–20 °C, is more suitable for hexaploid wheat, triticale, rye and perennial ryegrass (Flehinghaus-Roux et al. 1995; Madsen et al. 1995; Redha et al. 1998; Immonen and Robinson 2000), although Orshinsky and Sadasivaiah (1997) reported more embryos and green shoots when wheat donor plants were grown at high day/night temperature (25/18 °C) or when transferred from low (15/12 °C) to high temperature. The highest temperatures are most appropriate for maize (25–28/19–22 °C) (Afele and Kannenberg 1990; Saisingtong et al. 1996) and rice (28–31/18–20 °C) (Lentini et al. 1995). A 16-h day length regime is considered optimum for donor plant growth, with an irradiance ranging from 200 $\mu E\, m^{-2} s^{-1}$ for wheat (Henry et al. 1993) up to 600 $\mu E\, m^{-2} s^{-1}$ for maize (Martin and Widholm 1996).

2.1.2 Spike Sampling and Pretreatment

In most species, spikes were sampled just before the first microspore mitosis, i.e., from mid- to late-uninucleate stage, while maize tassels were collected when microspores were more advanced, from late uninucleate to early binucleate (Martin and Widholm 1996). The stage of microspores can easily be checked by squashing an anther taken from the middle of the spike in acetocarmine for microscopic examination. In the early uninucleate stage, the nucleus is near the pore and by late uninucleate it is opposite the pore (Kasha et al. 2001a). Pretreatment of whole spikes or isolated anthers is often required to switch the gametophytic pathway into a sporophytic development of the microspores (for review, see Reynolds 1997). The pretreatment induces stress to the microspore at a critical stage and can be applied as low or high temperature, osmotic shock, or chemicals to induce starvation and microtubule disruption chemicals. In cereals, a cold pretreatment has often been favored (Jähne and Lörz 1995) as it provides more flexibility and is less labor-intensive for routine production than some of the other pretreatments. Temperatures of about 4 °C have been used for barley, wheat, triticale and rye, while higher temperatures ranging from 7 to 10 °C are more desirable for maize and rice. However, for a hybrid between *Lolium multiflorum* × *Festuca arundinacea*, there is no beneficial effect of cold pretreatment on androgenesis response (Zare et al. 2002). The length of pretreatment also depends upon the species, with the longest time for barley, usually 28 days, intermediate for maize, triticale and wheat, and the lowest, for example 8 days and less, for rice and rye. Starvation on a 0.3- to 0.37-M mannitol solution was first reported for barley (Roberts-Oehlschlager and Dunwell 1990) and was progressively adopted by several laboratories (Ziauddin et al. 1992; Hoekstra et al. 1997; Caredda et al. 1999). Improvements were then achieved by incorporating culture media macronutrients in the mannitol solution (Hoekstra et al. 1997; van Bergen et al. 1999), by increasing the concentration of mannitol up to 0.7 M for low-responsive cultivars (Cistué et al. 1994, 1999; Castillo et al. 2001a) or by combining cold and starvation (Wojnarowiez et al. 2002). The effect of the hyperosmotic stress on the programmed cell death or apoptosis in the anther tissue of barley has been investigated by Wang et al. (1999) who found a more pronounced intranucleosomal cleavage of DNA in the pretreated anthers than in the control. Using the TUNEL reaction, electron microscopy and RNA quantification, they described how the loculus wall and loculus tapetum cells were mainly affected by the treatment. The authors suggested that the stress conditions stimulated abscisic acid (ABA) production, which might inhibit apoptosis in ABA-susceptible microspores. Van Bergen et al. (1999) confirmed that high ABA content during anther pretreatment was correlated with subsequent high regeneration efficiency. Other chemicals such as colchicine added during the first days of anther culture have also improved microspore-derived embryo frequency in wheat (Szakács and Barnabás 1995) and maize (Saisingtong et al. 1996).

2.1.3 Induction Media and Culture Conditions

In barley, the most popular medium has been FHG (Hunter 1987), which differs from the original Linsmaier and Skoog (LS) medium (1965) by a tenfold reduced concentration of the ammonium nitrate, by omitting cobalt chloride and by adding 750 mg l^{-1} L-glutamine. The basic Potato-2 medium (Chuang et al. 1978) is popular for wheat (Henry and de Buyser 1990), triticale (Wang and Hu 1984), and ryegrass (Olesen et al. 1988) anther culture. To ensure more consistent results some researchers prefer synthetic media such as N6 (Orshinsky and Sadasivaiah 1994; González et al. 1997), FHG (Lashermes 1992), 190–2 (Brisibe et al. 1997), W14 (Immonen and Robinson 2000; Puolimatka and Pauk 2000; Zamani et al. 2000) and C17 (Machii et al. 1998; Arzani and Darvey 2001). However, the modified Potato-2 medium outperformed all synthetic media for wheat anther culture (Henry and de Buyser 1990). In maize, the basal YP medium of Ku et al. (1978), modified by Genovesi and Collins (1982), has been used by many researchers (Büter 1997), while in rice the N6 medium of Chu (1978) remains popular for japonica and the SK3 medium more suitable for indica/japonica hybrids (He et al. 1998). Several carbohydrates have been used depending upon the species. In barley, the most significant development for improving success rates has been the use of maltose (Hunter 1987), which has sometimes replaced sucrose in tetraploid and hexaploid wheat (Navarro-Alvarez et al. 1994; Otani and Shimada 1994, 1995; Stober and Hess 1997), triticale (Immonen and Robinson 2000), rice (Lentini et al. 1995), rye (Flehinghaus-Roux et al. 1995), and ryegrass (Opsahl-Ferstad et al. 1994). However, in maize, no other carbohydrate has been superior to sucrose for embryo and plant production (Büter 1997). Disaccharide concentration in induction media varies from 6% for barley and rice to 9–12% for wheat, triticale and maize (Barnabás et al. 1999). The effects of plant growth regulators in anther culture have been controversial, but there are some trends. Four main auxins have been specified, IAA, NAA, PAA and 2,4-D, at concentrations ranging from 0.5–2 mg l^{-1}; cytokinins are also added such as kinetin or 6-benzylaminopurine (BAP) at 0.1–1 mg l^{-1} (Cai et al. 1992). In general, weaker auxins at low concentration have been used for barley more than in any other species. In some cases, the auxin has been omitted, leaving BAP as the sole growth substance (Kihara et al. 1994). Indeed, the anti-auxin 2,3,5-triiodobenzoic acid (TIBA) at low concentration (0.1 mg l^{-1}) may be suitable for maize (Dieu and Beckert 1986; Barnabás et al. 1999). Several other components have occasionally been added to the induction media. These include the ethylene antagonists silver thiosulfate or silver nitrate (Lashermes 1992; Evans and Batty 1994; Lentini et al. 1995), activated charcoal (Saisingtong et al. 1996) and L-proline to increase embryogenesis (Redha et al. 1998). Liquid, semi-liquid, and solid media have all been employed, and although Ficoll added to liquid medium enhanced embryo and plant production (Devaux 1992; Cistué et al. 1999; Immonen and Robinson 2000), solid media are easier to handle and cheaper and, therefore, they are preferred. Several gelling agents are available ranging from basic agar (0.7–0.8%) to high-grade agarose (0.6%), but gelrite (0.1–0.3%) now usually replaces agar in the induction media as it yields better results. Membrane rafts have been proposed in conjunction with liquid media, but adverse effects have been obtained (Luckett and Smithard 1995). Anther cultures have been usually maintained at a range of temperatures between 21 and 29°C often in darkness or with a 16-h photoperiod under dim light as embryogenic induction of microspore is inhibited by high-intensity white light (Reynolds and Crawford 1997). An increased (32°C) (Brisibe et al. 1997) or reduced 14°C temperature (Saisingtong et al. 1996; Redha et al. 1998) for the first 3–7 days may improve anther culture response.

2.1.4 Regeneration

Plantlet regeneration has been achieved by transferring the embryo-like-structures (ELS) and calli >1 mm in size to a fresh medium. The basal medium for regeneration may resemble or differ from the induction medium. A variety of similar synthetic media to those used for induction have been adopted for regeneration and solidified by agar or, more usually, gelrite. The most common carbohydrate used for regeneration is sucrose at 2–3%, but some laboratories prefer maltose (Castillo et al. 2001a). Growth substances may be omitted, but an auxin (IAA, NAA or 2,4-D), a cytokinin (kinetin, BAP) or a mixture of the two has been used for regeneration at concentrations ranging from 0.05–1 mg l^{-1}. After transfer to a regeneration medium, the cultures are kept in growth rooms at temperatures of about 25–26°C either in the dark or under a 16-h dim to mid-light regime. In most cases, the regenerated green plantlets are then transferred to the same solid medium but without any growth substances or with 1 mg l^{-1} IAA and 2% sucrose before transplanting to potting compost.

2.1.5 Albinism

Almost every species in the Poaceae has been affected by albinism. In barley and rice, which have been more severely affected by albinism, green plant:albino plantlet ratio (G:A) is usually <1 (Caredda and Clément 1999) but fluctuates according to genotype, pretreatment and culture conditions. In wheat, the growth conditions of the donor plants are an important factor, too (Orshinski and Sadasivaiah 1997; Dogramaci-Altuntepe et al. 2001). Some genotypes regenerate few green plants (Tuvesson et al. 1989; Jähne et al. 1991; Lentini et al. 1995; Moieni and Sarrafi 1995; Stober and Hess 1997), which precludes the use of androgenesis when these genotypes are used in breeding programs. Plastid differentiation during anther culture was investigated in two barley cultivars, 'Cork' and 'Igri', which produce high and low proportions of albino plants, respectively (Caredda et al. 2000). Differences in proplastid differentiation, thylakoid development, and the ability of cells to divide were observed and the fate of microspore plastids could be predetermined as early as the microspore stage (Caredda et al. 2004). In contrast to Cork, plastids in Igri contained DNA at the time of culture, indicating that perhaps DNA degradation occurred earlier in Cork than in Igri. Deletions of genes related to chlorophyll and photosynthesis in the nuclear and plastid genomes occur in albino plantlets of Igri (Dunford and Walden 1991) and rice (Harada et al. 1991; Yamagishi 2002) and probably arise during regeneration (Mouritzen and Holm 1994). Pretreatment and culture conditions are also critical factors influencing green plant regeneration. In barley, Caredda et al. (1999) observed that organelle structure and G:A ratios were both improved by mannitol pretreatment rather than a period of cold. The ratio was further increased on an induction medium with mannitol as an osmotic pressure regulator (Wojnarowiez et al. 2004). However, in rye and in triticale, cold pretreatment seems optimal for green plant production (Immonen 1999; Immonen and Robinson 2000). G:A ratios for wheat were significantly improved when colchicine was added to the medium at the start of microspore induction (Zamani et al. 2000) and by replacing sucrose with maltose in the induction media for barley (Finnie et al. 1989), wheat (Navarro-Alvarez et al. 1994) and rice (Lentini et al. 1995).

2.1.6 Ploidy Level of Regenerants and Chromosome Doubling

The distribution of regenerated plantlets from anther culture according to their ploidy level differs among species. The rate of spontaneous chromosome doubling is on average 60% in barley, 27% in wheat, 17% in triticale (Devaux 1992), 10% in maize (Büter 1997), 50% in rice (Guiderdoni et al. 1991) and 70% in rye (Immonen 1999). Moreover, a relatively high proportion of polyploid, mainly tetraploid plants (8%) are recovered from barley anther culture, the remaining plants being mainly haploid (Devaux 1992).

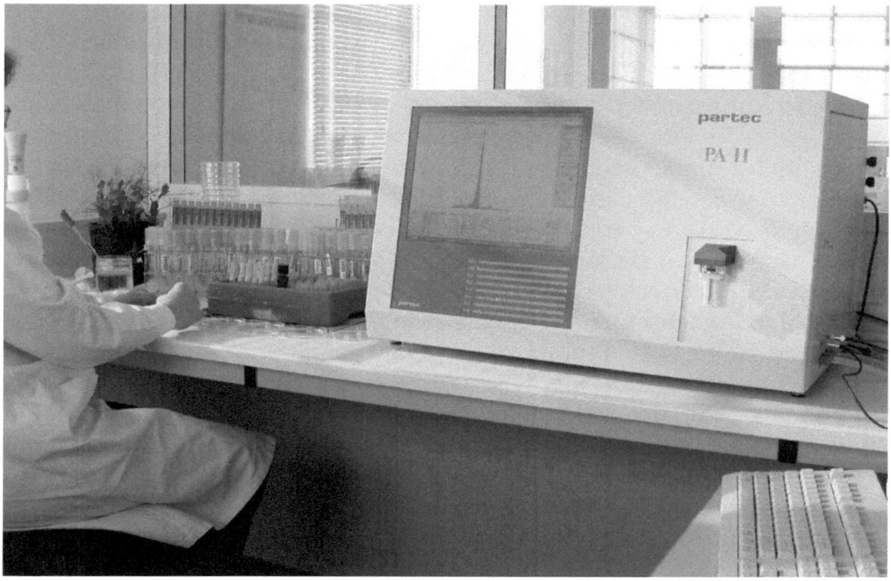

Fig. 1. Ploidy level determination of anther culture-derived plantlets by flow cytometry. Thirty plants can be checked in 1 h to identify spontaneous doubled haploid plantlets and to ensure an efficient chromosome doubling of haploid plants at an early stage

There are reports within a species of variations in spontaneous doubling according to the genotype (Alemanno and Guiderdoni 1994; Stober and Hess 1997), but ploidy level can be easily determined by flow cytometry (Coba de la Peña and Brown 2001) at a speed of 30 plants h^{-1} (Fig. 1). As an alternative, an indirect ploidy determination method, such as stomatal guard cell length (Borrino and Powell 1988) or stomatal chloroplast number (Ho et al. 1990), could be used. Following the analyses, polyploid plants can be discarded, diploid ones directly planted in the greenhouse and haploid plants treated with colchicine for chromosome doubling. Several factors influence the rate of spontaneous doubling in anther culture. These include the pretreatment stringency (Immonen and Robinson 2000), the carbohydrate source in the culture media (Navarro-Alvarez et al. 1994) and supplementing the medium with colchicine or antimicrotubule agents (herbicides) either during pretreatment (Antoine-Michard and Beckert 1997) or during incubation (Wan et al. 1991; Saisingtong et al. 1996; Redha et al. 1998; Barnabás et al. 1999).

2.2 Isolated Microspore Culture

Isolated microspore culture (IMC) is a development from anther culture in which the microspores are removed mechanically from the anther prior to in vitro culture. As haploid single cells, microspores are ideal for selection, mutation, transformation and biochemical analysis. Furthermore, recent improvements in IMC efficiency enable the technique to be considered for routine production of DH, especially barley (Kasha et al. 2001a). IMC differs technically from anther culture in the following respects. Donor plants are raised either in controlled environments such as greenhouses or in growth rooms, under similar conditions to those for anther culture, although higher temperatures can be used especially for spring types (Kasha et al. 2001a; Ritala et al. 2001). In barley and wheat, spikes may be collected when microspores have reached the late-uninucleate to early-binucleate stage (Gustafson et al. 1995; Ritala et al. 2001). Excised anthers are usually pretreated in 0.3–0.7 M mannitol solution for 3–7 days in the dark either at 25–33 °C (Mouritzen and Holm 1994; Touraev et al. 1996; Castillo et al. 2000) or at cooler temperatures (Guo and Pulli 2000), while entire spikes are cold-pretreated (Mordhorst and Lörz 1993; Puolimatka and Pauk 1999) with starvation (Kasha et al. 2001a). A beneficial effect of macronutrients in the mannitol solution has also been reported (Hu et al. 1995; Li and Devaux 2001). In wheat, pretreatment of spikes with an inducing agent consisting of 0.1 $g\,l^{-1}$ of 2-hydroxynicotinic acid (2-HNA), 10^{-6} $mol\,l^{-1}$ of 2,4-D and 10^{-6} $mol\,l^{-1}$ of BAP converted up to 50% of the microspores to the sporophytic pathway and resulted in high green plant production (Liu et al. 2002). Alternatively, a 2-HNA treatment can be applied to freshly isolated wheat microspores prior to culture (Zheng et al. 2001). Following pretreatment, microspores are released from anthers by blending (Mouritzen and Holm 1994; Gustafson et al. 1995), vortexing (Hu et al. 1995), stirring (Touraev et al. 1996), pestle, glass or Teflon rod maceration (Hoekstra et al. 1993; Cistué et al. 1995; Ritala et al. 2001) either in 0.3–0.4 M mannitol solution, washing solution or culture medium (Mejza et al. 1993; Salmenkallio-Marttila et al. 1995). Washing solutions differ from the induction medium by a reduced concentration of nutrients and by replacing maltose by sucrose (Kunz et al. 2000) or by adding 10 mM calcium chloride to the mannitol solution (Li and Devaux 2001). To save time, microspores can be isolated by blending wheat spikelets (Mejza et al. 1993) or barley spike segments (Mordhorst and Lörz 1993; Scott and Lyne 1994). The resulting microspore-containing slurry is filtered through a nylon mesh and subjected to several cycles of washing and centrifugation. To obtain consistent and viable cultures, dead cells and small debris can be removed by a density gradient of 18–21% maltose (Ritala et al. 2001) or by a discontinuous Percoll gradient (Gaillard et al. 1991; Touraev et al. 1996) followed by centrifugation. In liquid media, microspores are cultured at densities ranging from 7×10^3 to 2×10^5 microspores ml^{-1} (Gustafson et al. 1995; Zheng et al. 2002), while between 3×10^3 and 6×10^5 microspores are deposited on filter papers (Hoekstra et al. 1996; Kasha et al. 2001a). Microspore culture media are basically the same as those used for anther culture except that some are conditioned with ovary pre- or co-culture, resulting in higher efficiencies and less genotypic influence (Hu and Kasha 1997; Zheng et al. 2002). Arabinogalactan protein (AGP) which has been identified in conditioned media with barley

IMC (Paire et al. 2003) or related hydroxyproline glycoproteins can be added to the induction medium to improve microspore response (Kasha and Simion 2001). The use of the auxin PAA in the induction medium has improved plant regeneration from barley microspore culture (Ziauddin et al. 1992). When liquid media are employed, 0.5–2 ml of fresh medium can be added 1–2 weeks after the beginning of the culture. Cultures are kept stationary and put onto a rotary shaker at ca. 70 rpm after 1–2 weeks (Salmenkallio-Marttila et al. 1995; Li and Devaux 2001) or continuously shaken (Scott and Lyne 1994). In a comparative study, Davies and Morton (1998) showed that IMC was much more efficient than anther culture with the barley cultivar Igri and to a lesser extent with an Australian spring F_1 hybrid. Although three of the 17 F_1 hybrids investigated by Castillo et al. (2000) responded better with IMC than anther culture, the average yield of anther culture was twice that of IMC. From the results of many studies, high yields of plant production, e.g., 50 green plants per anther, can be achieved by IMC with model genotypes, e.g., Igri (Hoekstra et al. 1996), but if the critical parts of the protocol have been followed similar yields can be expected from any other genotype (Kasha et al. 2001a). The rate of spontaneous chromosome doubling from IMC ranged from 5–83 % (Pauk et al. 2000; Kasha et al. 2001b). Hu and Kasha (1999) and Kasha et al. (2001b) observed that following the first mitotic division during pretreatment, the two daughter nuclei fused to form a diploid nucleus, which then undergoes rapid divisions. Increases in spontaneous chromosome doubling have been achieved by combining cold pretreatment of spikes with mannitol treatment of microspores (Li and Devaux 2003) or by adding colchicine or antimicrotubule agents to the culture medium (Hansen and Andersen 1998a,b).

2.3 Interspecific and Intergeneric Hybridizations

2.3.1 Barley × H. bulbosum

This cross was the preferred method for producing barley DHs before great improvements in androgenesis caused interspecific hybridizations to be superseded by anther and microspore culture. However, the interspecific cross is still used as an alternative method to obtain barley DHs from hybrids that are recalcitrant to androgenesis and also where an unbiased random sample of gametes is required for a mapping population. In general, 10–15 DHs per 100 pollinated florets can be achieved. The interspecific method was first described by Kasha and Kao (1970) who elucidated the mechanism by which barley haploid embryo formation occurred. Following fertilization of the *H. vulgare* egg by the *H. bulbosum* male gamete, the complete genome of the wild species is rapidly eliminated in the first few days after fertilization. The resultant haploid embryos must be rescued aseptically to a defined medium (for example, Gamborg's B5) prior to endosperm degeneration, usually about 12–14 days after pollination (d.a.p.). Haploid plantlets that develop can be treated with an aqueous colchicine solution (0.05%) + 2% dimethyl sulfoxide to restore the fertility by doubling the chromosome number. Chromosome elimination may not always take place and depends on parental genotype (Simpson et al. 1980; Pickering 1983; Pickering and Rennie 1990) and temperature during the first few days after pollination (Pickering 1984). To prolong seed development to the time of embryo culture, gibberellic acid (GA_3) at 75 mg l^{-1} with a wetting agent (Tween 20) is applied to florets 1 or 2 d.a.p. Occasionally, some cultivars do not respond well, for example, 'Magnum', and a combination of GA_3 and 2,4-D was developed as a postpollination spray for these cultivars (Pickering and Wallace 1994). Since then, GA_3 +

Fig. 2. Emasculated barley spikes 13 days after pollinating florets with *Hordeum bulbosum* L. *Left* Postpollination spray treatment 1 day after pollination with 75 mg l^{-1} GA_3 + 2 mg l^{-1} 2,4-D + 1 mg l^{-1} dicamba; *right* postpollination spray treatment 1 day after pollination with 75 mg l^{-1} GA_3. Note differences in seed size and unfertilized florets (*arrowed*) and the selfed seed (*lowermost floret on the left-hand spike*). *Bar* 20 mm

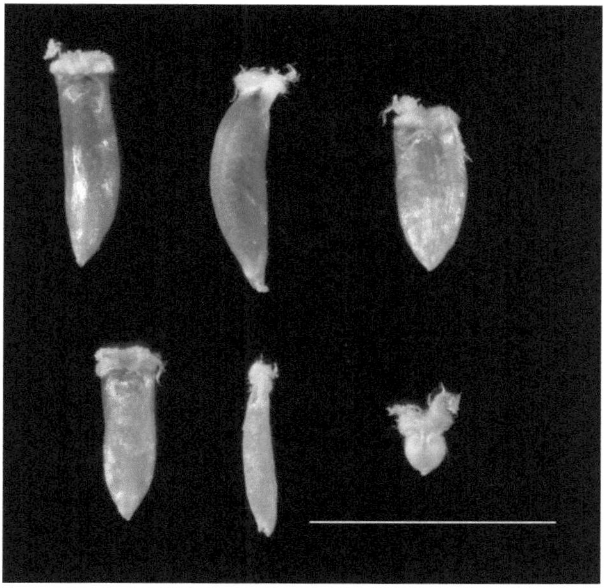

Fig. 3. Seeds and unfertilized florets removed from spikes illustrated in Fig. 1. *Top row, left* and *center* Seed development after postpollination spray treatment 1 day after pollination with 75 mgl^{-1} GA_3 + 2 mgl^{-1} 2,4-D + 1 mgl^{-1} dicamba; *right* unfertilized floret with the same postpollination treatment. *Bottom row* as for top row but spray treatment 75 mgl^{-1} GA_3 only. Note differences in seed development promoted by the addition of 2,4-D and dicamba. *Bar* 10 mm

2,4-D + dicamba at 75, 2 and 1 mgl^{-1}, respectively, has been used routinely 1 d.a.p. to reduce seed shrivelling (Figs. 2 and 3) without causing adverse effects on plant regeneration (Pickering, unpubl.). Apart from these minor refinements to the protocol since our previous review (Pickering and Devaux 1992), there have been no major reported improvements to the technique, so we refer the reader to this publication and to Devaux (2003) for more complete details.

2.3.2 Wheat × Maize; Triticale × Maize

To obtain DHs from tetraploid and hexaploid wheat via androgenesis has not been as successful as with barley. Initially, it was hoped that wide hybridizations between wheat and *H. bulbosum* might have been successful. However, unfortunately, because of a series of incompatibility (*Kr*) loci in most wheat cultivars, crossability has usually been very low (Laurie and Bennett 1987). Nevertheless, following the pioneering research of Laurie and associates (Laurie and Bennett 1988), wheat crossed with maize has been the method of choice for producing DHs and details of the many factors involved in DH production have been presented by Knox et al. (2000). In brief, a similar protocol is adopted as for barley × *H. bulbosum* although a higher concentration of postpollination plant growth regulators (PGRs) is usually applied either as a spray or injected into the last internode; GA_3 can be omitted as it appears to be ineffective on wheat (Knox et al. 2000). Some of the other factors that are important in achieving successful production are growth conditions for donor plants, including light (Campbell et al. 2001), emasculation method and PGR applications (Knox et al. 2000). Genotypic influences have been reported, but these are generally not as important a factor as for androgenesis. Success rates with hexaploid wheat are lower than those obtained from barley × *H. bulbosum* (around 6 DHs per 100 pollinated florets; Lefebvre and Devaux 1996) and tetraploid wheat DH success rates are very much lower (<1 DH per 100 pollinated florets; Knox et al. 2000; P. Devaux, unpubl.). There have been no major improvements to the success rates for tetraploid wheats, although some minor modifications to the protocol, such as optimum PGR applications especially dicamba (Knox et al. 2000), have raised DH yields. Regarding DH production from hexaploid triticale using the wide hybridization technique, very little research has been undertaken to improve success rates, which are generally similar to those obtained with durum wheat (around 1–2 DHs per 100 pollinated florets; P. Devaux, unpubl.) and can be attributed to the absence of the D genome in durum wheat and hexaploid triticale (Inagaki and Hash 1998). Increases in DH efficiency were gained by modifying the PGR composition, especially dicamba, and concentrations of the components (Wedzony et al. 1998). For the future, therefore, optimizing PGR composition and concentrations should result in further improvements for both durum and triticale DH production.

2.3.3 Oat × Maize

There was some optimism that oat × maize hybridizations might have resulted in haploid oat embryo formation, but although very small numbers of oat haploids were obtained (Rines and Dahleen 1990) it later became clear that there were some anomalous plants formed after embryo rescue that contained additional chromosomes (Riera-Lizarazu et al. 1996). These were identified as maize chromosomes and since then the complete series of oat–maize chromosome addition lines have been obtained (Kynast et al. 2001). They are proving invaluable for research into species relationships and mapping experiments. The problem still remains of consistently obtaining oat DHs, and screening maize parental genotypes would be appropriate. However, since the oat crop is relatively minor compared with wheat and barley, the amount of research undertaken is consequently much less and funding harder to procure.

3 Use of Doubled Haploids

3.1 Breeding

DHs have long been used for breeding since they can easily be produced from a wide range of crosses. The advantages and limitations of doubled haploidy in breeding have been discussed many times (see, for example, Kasha and Reinbergs 1982; Devaux 1995; Steffenson et al. 1995). Gallais (1990) proposed the use of DHs for recurrent selection and line development as being one of the most efficient methods for low heritable traits. Thomas et al. (2003) pointed out that many factors influence the deployment of the DH method in breeding. Although DHs can be used at different stages of a breeding scheme and for several purposes, an appropriate way to assess the value of doubled haploidy in practical breeding is the number of cultivars released as DHs. In barley, wheat, rice and triticale nearly 150 cultivars were produced by doubled haploidy (www.scri.sari.ac.uk/assoc/COST851/) and the number is continuously increasing. With the development of linkage maps based on DHs, segregation distortions at marker loci were often detected (Devaux et al. 1995; Yamagishi et al. 1996; Dufour et al. 2001). Although skewed segregations were also reported in F_2 (Yamagishi et al. 1996) and single-seed descent populations (Bjørnstad et al. 1993), the complete occurrence of one parental allele has never been reported in any DH progeny derived from polymorphic parents. For example, the strongest distortion reported by Devaux et al. (1995) was 21 DHs with the Steptoe allele and 80 DHs with the Morex allele. Therefore if the DH production technique is efficient enough many DHs with the less frequent alleles can be produced and developed for breeding. When reliable molecular markers for the desired traits have been identified, selection pressure can be performed at an early stage, for example in vitro regeneration to eliminate undesirable recombinants among progenies of a breeder's crosses. Time and space can, therefore, be saved, enabling the identification of superior agronomic types very quickly. Furthermore, the molecular characterization of individual alleles at markers is simple and accurate with haploid and doubled-haploid tissues and will be discussed in more detail in the following section. By scoring the presence vs. absence of particular fragments, haploid and doubled-haploid tissues can be genotyped using PCR-dominant markers (Devaux 1995) such as sequence tagged sites (STSs) and amplified fragment length polymorphisms (AFLPs). Another application of molecular markers with regard to DHs as well as traditional lines is fingerprinting. Commercially promising DHs or lines can be genotyped with a set of molecular markers, most commonly simple sequence repeats (SSR), dispersed on each chromosome along with their parents and the most commonly grown cultivars. The genotyping data along with phenotype records help breeders in the choice of potential crosses by strengthening the characterization of new lines as well as determining their relative genetic distance from the current cultivars.

3.2 Molecular Genetics and Genomics

3.2.1 QTLs Influencing in Vitro Response

In the early 1970s when anther culture was in its infancy, genotypic differences in anther culture response were reported and well-designed experiments confirmed the occurrence of genes controlling this trait (see, for example, Charmet and Bernard 1984; Lazar et al. 1984; Powell 1988; Afele and Kannenberg 1990; Quimio and Zapata 1990). Consequently, those genes became targets for mapping. In maize, Cowen et al. (1992) detected two major and two minor QTLs accounting for 57% of the genetic variability for embryo-like structure (ELS) production. While investigating three crosses, Murigneux et al. (1994) found three to four QTLs related to percentage of responding anthers or embryo production specific to each cross, explaining 30–40% of the phenotypic variation. Using different material Beaumont et al. (1995) confirmed the perfect match of at least three QTLs identified by Murigneux et al. (1994) on chromosomes 5, 7 and 8. From a linkage map based on an anther culture-derived DH population of barley (Zivy et al. 1992), Devaux and Zivy (1994) hypothesized that their protein markers, which deviated from an expected 1:1 segregation ratio, could be linked to genes involved in anther culture response. They established that two QTLs were linked to genes involved in both ELS production and green plant regeneration, whereas the other two QTLs were linked to genes involved only in green plant regeneration. In a similar test, Manninen (2000) described 10 QTLs associated with percentage of responsive anthers, plants per responsive anther and spontaneous diploidization located on chromosomes 2H, 3H and 4H. Furthermore, not all regions associated with anther culture response matched regions where distorted segregation occurred. Using a DH population derived from an indica/japonica rice hybrid, He et al. (1998) identified five QTLs for callus induction frequency on chromosomes 6, 7, 8, 10 and 12, two QTLs for green plantlet differentiation frequency on chromosomes 1 and 9 and a major QTL for albino plantlet differentiation frequency on chromosome 9. Also in rice, a QTL associated with green plant regeneration on chromosome 10 was detected over three culture methods (Kwon et al. 2002). Hence, one marker identified genotypes with good and poor regenerability across many cultivars. Four QTLs explaining 80% of the genotypic variation for green plant regeneration were detected in wheat anther culture (Torp et al. 2001).

3.2.2 Mapping with Molecular Markers

Using genetic markers to manipulate loci controlling traits of interest and to understand gene organization in complex genomes represents a major breakthrough for plant geneticists and breeders. Advances in methods for assaying DNA polymorphisms have produced hundreds of segregating genetic markers in many species. In traditional linkage analysis the genetic markers have been arranged into dense genetic linkage maps (http://wheat.pw.usda.gov/ggpages/maps.shtml), enabling them to be used as selection criteria when they have been mapped near trait loci. Different types of segregating populations have been used in genetic mapping including F_2, F_3, backcross, recombinant inbred lines (RILs) and DHs. RILs and DHs become immortal populations once each individual has been selfed and the seeds stored properly. These populations remain stable for different purposes such as phenotypic evaluation under different selection pressures and for continuously mapping new markers and traits (Guyomarc'h et al. 2002; Thiel et al. 2003). Therefore, RILs and DHs represent an invaluable tool to identify QTLs and to assess their variability across environments and years. For example, the barley DH population derived from the F_1(Dicktoo × Morex) has been repeatedly phenotyped and genotyped over several years (Hayes et al. 1996). In some species, recombinant DH populations are easy to produce and have the important advantage over RILs that they are faster to generate. Consequently, in the four species, barley, wheat, rice and maize, at least 43 DH populations have been used for map construction and QTL mapping (Forster and Thomas 2003). From simulation studies, Martinez et al. (2002) showed that DH design is a very useful tool for QTL mapping, particularly when the effect of the QTL is low and the residual genetic variation from other chromosomes can be controlled by using multiple markers. Wu (1999) reported a new method for mapping segregating QTLs in an open-pollinated progeny population using dominant markers derived from haploid tissues. Using DH populations, many morphological, qualitative single locus and QTLs including genes conferring resistance to pests and diseases (Chen et al. 1994; Kicherer et al. 2000; Toojinda et al. 2000; Scheurer et al. 2001) and tolerance to abiotic stresses (Pan et al. 1994; Ellis et al. 2002; This et al. 2003), associated with heading time, photoperiod response (Sourdille et al. 2000), quality (Marquez-Cedillo et al. 2000; Perretant et al. 2000) and yield (Backes et al. 1995; Marquez-Cedillo et al. 2001) have been mapped. When no genetic linkage map is available or when a gene or genes controlling a trait have not yet been mapped, bulk segregant analysis (BSA) can be applied to identify markers for the genes (Michelmore et al. 1991). Precise phenotype characterization of each individual within the bulks has to be performed while markers are much easier to analyze when DH vs. F_2 lines are used in the bulks. BSA and DH quickly identified SSR markers linked with a new gene conferring resistance to barley mild mosaic virus (BaMMV) from the Japanese cv. Chikurin Ibaraki 1 and the locus was then mapped on the chromosome 6H (Le Gouis et al. 2004). Borovkova et al. (1995), using similar methods, identified RAPD and RFLP markers linked to the stem rust resistance gene *rpg4* in barley.

3.2.3 DHs and Expressed Sequence Tags (ESTs)

The development of expressed sequence tag (EST) sequencing projects has generated a large amount of sequence information. In wheat, the recent deposit of 200,000 ESTs into GenBank by DuPont has led to the availability of more than 500,000 genome data in the species (http://www.ncbi.nlm.nih.gov/dbEST/). From these sequences, molecular markers, such as SSR (Holton et al. 2002; Kantety et al. 2002) and single nucleotide polymorphisms (SNP), can be defined. SNPs are very powerful and abundant and have contributed greatly to allele diversity through evolution. They are usually discovered by sequencing PCR products generated from different individuals (Dietrich et al. 1999). Using this approach, Kota et al. (2001) identified SNPs between two barley accessions and subsequently mapped them using a DH population derived from the cross between the two accessions and denaturing high-performance liquid chromatography (DHPLC). The homozygous state of the DHs allowed simplified profile analyses; ESTs that were monomorphic at the RFLP level were then mapped.

3.2.4 Gene Cloning

Isolating important genes is a goal of many genomic projects, but progress has been slower in species with large genomes such as barley and wheat compared with rice, which is the model genome for grasses (Devos and Gale 2000). As a demonstration of this goal, Horvath et al. (2003) genetically engineered barley with the cloned *Rpg1* gene. DNA sequence comparisons in grass genomes have shown that coding regions are usually well conserved, but the distances between the genes seem to be correlated with genome size (Bennetzen 2000), hindering isolation of specific genes in large grass genomes. While it is possible to isolate a single gene using an F_2 population followed by inbreeding to obtain the homozygous recombinants, a large DH population is preferred when multiple gene isolations have to be carried out (A. Kleinhofs, pers. comm., 2003). Progress could be made in map-based cloning by further increasing the number of recombinant DHs in a population for fine mapping (Kilian et al. 1997). At this level, the complete homozygosity and immortality of the DH population are tremendous advantages for facilitating the strategy. Few genes related to microspore embryogenesis have so far been isolated. Early studies in functional proteomics had shown the involvement of extracellular proteins in the initiation of somatic embryogenesis in carrot (van Engelen and de Vries 1992). In maize, Vergne et al. (1993) found that a 32-kDa protein named MAR32 was induced and accumulated in the anthers during cold pretreatment of the tassels and that the amount of MAR32 was positively correlated with the proportion of responding anthers and the production of ELS. In androgenetic embryos of barley, the expression of two embryo-specific genes was detected more intensively at the globular stage of the proembryos (Stirn et al. 1995). Reynolds and Crawford (1996) identified an ABA-responsive metallothionein (EcMt) gene expressed strongly in early stages of wheat anther culture, but the gene transcript was not detected in mature zygotic embryos, vegetative tissues or developing pollen. Five of the QTLs associated with anther culture in maize mapped near viviparous mutant loci, which are related to ABA production and regulation (Beaumont et al. 1995), confirming the importance of ABA for androgenesis. By differential screening, Vrinten et al. (1999) isolated three cDNAs in barley microspore culture that represented genes not previously identified in barley. Two of them showed homology with glutathione-S-transferases and lipid transfer protein genes, while the third had no homology to any isolated gene.

3.3 Mutation and Genetic Transformation

Increasing the genetic variability of crop species has been a goal of scientists and breeders in order to generate new and superior recombinants. Genetic modification of single haploid cells followed by regeneration enables the direct observation of recessive genes on the phenotype of a non-chimeric plant and can be induced by in vitro culture, mutagenic agents and by genetic transformation. In vitro culture has long been known to induce genetic changes (for review, see Karp 1991). Phenotypic as well as molecular changes have been occasionally reported in plants regenerated from haploid tissues (Snape et al. 1988; Devaux et al. 1993a; Wan and Widholm 1993). Unfortunately, most of these so-called gametoclonal variations negatively affect the agronomic performance of DHs (Powell et al. 1984; Snape et al. 1988). Although selection agents such as Na_2SO_3 or Al can be added to anther or microspore culture media to enhance the recovery of plants with high tolerance to the agent, there are indications that these plants can result from either recombination of genes (Ye et al. 1987) or from mutation (Barnabás et al. 2000). To further increase the rate of mutations, mutagenic agents can be applied to anther or microspore cultures (Castillo et al. 2001b) or to seeds from which plants are used for anther culture (Szarejko et al. 1995). In barley and wheat, both microspore-derived callus and isolated microspores have been used as explants for genetic transformation by particle bombardment (Jähne et al. 1994; Yao et al. 1997; Folling and Olesen 2001), yielding both homozygous (Jähne et al. 1994) and heterozygous transgenic plants (Yao et al. 1996) with improved phenotype (Leckband and Lörz 1998).

3.4 Other Research on Wide Hybrids in the Poaceae

Chromosome-engineered plants and introgressions obtained from wide hybrids via androgenesis is an important adjunct to DH production in the Poaceae, and although it is beyond the scope of this chapter to devote much space to these topics, we will describe some recent developments. Anther culture has been successfully employed in the Poaceae to obtain novel genetic combinations derived from the parental genomes of interspecific and intergeneric hybrids. The aim was to regenerate plants with novel chromosomal and genetic constitutions that could not be obtained from conventional crossing procedures. Regenerants from cultured anthers of barley × *H. bulbosum* comprised a small number of chromosomally engineered derivatives (multiple chromosome substitution lines), which have proved useful in mapping introgressions from *H. bulbosum* into barley (Pickering and Fautrier 1993). A subsequent larger-scale experiment was carried out (Gilpin et al. 1997) and similar results obtained, but the number of novel plants obtained from androgenesis was far fewer than could be produced by conventional crosses between the two species (Pickering 1992; Zhang et al. 2001). Anther culture has also been used effectively in hybrids involving hexaploid and octoploid triticale × wheat (Wang and Hu 1985; Wang et al. 1996) and *Triticum–Agropyron* × wheat (Miao et al. 1988). Chromosome addition lines as well as translocations were obtained from these hybrids and this is an efficient way of obtaining novel chromosomally engineered plants. Similar results have been obtained with intergeneric hybrids involving the forage grass species *Lolium* and *Festuca*, and the results from these crosses have been recently extensively reviewed (Humphreys et al. 2000).

4 Conclusion

To summarize, in a relatively short time DH production in the Poaceae has reached the point where it is a routine procedure for several of the cereals. Improvements are still awaited in some of the less widely grown cereals, such as oats, and it is speculative whether such improvements will be forthcoming given the minor importance and lower research funding for such crops compared with the major cereals. The uses of DHs have lately expanded from being merely a breeding tool to achieve homozygosity from early generation hybrid material to playing a crucial role in marker-assisted selection, molecular mapping, and gene cloning. Haploids can also be used as a source of explants for mutation breeding and are eminently suitable in this role since there are no masking effects in haploid tissue that would confound and slow the pace of screening for recessive mutations. The future is exciting for breeders and geneticists alike, since further technical refinements can be expected and DH production will become even more of a routine procedure than it currently is. There will, of course, be expansion into the area of genomics and the possible use of haploid tissue in conjunction, for example, with microarrays, which will increase our knowledge of gene expression in plant ontogeny.

References

Afele JC, Kannenberg LW (1990) Genetic studies of corn (*Zea mays* L.) anther culture response. Theor Appl Genet 80:459–464

Alemanno L, Guiderdoni E (1994) Increased doubled haploid plant regeneration from rice (*Oryza sativa* L.) anthers cultured on colchicine-supplemented media. Plant Cell Rep 13:432–436

Antoine-Michard S, Beckert M (1997) Spontaneous versus colchicine-induced chromosome doubling in maize anther culture. Plant Cell Tissue Organ Cult 48:203–207

Arzani A, Darvey N (2001) The effect of colchicine on triticale anther-derived plants: microspore pre-treatment and haploid-plant treatment using a hydroponic recovery system. Euphytica 122:235–241

Backes G, Graner A, Foroughi-Wehr B, Fischbeck G, Wenzel G, Jahoor A (1995) Localization of quantitative trait loci (QTL) for agronomic important characters by the use of a RFLP map in barley (*Hordeum vulgare* L.). Theor Appl Genet 90:294–302

Barceló P, Cabrera A, Hagel C, Lörz H (1994) Production of doubled-haploid plants from tritordeum anther culture. Theor Appl Genet 87:741–745

Barnabás B, Obert B, Kovács G (1999) Colchicine, an efficient genome-doubling agent for maize (*Zea mays* L.) microspores cultured in anthero. Plant Cell Rep 18:858–862

Barnabás B, Kovács G, Hegedus A, Erdei S, Horváth G (2000) Regeneration of doubled haploid plants from in vitro selected microspores to improve aluminium tolerance in wheat. J Plant Physiol 156:217–222

Beaumont VH, Rocheford TR, Widholm JM (1995) Mapping the anther culture response genes in maize (*Zea mays* L.). Genome 38:968–975

Bennetzen JL (2000) Comparative sequence analysis of plant nuclear genomes: microcolinearity and its many exceptions. Plant Cell 12:1021–1029

Bjørnstad Å, Skinnes H, Uhlen AK, Marum P, Maroy AG (1993) Genetic-marker segregations in doubled haploids in spring wheat crosses. Hereditas 118:55–62

Borovkova IG, Steffenson BJ, Jin Y, Rasmussen JB, Kilian A, Kleinhofs A, Rossnagel BG, Kao KN (1995) Identification of molecular markers linked to the stem rust resistance gene *rpg4* in barley. Phytopathology 85:181–185

Borrino EM, Powell W (1988) Stomatal guard cell length as an indicator of ploidy in microspore-derived plants of barley. Genome 30:158–160

Brisibe EA, Olesen A, Andersen SB (1997) Characterization of anther culture-derived cell suspensions exclusively regenerating green plantlets in wheat (*Triticum aestivum* L.). Euphytica 93:321–329

Büter B (1997) In vitro haploid production in maize. In: Jain SM, Sopory SK, Veilleux RE (eds) In vitro haploid production in higher plants, vol 4. Cereals. Kluwer, Dordrecht, pp 37–71

Cai Q, Szarejko I, Polok K, Maluszynski M (1992) The effect of sugars and growth regulators on embryoid formation and plant regeneration from barley anther culture. Plant Breed 109:218–226

Campbell AW, Griffin WB, Burritt DJ, Conner AJ (2001) The importance of light intensity for pollen tube growth and embryo survival in wheat × maize crosses. Ann Bot 87:517–522

Caredda S, Clément C (1999) Androgenesis and albinism in Poaceae: influence of genotype and carbohydrates. In: Clément C, Pacini E, Audran JC (eds) Anther and pollen: from biology to biotechnology. Springer, Berlin Heidelberg New York, pp 211–228

Caredda S, Devaux P, Sangwan RS, Clément C (1999) Differential development of plastids during embryogenesis in barley. Protoplasma 208:248–256

Caredda S, Doncoeur C, Devaux P, Sangwan RS, Clément C (2000) Plastid differentiation during androgenesis in albino and non-albino producing cultivars of barley (*Hordeum vulgare* L.). Sex Plant Reprod 13:95–104

Caredda S, Devaux P, Sangwan RS, Proult I, Clément C (2004) Plastid ultrastructure and DNA related to albinism in androgenetic embryos of various barley (*Hordeum vulgare* L.) cultivars. Plant Cell Tissue Organ Cult 76:35–43

Castillo AM, Vallés MP, Cistué L (2000) Comparison of anther and isolated microspore culture in barley. Effects of culture density and regeneration medium. Euphytica 113:1–8

Castillo AM, Cistué L, Romagosa I, Vallés MP (2001a) Low responsiveness of six-rowed genotypes to androgenesis in barley does not have a pleiotropic basis. Genome 44:936–940

Castillo AM, Cistué L, Vallés MP, Sanz JM, Romagosa I, Molina-Cano JL (2001b) Efficient production of androgenetic doubled-haploid mutants in barley by the application of sodium azide to anther and microspore cultures. Plant Cell Rep 20:105–111

Charmet G, Bernard S (1984) Diallel analysis of androgenetic plant production in hexaploid triticale (× *Triticosecale*, Wittmack). Theor Appl Genet 69:55–61

Chen FQ, Prehn D, Hayes PM, Mulrooney D, Corey A, Vivar H (1994) Mapping genes for resistance to barley stripe rust (*Puccinia striiformis* f. sp. *hordei*). Theor Appl Genet 88:215–219

Chu CC (1978) The N6 medium and its application to anther culture of cereal crops. In: Proc Symp Plant Tissue Cultivation. Science Press, Peking, pp 43–50

Chuang CC, Ouyang TW, Chia H, Chou SM, Ching CK (1978) A set of potato media for wheat anther culture. In: Proc Symp Plant Tissue Cultivation. Science Press, Peking, pp 51–56

Chupeau Y, Caboche M, Henry Y (eds) (1998) Androgenesis and haploid plants. Springer, Berlin Heidelberg New York

Cistué L, Ramos A, Castillo AM, Romagosa I (1994) Production of large number of doubled haploid plants from barley anthers pretreated with high concentrations of mannitol. Plant Cell Rep 13:709–712

Cistué L, Ziauddin A, Simion E, Kasha KJ (1995) Effects of culture conditions on isolated microspore response of barley cultivar Igri. Plant Cell Tissue Organ Cult 42:163–169

Cistué L, Ramos A, Castillo AM (1999) Influence of anther pre-treatment and culture medium composition on the production of barley doubled haploids from model and low responding cultivars. Plant Cell Tissue Organ Cult 55:159–166

Coba de la Peña T, Brown S (2001) Flow cytometry. In: Hawes C, Satiat-Jeunemaitre B (eds) Plant cell biology, 2nd edn. Oxford University Press, Oxford, pp 85–106

Cowen NM, Johnson CD, Armstrong K, Miller M, Woosley A, Pescitelli S, Skokut M, Belmar S, Petolino JF (1992) Mapping genes conditioning in vitro androgenesis in maize using RFLP analysis. Theor Appl Genet 84:720–724

Davies PA, Morton S (1998) A comparison of barley isolated microspore and anther culture and the influence of cell culture density. Plant Cell Rep 17:206–210

Devaux P (1992) Haploidy in barley and wheat improvement. In: Dattée Y, Dumas C, Gallais A (eds) Reproductive biology and plant breeding. In: Proc 13th Eucarpia Congr, Angers. Springer, Berlin Heidelberg New York, pp 139–151

Devaux P (1995) Production and use of doubled haploids for breeding barley. In: Proc 7th Australian Barley Tech Symp The Grain Pool of Western Australia, Perth, pp 195–199

Devaux P (2003) The *Hordeum bulbosum* (L.) method. In: Maluszynski M, Kasha KJ, Forster BP, Szarejko I (eds) Doubled haploid production in crop plants. A manual. Kluwer, Dordrecht, pp 15–19

Devaux P, Zivy M (1994) Protein markers for anther culturability in barley. Theor Appl Genet 88:701–706

Devaux P, Kilian A, Kleinhofs A (1993a) Anther culture and *Hordeum bulbosum*-derived barley doubled haploids: mutations and methylation. Mol Gen Genet 241:674–679

Devaux P, Hou L, Ullrich S, Huang Z, Kleinhofs A (1993b) Factors affecting anther culturability of recalcitrant barley genotypes. Plant Cell Rep 13:32–36

Devaux P, Kilian A, Kleinhofs A (1995) Comparative mapping of the barley genome with male and female recombination-derived, doubled haploid populations. Mol Gen Genet 249:600–608

Devos K, Gale MD (2000) Genome relationships: the grass model in current research. Plant Cell 12:637–646

Dietrich WF, Weber JL, Nickerson DA, Kwok PY (1999) Identification and analysis of DNA polymorphisms. In: Birren B, Green ED, Hieter P, Klapholz S, Myers RM, Riethman H, Roskams

J (eds) Genome analysis: a laboratory manual, vol 4. Mapping genomes. CSHL Press, New York, pp 135–186
Dieu P, Beckert M (1986) Further studies of androgenetic embryo production and plant regeneration from in vitro cultured anthers of maize (*Zea mays* L.). Maydica 31:245–259
Dogramaci-Altuntepe M, Peterson TS, Jauhar PP (2001) Anther culture-derived regenerants of durum wheat and their cytological characterization. J Hered 92:56–64
Dufour P, Johnsson C, Antoine-Michard S, Cheng R, Murigneux A, Beckert M (2001) Segregation distortion at marker loci: variation during microspore embryogenesis in maize. Theor Appl Genet 102:993–1001
Dunford RP, Walden RM (1991) Plastid genome structure and plastid-related transcript levels in albino barley plants derived from anther culture. Curr Genet 20:339–347
Ellis RP, Forster BP, Gordon DC, Handley LL, Keith RP, Lawrence P, Meyer R, Powell W, Robinson D, Scrimgeour CM, Young G, Thomas WTB (2002) Phenotype/genotype associations for yield and salt tolerance in a barley mapping population segregating for two dwarfing genes. J Exp Bot 53:1163–1176
Evans JM, Batty NP (1994) Ethylene precursors and antagonists increase embryogenesis of *Hordeum vulgare* L. anther culture. Plant Cell Rep 13:676–678
Finnie SJ, Powell W, Dyer AF (1989) The effect of carbohydrate composition and concentration on anther culture response in barley (*Hordeum vulgare* L.). Plant Breed 103:110–118
Flehinghaus-Roux T, Deimling S, Geiger HH (1995) Anther-culture ability in *Secale cereale* L. Plant Breed 114:259–261
Folling L, Olesen A (2001) Transformation of wheat (*Triticum aestivum* L.) microspore-derived callus and microspores by particle bombardment. Plant Cell Rep 20:629–636
Forster BP, Thomas WTB (2003) Doubled haploids in genetic mapping and genomics. In: Maluszynski M, Kasha KJ, Forster BP, Szarejko I (eds) Doubled haploid production in crop plants. A manual. Kluwer, Dordrecht, pp 367–390
Gaillard A, Vergne P, Beckert M (1991) Optimization of maize microspore isolation and culture conditions for reliable plant regeneration. Plant Cell Rep 10:55–58
Gallais A (1990) Quantitative genetics of doubled haploid populations and application to the theory of line development. Genetics 124:199–206
Genovesi AD, Collins GB (1982) In vitro production of haploid plants of corn via anther culture. Crop Sci 22:1137–1144
Gilpin MJ, Pickering RA, Fautrier AG, Mcneil DL, Szigat G, Hill AM, Kynast RG (1997) Morphological and molecular analysis of androgenetic, selfed and backcrossed plants produced from a *Hordeum vulgare* L. (barley) × *H. bulbosum* L. hybrid. Plant Breed 116:505–510
González M, Hernández I, Jouve N (1997) Analysis of anther culture response in hexaploid triticale. Plant Breed 116:302–304
Guiderdoni E, Courtois B, Boissot N, Valdez M (1991) Rice somatic tissue and anther cultures: current status in France. In: Bajaj YPS (ed) Biotechnology in agriculture and forestry, vol 14. Rice. Springer, Berlin Heidelberg New York, pp 591–618
Guo YD, Pulli S (2000) Isolated microspore culture and plant regeneration in rye (*Secale cereale* L.). Plant Cell Rep 19:875–880
Gustafson VD, Baenziger PS, Wright MS, Stroup WW, Yen Y (1995) Isolated wheat microspore culture. Plant Cell Tissue Organ Cult 42:207–213
Guyomarc'h H, Sourdille P, Charmet G, Edwards KJ, Bernard M (2002) Characterisation of polymorphic microsatellite markers from *Aegilops tauschii* and transferability to the D-genome of bread wheat. Theor Appl Genet 104:1164–1172
Hallauer AR, Miranda JB (eds) (1981) Quantitative genetics in maize breeding. Iowa State University Press, Ames
Hansen NJP, Andersen SB (1998a) In vitro chromosome doubling with colchicine during microspore culture in wheat (*Triticum aestivum* L.). Euphytica 102:101–108
Hansen NJP, Andersen SB (1998b) Efficient production of doubled haploid wheat plants by in vitro treatment of microspores with trifluralin or APM. Plant Breed 117:401–405

Harada T, Sato T, Asaka D, Mitsukawa I (1991) Large-scale deletions of rice plastid DNA in anther culture. Theor Appl Genet 81:157–161

Hayes PM, Chen FQ, Kleinhofs A, Kilian A, Mather DE (1996) Barley genome mapping and its application. In: Jauhar PP (ed) Methods of genome analysis in plants. CRC Press, Boca Raton, pp 229–249

He P, Shen L, Lu C, Chen Y, Zhu L (1998) Analysis of quantitative trait loci which contribute to anther culturability in rice (*Oryza sativa* L.). Mol Breed 4:165–172

Henry Y, de Buyser J (1990) Wheat anther culture: agronomic performance of doubled haploid lines and the release of a new variety „Florin". In: Bajaj YPS (ed) Biotechnology in agriculture and forestry, vol 13. Wheat. Springer, Berlin Heidelberg New York, pp 285–352

Henry Y, Bernard S, Bernard M, Gay G, Marcotte JL, de Buyser J (1993) Nuclear gametophytic genes from chromosome arm 1RS improve regeneration of wheat microspore-derived embryos. Genome 36:808–814

Ho I, Wan Y, Widholm JM, Rayburn AL (1990) The use of stomatal chloroplast number for rapid determination of ploidy level in maize. Plant Breed 105:203–210

Hoekstra S, van Zijderveld MH, Louwerse JD, Heidekamp F, van der Mark F (1992) Anther and microspore culture of *Hordeum vulgare* L. cv. Igri. Plant Sci 86:89–96

Hoekstra S, van Zijderveld MH, Heidekamp F, van der Mark F (1993) Microspore culture of *Hordeum vulgare* L.: the influence of density and osmolality. Plant Cell Rep 12:661–665

Hoekstra S, van Bergen S, van Brouwershaven IR, Schilperoort RA, Heidekamp F (1996) The interaction of 2,4-D application and mannitol pretreatment in anther and microspore culture of *Hordeum vulgare* L. cv. Igri. J Plant Physiol 148:696–700

Hoekstra S, van Bergen S, van Brouwershaven IR, Schilperoort RA, Wang M (1997) Androgenesis in *Hordeum vulgare* L.: effects of mannitol, calcium and abscisic acid on anther pretreatment. Plant Sci 126:211–218

Holton TA, Christopher JT, McClure L, Harker N, Henry RJ (2002) Identification and mapping of polymorphic SSR markers from expressed gene sequences of barley and wheat. Mol Breed 9:63–71

Horvath H, Rostoks N, Brueggeman R, Steffenson B, von Wettstein D, Kleinhofs A (2003) Genetically engineered stem rust resistance in barley using the *Rpg1* gene. Proc Natl Acad Sci USA 100:364–369

Hou L, Ullrich SE, Kleinhofs A (1994) Inheritance of anther culture traits in barley. Crop Sci 34:1243–1247

Hu T, Kasha KJ (1997) Improvement of isolated microspore culture of wheat (*Triticum aestivum* L.) through ovary co-culture. Plant Cell Rep 16:520–525

Hu T, Kasha KJ (1999) A cytological study of pretreatments used to improve isolated microspore cultures of wheat (*Triticum aestivum* L.) cv Chris. Genome 42:432–441

Hu TC, Ziauddin A, Simion E, Kasha KJ (1995) Isolated microspore culture of wheat (*Triticum aestivum* L.) in a defined media. I. Effects of pre-treatment, isolation methods, and hormones. In Vitro Cell Dev Biol 31:79–83

Humphreys MW, Zwierzykowski Z, Collin HA, Rogers WJ, Zare AG, Lesniewska A (2000) Androgenesis in grasses – methods and aspects for future breeding. Biotechnological approaches for utilization of gametic cells. In: Proc COST 824 Final Meeting, Slovenia, pp 1–5

Hunter CP (1987) Plant generation method. European Patent Office application no 87200773.7

Immonen S (1999) Androgenetic green plants from winter rye, *Secale cereale* L., of diverse origin. Plant Breed 118:319–322

Immonen S, Robinson J (2000) Stress treatments and ficoll for improving green plant regeneration in triticale anther culture. Plant Sci 150:77–84

Inagaki MN, Hash CT (1998) Production of haploids in bread wheat, durum wheat and hexaploid triticale crossed with pearl millet. Plant Breed 117:485–487

Jähne A, Lörz H (1995) Cereal microspore culture. Plant Sci 109:1–12

Jähne A, Lazzeri PA, Jäger-Gussen M, Lörz H (1991) Plant regeneration from embryogenic cell suspensions derived from anther cultures of barley (*Hordeum vulgare* L.). Theor Appl Genet 82:74–80

Jähne A, Becker D, Brettschneider R, Lörz H (1994) Regeneration of transgenic microspore-derived, fertile barley. Theor Appl Genet 89:525–533
Jain SM, Sopory SK, Veilleux RE (eds) (1996/1997) In vitro haploid production in higher plants, vol 1–5. Kluwer, Dordrecht
Kantety RV, La Rota M, Matthews DE, Sorrells ME (2002) Data mining for simple sequence repeats in expressed sequence tags from barley, maize, rice, sorghum and wheat. Plant Mol Biol 48:501–510
Karp A (1991) On the current understanding of somaclonal variation. Oxf Surv Plant Mol Cell Biol 7:1–58
Karsai I, Bedö Z, Hayes P (1994) Effect of induction medium pH and maltose concentration on in vitro androgenesis of hexaploid winter triticale and wheat. Plant Cell Tissue Organ Cult 39:49–53
Kasha KJ, Kao KN (1970) High frequency haploid production in barley (*Hordeum vulgare* L.). Nature 225:874–876
Kasha KJ, Reinbergs E (1982) Recent developments in the production and utilization of haploids in barley. In: Asher MJC, Ellis RP, Whitehouse RNH (eds) Proc 4th Int Barley Genet Symp. Edinburgh University Press, Edinburgh, pp 655–665
Kasha KJ, Simion E (2001) Embryogenesis and plant regeneration from microspores. World Intellectual Property Organization, International patent public no W0 01/41557, 14 June 2001
Kasha KJ, Simion E, Oro R, Yao QA, Hu TC, Carlson AR (2001a) An improved in vitro technique for isolated microspore culture of barley. Euphytica 120:379–385
Kasha KJ, Hu TC, Oro R, Simion E, Shim YS (2001b) Nuclear fusion leads to chromosome doubling during mannitol pre-treatment of barley (*Hordeum vulgare* L.) microspores. J Exp Bot 52:1227–1238
Kicherer S, Backes G, Walther U, Jahoor A (2000) Localising QTLs for leaf rust resistance and agronomic traits in barley (*Hordeum vulgare* L.). Theor Appl Genet 100:881–888
Kihara M, Fukuda K, Funatsuki H, Kishinami I, Aida Y (1994) Plant regeneration through anther culture of three wild species of *Hordeum* (*H. murinum*, *H. marinum* and *H. bulbosum*). Plant Breed 112:244–247
Kilian A, Chen J, Han F, Steffenson B, Kleinhofs A (1997) Towards map-based cloning of the barley stem rust resistance genes *Rpg1* and *rpg4* using rice as an intergenomic cloning vehicle. Plant Mol Biol 35:187–195
Kintzios S, Fischbeck G (1994) Anther culture response of *Hordeum spontaneum*-derived winter barley lines. Plant Cell Tissue Organ Cult 37:165–170
Knox RE, Clarke JM, DePauw RM (2000) Dicamba and growth condition effects on doubled haploid production in durum wheat crossed with maize. Plant Breed 119:289–298
Kota R, Wolf M, Michalek W, Graner A (2001) Application of denaturing high-performance liquid chromatography for mapping of single nucleotide polymorphisms in barley (*Hordeum vulgare* L.). Genome 44:523–528
Ku MK, Cheng WC, Kuo LC, Kuan YL, An HP, Huang CH (1978) Induction factors and morphocytological characteristics of pollen-derived plants in maize (*Zea mays*). In: Proc Symp Plant Tissue Cultivation. Science Press, Peking, pp 35–42
Kunz C, Islam SMS, Berberat J, Peter SO, Büter B, Stamp P, Schmid JE (2000) Assessment and improvement of wheat microspore derived embryo induction and regeneration. J Plant Physiol 156:190–196
Kwon YS, Kim KM, Eun MY, Sohn JK (2002) QTL mapping and associated marker selection for the efficacy of green plant regeneration in anther culture of rice. Plant Breed 121:10–16
Kynast RG, Riera-Lizarazu O, Vales MI, Okagaki RJ, Maquieira SB, Chen G, Ananiev EV, Odland WE, Russell CD, Stec AO, Livingston SM, Zaia HA, Rines HW, Philips RL (2001) A complete set of maize individual chromosome additions to the oat genome. Plant Physiol 125:1216–1227
Lashermes P (1992) Improved anther culture method for obtaining direct regeneration in wheat (*Triticum aestivum* L.). J Genet Breed 46:99–102

Laurie DA, Bennett MD (1987) The effect of the crossability loci *Kr1* and *Kr2* on fertilization frequency in hexaploid wheat × maize crosses. Theor Appl Genet 73:403–409

Laurie DA, Bennett MD (1988) The production of haploid wheat plants from wheat × maize crosses. Theor Appl Genet 76:393–397

Lazar MD, Baenziger PS, Schaeffer GW (1984) Combining abilities and heritability of callus formation and plantlet regeneration in wheat (*Triticum aestivum* L.) anther culture. Theor Appl Genet 68:131–134

Leckband G, Lörz H (1998) Transformation and expression of a stilbene synthase gene of *Vitis vinifera* L. in barley and wheat for increased fungal resistance. Theor Appl Genet 96:1004–1012

Lefebvre D, Devaux P (1996) Doubled haploids of wheat from wheat × maize crosses: genotypic influence, fertility and inheritance of the 1BL-1RS chromosome. Theor Appl Genet 93:1267–1273

Le Gouis J, Devaux P, Werner K, Hariri D, Bahrman N, Béghin D, Ordon F (2004) *rym15* from the Japanese cultivar 'Chikurin Ibaraki 1' is a new barley mild mosaic virus (BaMMV) resistance gene mapped on chromosome 6H. Theor Appl Genet 198:1521–1525

Lentini Z, Reyes P, Martinez CP, Roca WM (1995) Androgenesis of highly recalcitrant rice genotypes with maltose and silver nitrate. Plant Sci 110:127–138

Li H, Devaux P (2001) Enhancement of microspore culture efficiency of recalcitrant barley genotypes. Plant Cell Rep 20:475–481

Li H, Devaux P (2003) High frequency regeneration of barley doubled haploid plants from isolated microspore culture. Plant Sci 164:379–386

Linsmaier EM, Skoog F (1965) Organic growth factor requirements of tobacco tissue cultures. Physiol Plant 18:100–127

Liu W, Zheng MY, Polle EA, Konzak CF (2002) Highly efficient doubled-haploid production in wheat (*Triticum aestivum* L.) via induced microspore embryogenesis. Crop Sci 42:686–692

Luckett DJ, Smithard RA (1995) Barley anther culture using membrane rafts. Plant Cell Tissue Organ Cult 42:287–290

Machii H, Mizuno H, Hirabayashi T, Li H, Hagio T (1998) Screening wheat genotypes for high callus induction and regeneration capability from anther and immature embryo cultures. Plant Cell Tissue Organ Cult 53:67–74

Madsen S, Olesen A, Dennis B, Andersen SB (1995) Inheritance of anther-culture response in perennial ryegrass (*Lolium perenne* L.). Plant Breed 114:165–168

Manninen OM (2000) Associations between anther-culture response and molecular markers on chromosomes 2H, 3H and 4H of barley (*Hordeum vulgare* L.). Theor Appl Genet 100:57–62

Marquez-Cedillo LA, Hayes PM, Jones BL, Kleinhofs A, Legge WG, Rossnagel BG, Sato K, Ullrich SE, Wesenberg DM (2000) QTL analysis of malting quality in barley based on the doubled-haploid progeny of two elite North American varieties representing different germplasm groups. Theor Appl Genet 101:173–184

Marquez-Cedillo LA, Hayes PM, Kleinhofs A, Legge WG, Rossnagel BG, Sato K, Ullrich SE, Wesenberg DM (2001) QTL analysis of agronomic traits in barley based on the doubled haploid progeny of two elite North American varieties representing different germplasm groups. Theor Appl Genet 103:625–637

Martin B, Widholm JM (1996) Ploidy of small individual embryo-like structures from maize anther cultures treated with chromosome doubling agents and calli derived from them. Plant Cell Rep 15:781–785

Martinez VA, Hill WG, Knott SA (2002) On the use of double haploids for detecting QTL in outbred populations. Heredity 88:423–431

Mejza SJ, Morgant V, DiBona DE, Wong JR (1993) Plant regeneration from isolated microspores of *Triticum aestivum*. Plant Cell Rep 12:149–153

Miao Z, Zhuang J, Hu H (1988) Expression of various gametic types in pollen plants regenerated from hybrids between *Triticum–Agropyron* and wheat. Theor Appl Genet 75:485–491

Michelmore RW, Paran I, Kesseli RV (1991) Identification of markers linked to disease-resistance genes by bulked segregant analysis: a rapid method to detect markers in specific genomic regions by using segregating populations. Proc Natl Acad Sci USA 88:9828–9832
Moieni A, Sarrafi A (1995) Genetic analysis for haploid-regeneration responses of hexaploid-wheat anther cultures. Plant Breed 114:247–249
Mordhorst AP, Lörz H (1993) Embryogenesis and development of isolated barley (*Hordeum vulgare* L.) microspores are influenced by the amount and composition of nitrogen sources in culture media. J Plant Physiol 142:485–492
Mouritzen P, Holm PB (1994) Chloroplast genome breakdown in microspore cultures of barley (*Hordeum vulgare* L.) occurs primarily during regeneration. J Plant Physiol 144:586–593
Murigneux A, Bentolila S, Hardy T, Baud S, Guitton C, Jullien H, Ben Tahar S, Freyssinet G, Beckert M (1994) Genotypic variation of quantitative trait loci controlling in vitro androgenesis in maize. Genome 37:970–976
Navarro-Alvarez W, Baenziger PS, Eskridge KM, Shelton DR, Gustafson VD, Hugo M (1994) Effect of sugars in wheat anther culture media. Plant Breed 112:53–62
Olesen A, Andersen SB, Due IK (1988) Anther culture response in perennial ryegrass (*Lolium perenne* L.). Plant Breed 101:60–65
Opsahl-Ferstad HG, Bjørnstad Å, Rognli OA (1994) Genetic control of androgenetic response in *Lolium perenne* L. Theor Appl Genet 89:133–138
Orshinsky BR, Sadasivaiah RS (1994) Effects of media on embryoid induction and plant regeneration from cultured anthers of soft white spring wheats (*Triticum aestivum* L.). Plant Sci 102:99–107
Orshinsky BR, Sadasivaiah RS (1997) Effect of plant growth conditions, plating density, and genotype on the anther culture response of soft white spring wheat hybrids. Plant Cell Rep 16:758–762
Otani M, Shimada T (1994) Pollen embryo formation and plant regeneration from cultured anthers of tetraploid wheat. J Genet Breed 48:103–106
Otani M, Shimada T (1995) Effect of synthetic medium on pollen embryo formation of common wheat and tetraploid wheat species. Bull RIAR Ishikawa Agric Coll 4:45–51
Paire A, Devaux P, Lafitte C, Dumas C, Matthys-Rochon E (2003) Proteins produced by barley microspores and their derived androgenic structures promote in vitro zygotic maize embryo formation. Plant Cell Tissue Organ Cult 73:167–176
Pan A, Hayes PM, Chen F, Chen THH, Blake T, Wright S, Karsai I, Bedö Z (1994) Genetic analysis of the components of winter hardiness in barley (*Hordeum vulgare* L.). Theor Appl Genet 89:900–910
Pauk J, Puolimatka M, Tóth KL, Monostori T (2000) In vitro androgenesis of triticale in isolated microspore culture. Plant Cell Tissue Organ Cult 61:221–229
Perretant MR, Cadalen T, Charmet G, Sourdille P, Nicolas P, Boeuf C, Tixier MH, Branlard G, Bernard S, Bernard M (2000) QTL analysis of bread-making quality in wheat using a doubled haploid population. Theor Appl Genet 100:1167–1175
Pickering RA (1983) The influence of genotype on doubled haploid barley production. Euphytica 32:863–876
Pickering RA (1984) The influence of genotype and environment on chromosome elimination in crosses between *Hordeum vulgare* L. × *H. bulbosum* L. Plant Sci Lett 34:153–164
Pickering RA (1992) Monosomic and double monosomic substitutions of *Hordeum bulbosum* L. chromosomes into *H. vulgare* L. Theor Appl Genet 84:466–472
Pickering RA, Devaux P (1992) Haploid production: approaches and use in plant breeding. In: Shewry PR (ed) Barley: genetics, molecular biology and biotechnology. CAB International, Wallingford, pp 511–539
Pickering RA, Fautrier AG (1993) Anther culture-derived regenerants from *Hordeum vulgare* × *H. bulbosum* crosses. Plant Breed 110:41–47
Pickering RA, Rennie WF (1990) The evaluation of superior *Hordeum bulbosum* L. genotypes for use in a doubled haploid barley breeding programme. Euphytica 45:251–255

Pickering RA, Wallace AR (1994) Gibberellic acid + 2,4-D improves seed quality in *Hordeum vulgare* L. × *H. bulbosum* L. crosses. Plant Breed 113:174–176

Powell W (1988) Diallel analysis of barley anther culture response. Genome 30:152–157

Powell W, Hayter AM, Wood W, Dunwell JM, Huang B (1984) Variation in the agronomic characters of microspore-derived plants of *Hordeum vulgare* cv. Sabarlis. Heredity 52:19–23

Puolimatka M, Pauk J (1999) Impact of explant type, duration and initiation time on the co-culture effect in isolated microspore culture of wheat (*Triticum aestivum* L.). J Plant Physiol 154:367–373

Puolimatka M, Pauk J (2000) Effect of induction duration and medium composition on plant regeneration in wheat (*Triticum aestivum* L.) anther culture. J Plant Physiol 156:197–203

Quimio CA, Zapata FJ (1990) Diallel analysis of callus induction and green plant regeneration in rice anther culture. Crop Sci 30:188–192

Redha A, Attia T, Büter B, Stamp P, Schmid JE (1998) Single and combined effects of colchicine, L-proline and post-inoculation low temperature on anther culture of wheat, *Triticum aestivum* L. Plant Breed 117:335–340

Reynolds TL (1997) Pollen embryogenesis. Plant Mol Biol 33:1–10

Reynolds TL, Crawford RL (1996) Changes in abundance of an abscisic acid-responsive, early cysteine-labeled metallothionein transcript during pollen embryogenesis in bread wheat (*Triticum aestivum*). Plant Mol Biol 32:823–829

Reynolds TL, Crawford RL (1997) Effects of light on the accumulation of abscisic acid and expression of an early cysteine-labeled metallothionein gene in microspores of *Triticum aestivum* during induced embryogenic development. Plant Cell Rep 16:458–463

Riera-Lizarazu O, Rines HW, Phillips RL (1996) Cytological and molecular characterization of oat × maize partial hybrids Theor Appl Genet 93:123–135

Rines HW, Dahleen LS (1990) Haploid oat plants produced by application of maize pollen to emasculated oat florets. Crop Sci 30:1073–1078

Ritala A, Mannonen L, Oksman-Caldentey KM (2001) Factors affecting the regeneration capacity of isolated barley microspores (*Hordeum vulgare* L.). Plant Cell Rep 20:403–407

Roberts-Oehlschlager SL, Dunwell JM (1990) Barley anther culture: pre-treatment on mannitol stimulates production of microspore-derived embryos. Plant Cell Tissue Organ Cult 20:235–240

Saisingtong S, Schmid JE, Stamp P, Büter B (1996) Colchicine-mediated chromosome doubling during anther culture of maize (*Zea mays* L.). Theor Appl Genet 92:1017–1023

Salmenkallio-Marttila M, Kurtén U, Kauppinen V (1995) Culture conditions for efficient induction of green plants from isolated microspores of barley. Plant Cell Tissue Organ Cult 43:79–81

Scheurer KS, Friedt W, Huth W, Waugh R, Ordon F (2001) QTL analysis of tolerance to a German strain of BYDV-PAV in barley (*Hordeum vulgare* L.). Theor Appl Genet 103:1074–1083

Scott P, Lyne RL (1994) Initiation of embryogenesis from cultured barley microspores: a further investigation into the toxic effects of sucrose and glucose. Plant Cell Tissue Organ Cult 37:61–65

Simpson E, Snape JW, Finch RA (1980) Variation between *Hordeum bulbosum* genotypes in their ability to produce haploids of barley, *Hordeum vulgare*. Z Pflanzenzüchtung 85:205–211

Snape JW, Sitch LA, Simpson E, Parker BB (1988) Tests for the presence of gametoclonal variation in barley and wheat doubled haploids produced using the *Hordeum bulbosum* system. Theor Appl Genet 75:509–513

Sourdille P, Snape JW, Cadalen T, Charmet G, Nakata N, Bernard S, Bernard M (2000) Detection of QTLs for heading time and photoperiod response in wheat using a doubled-haploid population. Genome 43:487–494

Steffenson BJ, Jin Y, Rossnagel BG, Rasmussen JB, Kao K (1995) Genetics of multiple resistance in a doubled-haploid population of barley. Plant Breed 114:50–54

Stirn S, Mordhorst AP, Fuchs S, Lörz H (1995) Molecular and biochemical markers for embryogenic potential and regenerative capacity of barley (*Hordeum vulgare* L.) cell cultures. Plant Sci 106:195–206

Stober A, Hess D (1997) Spike pretreatments, anther culture conditions, and anther culture response of 17 German varieties of spring wheat (*Triticum aestivum* L.). Plant Breed 116:443–447

Szakács É, Barnabás B (1995) The effect of colchicine treatment on microspore division and microspore-derived embryo differentiation in wheat (*Triticum aestivum* L.) anther culture. Euphytica 83:209–213

Szarejko I, Guzy J, Jimenez Davalos J, Roland Chavez A, Maluszynski M (1995) Production of mutants using barley DH systems. In: Induced mutations and molecular techniques for crop improvement. IAEA, Vienna, pp 517–530

Thiel T, Michalek W, Varshney RK, Graner A (2003) Exploiting EST databases for the development and characterization of gene-derived SSR-markers in barley (*Hordeum vulgare* L.). Theor Appl Genet 106:411–422

This D, Teulat B, Devaux P, Deleens E, Merah O (2003) QTLs for carbon isotope discrimination in barley. Plant and animal genome XI. Scherago Int, New York, 179 pp

Thomas WTB, Forster BP, Gertsson B (2003) Doubled haploids in breeding. In: Maluszynski M, Kasha KJ, Forster BP, Szarejko I (eds) Doubled haploid production in crop plants. A manual. Kluwer, Dordrecht, pp 337–349

Toojinda T, Broers LH, Chen XM, Hayes PM, Kleinhofs A, Korte J, Kudrna D, Leung H, Line RF, Powell W, Ramsay L, Vivar H, Waugh R (2000) Mapping quantitative and qualitative disease resistance genes in a doubled haploid population of barley (*Hordeum vulgare*). Theor Appl Genet 101:580–589

Torp AM, Hansen AL, Andersen SB (2001) Chromosomal regions associated with green plant regeneration in wheat (*Triticum aestivum* L.) anther culture. Euphytica 119:377–387

Touraev A, Indrianto A, Wratschko I, Vicente O, Heberle-Bors E (1996) Efficient microspore embryogenesis in wheat (*Triticum aestivum* L.) by starvation at high temperature. Sex Plant Reprod 9:209–215

Tuvesson IKD, Pedersen S, Andersen SB (1989) Nuclear genes affecting albinism in wheat (*Triticum aestivum* L.) anther culture. Theor Appl Genet 78:879–883

Tuvesson S, Ljungberg A, Johansson N, Karlsson KE, Suijs LW, Josset JP (2000) Large-scale production of wheat and triticale double haploids through the use of a single anther culture method. Plant Breed 119:455–459

Van Bergen S, Kottenhagen MJ, Van der Meulen RM, Wang M (1999) The role of abscisic acid in induction of androgenesis: a comparative study between *Hordeum vulgare* L. Cvs. Igri and Digger. Plant Growth Regul 18:135–143

Van Engelen FA, de Vries SC (1992) Extracellular proteins in plant embryogenesis. Trends Genet 8:66–70

Vergne P, Riccardi F, Beckert M, Dumas C (1993) Identification of a 32-kDa anther marker protein for androgenic response in maize, *Zea mays* L. Theor Appl Genet 86:843–850

Vrinten PL, Nakamura T, Kasha KJ (1999) Characterization of cDNAs expressed in the early stages of microspore embryogenesis in barley (*Hordeum vulgare* L.). Plant Mol Biol 41:455–463

Wan Y, Widholm JM (1993) Anther culture of maize. Plant Breed Rev 11:199–223

Wan Y, Duncan DR, Rayburn AL, Petolino JF, Widholm JM (1991) The use of antimicrotubule herbicides for the production of doubled haploid plants from anther-derived maize callus. Theor Appl Genet 81:205–211

Wang M, Hoekstra S, van Bergen S, Lamers GEM, Oppedijk BJ, van der Heijden MW, de Priester W, Schilperoort RA (1999) Apoptosis in developing anthers and the role of ABA in this process during androgenesis in *Hordeum vulgare* L. Plant Mol Biol 39:489–501

Wang X, Hu H (1984) The effect of potato II medium for triticale anther culture. Plant Sci Lett 36:237–239

Wang X, Hu H (1985) The chromosome constitution of plants derived from pollen of hexaploid triticale × common wheat F_1 hybrids. Theor Appl Genet 70:92–96

Wang YB, Hu H, Snape JW (1996) The genetic and molecular characterization of pollen-derived plant lines from octoploid triticale × wheat hybrids. Theor Appl Genet 92:811–816

Wedzony M, Marcinska I, Ponitka A, Slusarkiewicz-Jarzina A, Wozna J (1998) Production of doubled haploids in triticale (× *Triticosecale* Wittm.) by means of crosses with maize (*Zea mays* L.) using picloram and dicamba. Plant Breed 117:211–215

Wojnarowiez G, Jacquard C, Devaux P, Sangwan RS, Clément C (2002) Influence of copper sulfate on anther culture in barley (*Hordeum vulgare* L.). Plant Sci 162:843–847

Wojnarowiez G, Caredda S, Devaux P, Sangwan RS, Clément C (2004) Barley anther culture: assessment of carbohydrate effects on embryo yield, green plant production and differential plastid development in relation with albinism. J Plant Physiol 161(6):757–760

Wu RL (1999) Mapping quantitative trait loci by genotyping haploid tissues. Genetics 152:1741–1752

Yamagishi M (2002) Heterogeneous plastid genomes in anther culture-derived albino rice plants. Euphytica 123:67–74

Yamagishi M, Yano M, Fukuta Y, Fukui K, Otani M, Shimada T (1996) Distorted segregation of RFLP markers in regenerated plants derived from anther culture of an F_1 hybrid of rice. Genes Genet Syst 71:37–41

Yao QA, Simion E, William M, Krochko J, Kasha KJ (1997) Biolistic transformation of haploid isolated microspores of barley (*Hordeum vulgare* L.). Genome 40:570–581

Ye JM, Kao KN, Harvey BL, Rossnagel BG (1987) Screening salt-tolerance barley genotypes via F_1 anther culture in salt stress media. Theor Appl Genet 74:426–429

Zamani I, Kovács G, Gouli-Vavdinoudi E, Roupakias DG, Barnabás B (2000) Regeneration of fertile doubled haploid plants from colchicine-supplemented media in wheat anther culture. Plant Breed 119:461–465

Zare AG, Humphreys MW, Rogers JW, Mortimer AM, Collin HA (2002) Androgenesis in a *Lolium multiflorum* × *Festuca arundinaceae* hybrid to generate genotypic variation for drought resistance. Euphytica 125:1–11

Zhang L, Pickering RA, Murray BG (2001) A *Hordeum vulgare* × *H. bulbosum* tetraploid hybrid provides useful agronomic introgression lines for breeders. N Z J Crop Hortic Sci 29:239–246

Zheng MY, Liu W, Weng Y, Polle E, Konzak CF (2001) Culture of freshly isolated wheat (*Triticum aestivum* L.) microspores treated with inducer chemicals. Plant Cell Rep 20:685–690

Zheng MY, Weng Y, Liu W, Konzak CF (2002) The effect of ovary-conditioned medium on microspore embryogenesis in common wheat (*Triticum aestivum* L.). Plant Cell Rep 20:802–807

Ziauddin A, Marsolais A, Simion E, Kasha KJ (1992) Improved plant regeneration from wheat anther and barley microspore culture using phenylacetic acid (PAA). Plant Cell Rep 11:489–498

Zivy M, Devaux P, Blaisonneau J, Jean R, Thiellement H (1992) Segregation distortion and linkage studies in microspore derived doubled haploid lines of *Hordeum vulgare* L. Theor Appl Genet 83:919–924

II.4 Haploids in the Improvement of Woody Species

SVEN B. ANDERSEN[1]

1 Introduction

Woody species make up a very considerable part of our cultivated plants. Huge areas are cultivated as forests for production of timber, pulp and paper and other products. However, many woody plants are also important for parks and other recreational areas, where they constitute a major component of our living environment. Finally, many woody plants are grown on a large scale for production of fruits and special products like cork, rubber, syrup and oil. For areas cultivated with planted or sown material, selection of plant types with improved ability to grow and produce under environmentally friendly cultivation conditions is of prime importance.

For a number of important cultivated herbaceous species, methods of large-scale production of chromosome doubled haploid plants (DHs) and subsequently homozygous lines by selfing have been developed over the last 25 years. Notably for *Nicotiana tabacum*, *Hordeum vulgare*, *Brassica napus*, *Brassica oleracea* and *Triticum aestivum* such lines are now being produced and used in commercial breeding programmes. Several more species, e.g. *Oryza sativa* and *Zea mays*, are expected to join this line of cultivated crops with DH-based breeding in the near future. A common feature of this first wave of crops, which can be bred using DHs, is that they have a history of intensive improvement based on inbreeding. Traditional inbreeding by repeated selfings of the material produces more or less stable homozygous inbreds, which are either marketed directly as line cultivars or used as parental inbreds for hybrid cultivars. DHs in such programmes perform a simple substitution for inbreeding to save three to four generations of time and resources. However, their complete homozygosity also makes early selection on small plots more efficient than with traditionally inbred material because of higher within-line uniformity. For some highly cross-breeding crops, e.g. *Lolium perenne* and *Lolium multiflorum* (Andersen et al. 1997), quite efficient DH production systems have been developed. Regular use of DH for breeding in such allogamous species, however, is still hampered because of strong inbreeding depression and self-incompatibility. These genetic phenomena

[1] Department of Agricultural Sciences, The Royal Veterinary and Agricultural University, Thorvaldsensvej 40, 1871 Frederiksberg C, Denmark

Biotechnology in Agriculture and Forestry, Vol. 56
Haploids in Crop Improvement II (ed. by C.E. Palmer, W.A. Keller, and K.J. Kasha)

make the introduction of DH to traditional breeding programmes of the species more complex. For the same reasons, the benefit of DH for breeding of vegetatively propagated plants is still unclear.

Cultivated woody species are generally highly cross-breeding in nature. They normally have long life cycles from 5–50 years or more, and they are generally less intensively bred than most crop species. For these reasons, it is no surprise that most of these important species still have not experienced wide use of DH for their genetic improvement. There are several reasons for this delay: (1) most applications of DH for crop improvement demand large number of DHs, which should be available at a low price, because most plants or their offspring are discarded during the early selection process. Efficient DH production systems are still rare among woody plants; (2) the long generation time of most species makes investments in such use of DHs rather long term; and (3) the mode of application of completely homozygous material in the breeding process is far from traditional approaches in these species and still mostly unclarified. This chapter reviews the overall status and progress in haploid induction and use in the improvement of woody species, without details of media and methods, which can be found in the original literature or reviews.

2 Gametophyte Development in Woody Species

Woody species botanically divide into gymnosperms and angiosperms. For both types of woody plants, the main strategies for production of DH are clear. Haploids and subsequently chromosome doubled haploids may be achieved either from the female (gynogenesis) or the male (androgenesis) haploid phase of the life cycle. Detailed descriptions of development of both male and female gametophytes as well as plant development from these structures are given by Baldursson and Ahuja (1996a,b). These texts also present excellent reviews on the complete activities performed with DH induction in such species until 1996.

Most DH production systems efficient enough for plant improvement purposes of herbaceous species induce haploid or DH plants from microspores or immature male gametophytes cultured in vitro. Microspores of both gymnosperms and angiosperms are haploid single-celled products of meiosis. They pass through a development in vivo from initially cytoplasmically dense cells surrounded by thin walls to highly vacuolated cells with thickened walls and the haploid nucleus displaced toward the cell wall by vacuolation. Microspore development ends with the first pollen mitosis, which is a highly asymmetric division, to form a small generative cell and a large vegetative cell during the first pollen mitosis. Successful in vitro culture to develop plants from such structures generally initiates immediately before, during or soon after this first pollen mitosis. The two-celled structure formed by the asymmetric first pollen mitosis is the male gametophyte. Male gametophytes of angiosperm species in situ will mature directly and form two elongated sperm cells during a second pollen mitosis either during or before anthesis. In gymnosperms the male gametophyte undergoes two additional pollen mitoses to form a vegetative cell, two prothallus cells and a generative cell, before formation of two sperm cells (Favre-Duchartre 1956). Successful in vitro culture apparently disturbs normal differentiation of the two initial cells of male gametophytes. This disturbance of the tissue leads to extra cell divisions by one or both cells to form a multicelled structure inside the walls of the young pollen grain. This initial multicelled structure with 50 to several hundred small cells with dense cytoplasm has many features in common with early proembryos derived from fertilised egg cells. In many cases the multicelled structures from microspores (proembryos) can be further developed into complete plants in suitable in vitro culture systems.

3 Haploid Development in Woody Species

In spite of extensive basic and applied efforts to study the nature of the embryogenesis inducing change in cultured microspores or young pollen, the molecular nature of the change is still unclear. Subsequent development of proembryos into plants is apparently rather autonomous in the sense that external stimuli seem to be unnecessary for embryo differentiation. Balanced nutrition with inorganic salts and some organic compounds, notably a sugar carbon source, however, is needed for regular embryo development and germination into plants. In spite of extensive studies and wide application of DH development for several herbaceous plant species, basic understanding of nutritional demands for regular plant development is missing. In many in vitro culture systems, embryogenesis from the initial proembryo is distorted to form embryos incapable of normal germination into plants. In other cases, organisation of the developing embryo breaks down and apparently undifferentiated cell masses of callus are formed, from which plants may be subsequently obtained via adventitious shoots or secondary embryo formation.

Tulecke (1953) reported the first observation of haploid tissue induced from in vitro-cultured male gametophytes of the gymnosperm *Ginkgo biloba* and LaRue (1954) soon after for *Taxus brevifolia*. A considerable number of papers (approximately 20) (Baldursson and Ahuja 1996b) of such studies of in vitro culture of gymnosperm male gametophytes have indicated that formation of callus-like tissues from such structures in several species is a frequent phenomenon. Embryo development from in vitro-cultured microspores of *Ginkgo biloba* has been reported by Laurain et al. (1993), but plant regeneration from gymnosperm microspores has not been obtained.

Much more success has been achieved with in vitro cultures of microspores and immature male gametophytes of angiosperms, since Guha and

Maheshwari (1966) for the first time observed embryo formation from cultured anthers of *Datura innoxia* and Nitsch and Nitsch (1969) reported haploid plant formation from in vitro-cultured anthers of *Nicotiana tabacum*. These results introduced the phenomenon of anther culture where entire anthers with microspores inside are cultured in vitro. During anther culture, initial development into embryos or callus from microspores proceeds inside the anther loculus. The entire complex of somatic anther tissue and haploid microspores and developing embryogenic structures presumably generates a suitable physical and chemical environment for development of some of the plant initials. Anther cultures have been successful for generation of haploid or chromosome doubled plants from microspores in a large number of angiosperm species. More advanced systems of DH production use culture of isolated microspores in shallow layers of liquid nutrient media to obtain embryo development directly from the cultured microspores. Reports describing successful plant regeneration from in vitro-cultured microspores of angiosperm woody species are summarised in Table 1. For a more complete review of all reports with angiosperm woody species anther and microspore culture, see Baldursson and Ahuja (1996b).

In spite of the very considerable number of haploid or chromosome doubled haploid plants reported for 32 different woody species using anther culture (Table 1), efficient anther culture systems are still limited to a few species. This is no surprise since development of efficient systems for herbaceous species, e.g. wheat, barley and rapeseed, has involved huge efforts in many different laboratories worldwide over a period of 15–20 years. A comparable investment of resources for development of such technology in woody species has still not been realised.

Female macrospores and megagametophytes are derived from meiosis of megaspore mother cells in the female part of flowers. The reader is referred to Baldursson and Ahuja (1996b) and references therein for a detailed description of megagametophyte development. In both gymnosperms and angiosperms, only one of the initial four cells of meiosis from a megaspore mother cell will remain viable, while the other three cells degrade. In angiosperms development of the megagametophyte from the megaspore consists of only three cell divisions to form eight haploid cells in the ovule. Two of these cells normally fuse to form a diploid central nucleus. The remaining cells form one haploid egg cell and two synergids at the micropylar end of the ovule and three haploid antipodal cells at the other end of the ovule. In gymnosperms, the megagametophyte formation involves many cell divisions to form a highly specialised haploid tissue, which finally differentiates one or several archegonia in which specialised haploid egg cells are formed.

Fertilisation in angiosperms is a double fertilisation. The fertilising pollen delivers two sperm cells, one to fuse with the central nucleus and one to fuse with the egg cell. The fertilised central nucleus in angiosperms gives rise to the seed endosperm while the fertilised egg cell forms the embryo. In gymnosperms the fertilisation process involves only the egg cell to initiate an

Table 1. Induced haploid plant formation in angiosperm woody species

Species	Method of induction	Reference
Aesculus carnea	Anther culture	Radojevic et al. (1989); Marinkovic and Radojevic (1992)
Aesculus hippocastanum	Anther culture	Radojevic (1978); Jörgensen (1991)
Betula pendula	Anther culture	Huhtinen (1978)
Carica papaya	Anther culture	Tsay and Su (1985)
Poncirus trifoliata	Anther culture	Hidaka et al. (1979)
Citrus deliciosa, C. reticulum, C. sinensis	Anther culture	Geraci and Starrantino (1990)
Citrus clementina	Anther culture	Germanà et al. (1994)
Citrus clementina	Pollination from triploid	Oiyama and Kobayashi (1993); Germanà and Chiancone (2001)
Citrus madurensis	Anther culture	Chen (1985)
Citrus microcarpa	Anther culture	Chen et al. (1980)
Coffea arabica	Plants via anther or microspore culture	Sondahl and Sharp (1979); Carneiro (1993); Neuenschwander and Baumann (1995); Raghuramulu and Prakash (1996); Herrera et al. (2002)
Coffea canephora	Parthenogenesis in seeds	Lashermes et al. (1994)
Hevea brasiliensis	Anther culture	Chen et al. (1982); Chen (1990)
Litchi chienesis	Anther culture	Lianfang (1990)
Malus × domestica	Pollination-irradiated pollen	Zhang and Lespinasse (1991)
Malus × domestica	Anther culture	Höfer (1995); Kadota et al. (2002)
Malus prunifolia	Anther culture	Wu (1981)
Malus pumila	Anther culture	Fei and Xue (1981)
Pyrus communis, P. pyrifolia	Anther culture	Kadota et al. (2002)
Populus berolinensis	Anther culture	Anonymous (1975)
Populus deltoides	Anther culture	Ho and Raj (1985); Uddin et al. (1988); Kiss et al. (2001)
Populus maximowiczii	Anther culture	Stoehr and Zsuffa (1990a); Baldursson et al. (1993)
Populus nigra	Anther culture	Wang et al. (1975); Ho and Ray (1985); Kiss et al. (2001)
Populus trichocarpa	Anther culture	Baldursson et al. (1993)
Populus ussuriensis	Anther culture	Anonymous (1975)
Populus euphratica	Anther culture	Mofidabadi et al. (2001)
Populus × various hybrids	Anther culture	Anonymous (1975); Wang et al. (1975); Ho and Raj (1985); Wu and Nagarajan (1990)
Prunus cerasus	Anther culture	Seirlis et al. (1979)
Quercus petraea	Anther culture	Jörgensen (1988, 1991)
Quercus suber	Anther culture	Bueno et al. (1997)
Vitis rupestris	Anther culture	Cersosimo et al. (1990)
Vitis vinifera	Anther culture	Zou and Li (1981); Li and Cao (1993)

embryo. The remaining haploid megagametophyte of gymnosperms subsequently develops into a storage tissue of the seed with functions like the endosperm of angiosperm seeds.

Misfunctioning to prevent fertilisation of the egg cell can lead to formation of seeds with haploid embryos in both angiosperms and gymnosperms. Such haploids derived from the egg cell are generally termed gynogenic. In angiosperms, for an unfertilised egg cell to form a functional embryo the central cell of the embryo sac must still be fertilised to form a functional seed endosperm. Such gynogenetic plants can generally be found in low frequencies among ordinary seed offspring in most species including woody plants. There are reports of seeds with haploid embryos from two woody species: *Coffea arabica* (Sreenivasan et al. 1981) and *Pinus sylvestris* (Isakov et al. 1981). Such rare haploid plants among many normal diploid seed offspring are difficult to detect unless specific genetic markers in the pollinator are used for selection as exemplified for apple (Lespinasse and Godicheau 1980). However, gynogenic haploids in seeds of angiosperms are frequent among seeds germinating with two or more seedlings. Initial selection for polyembryonic seeds followed by detection of spontaneous haploids has been used for woody plants in *Coffea canephora* (Lashermes et al. 1994). Generally, however, frequencies of spontaneous haploid embryos in seeds are too low to be efficient for DH production for breeding purposes. Pollination of ordinary plants with low or non-functional pollen in some cases has been shown to drastically increase frequencies of seeds with haploid embryos in the offspring. Examples of such approaches to produce haploids in woody species comprise pollination with irradiated pollen to induce gynogenesis in poplars (Stettler and Bawa 1971) and *Malus domestica* (Zhang and Lespinasse 1991) or pollination with pollen from triploid pollinators in *Citrus clementina* (Oiyama and Kobayashi 1993; Germanà and Chiancone 2001).

In vitro methods to produce gynogenetic haploid or DH plants exploit the general totipotency of haploid cells in female gametophytes. For angiosperm species, ovules or ovaries containing the female gametophytic cells are cultured, with the culture substrate substituting for the nutritive functions of the seed endosperm. In successful cultures one or more of the cells in the embryo sac develop into an embryo or callus, which subsequently emerges from the inside of the cultured ovule. Such approaches have led to haploid plant formation in several woody species, e.g. *Populus* (Wu and Nagarajan 1990) and *Hevea* (Chen et al. 1987). However, isolation and culture of such explants in large numbers are quite resource demanding and frequency of plant regeneration has been generally low.

Gymnosperms comprise only woody species (about 1,000), which makes haploid induction from female megagametophytes special to woody plants. Considerable efforts have been devoted to in vitro culture of haploid megagametophytes with the purpose of haploid or DH plant regeneration since LaRue (1948) reported plant regeneration from such cultures of *Zamia floridana*. For a complete review comprising also non-plant-producing attempts

Table 2. Plant and shoot regeneration from gymnosperm megagametophyte in vitro cultures

Species	Response	Reference
Zamia floridana	Roots and shoots	La Rue (1948)
Zamia integrifolia	Roots and leaves	Norstog (1965)
Ceratozamia mexicana	Shoots	Chavez and Norstog (1992)
Ceratozamia hildae	Shoots	Chavez and Norstog (1992)
Ephedra foliata	Roots and shoots	Konar and Singh (1979); Singh et al. (1981); Bhatnagar and Singh (1984)
Larix decidua	Plants via embryos	von Aderkas and Bonga (1993)
Larix leptolepis	Plants via embryos	von Aderkas et al. (1990)
Larix × eurolepis	Plants via embryos	von Aderkas et al. (1990)
Picea abies	Plants from callus	Simola and Honkanen (1983)

with gymnosperm megagametophytes as well as culture media, see Baldursson and Ahuja (1996b). Results with successful plant regeneration from in vitro-cultured megagametophytes are summarised in Table 2. In spite of the very considerable efforts spent on studies of formation of cell, tissues and plants from in vitro-cultured gymnosperm megagametophytes, most such cultures are incapable of plant formation. The optimal developmental stage for culture of megametophytes for cell proliferation is less clear than for culture of microspores and may depend on the species (Baldursson and Ahuja 1996b). Studies of well-functioning cultures of *Larix* (von Aderkas and Bonga 1988; von Aderkas et al. 1991; von Aderkas and Dawkins 1993) have documented plant regeneration via distinct embryonal masses also known from somatic in vitro cultures of gymnosperms.

4 Examples of Woody Angiosperms

In a few notably angiosperm woody species, haploid induction systems efficient enough to enable studies of plant performance and use for breeding have been established. Status and results from *Coffea*, *Hevea*, *Populus* and *Citrus* are discussed below.

4.1 Coffee

Production and use of haploids in coffee have been reviewed in much detail by Raghuramulu and Prakash (1996). Most cultivated coffee (approx. 80%) is of the allotetraploid species *Coffea arabica*, which is self-fertile. The species tolerates inbreeding and can be made homozygous through repeated selfing for efficient seed multiplication. Such inbreeding, however, takes 15–18 years due to the prolonged life cycle of the material. Based on positive experience with herbaceous self-fertile species like barley and rape seed, DHs may be an efficient substitute for inbreeding to save time in future breeding of *Coffea arabica*. Such efficient DH technology for *Coffea arabica* based on anther and microspore culture has been under development (Sondahl and Sharp 1979; Carneiro 1993; Neuenschwander and Baumann 1995; Raghuramulu and Prakash 1996) and is approaching useful efficiency (Herrera et al. 2002).

Coffea arabica has a narrow genetic base and an important part of its improvement depends on successful hybridisation with diploid species of *Coffea* to enlarge the gene pool, particularly to gain new resistance toward several diseases, e.g. coffee leaf rust (*Hemileia vastatrix* Berk and Br.). Hybrids between the allotetraploid *Coffea arabica* and diploids like *Coffea canephora* are highly sterile, but can be multiplied vegetatively (Sreenivasan et al. 1993). Vegetative multiplication, however, is resource taking and sterility of the hybrids seriously hampers further genetic improvement to regain the superior quality of *Coffea arabica* types (Sondahl and Lauritis 1992). DHs derived from such sterile primary hybrids involving polyploids may directly generate fertile chromosome substitution, addition or translocation lines for further use in breeding as has been demonstrated in wheat and grass (Hu et al. 1982)

In the less widely cultivated *Coffea canephora*, known as Robusta coffee, an alternative haploid induction system has been developed based on spontaneous gynogenetic development of haploid embryos in seeds (Couturon 1982; Lashermes et al. 1994). The system uses selection of potential haploids among polyembryonic seeds and subsequent grafting of candidates on freshly germinated seedlings. Although the actual number of DHs obtainable with such approaches is limited they have nevertheless enabled production of several hundred completely homozygous clones in this highly cross-breeding species (Lashermes et al. 1994). If further developed the technology may permit future breeding of uniform F_1 hybrid cultivars of Robusta coffee.

4.2 *Hevea*

The rubber tree (*Hevea brasiliensis* Muell. Arg.) is the source of natural rubber used worldwide and therefore of considerable economic importance particularly for many developing countries. A rather efficient system for production of DHs in cultivated types of the species was developed by Chen and coworkers during the 1980s, since the first report by Chen et al. (1978). Excellent detailed reviews have been given previously, including protocols for in vitro culture (Chen et al. 1982; Chen 1990). The anther culture technique established induces plant development from microspores through a number of in vitro culture steps to form microspore-derived embryoids, which are subsequently germinated or regenerated into complete plants. The system has been efficient enough to establish hundreds of microspore-derived trees in the field and initiate studies of their breeding value (Chen et al. 1982; Chen 1990).

Cultivated *H. brasiliensis* is an allotetraploid ($2n=4x=36$) that is normally vegetatively propagated through grafting. Clones used for rubber production are highly heterozygous but also believed to have a narrow genetic base. Generation of new types of genetic variation within the species as well as possible transfer of disease resistance, quality characteristics and stress tolerance from relatives thus have interest for breeding (Chen 1990). Chen et al. (1982) and Chen (1990) documented that regenerated anther-derived trees have a wide variation in chromosome numbers mostly between the haploid number of 18 and the diploid number 36. There is good indication that originally haploid tissues during plant regeneration and further plant development gradually increase the number of chromosomes. Some trees reach the diploid level, but many trees apparently remain intermediate in chromosome number. Chen (1990) also demonstrated that some microspore-derived trees could out-yield their original donor clone when grafted material was evaluated for growth (stem girth) and latex production. It thus seems possible that this DH production system can produce offspring that show improved performance and that may be useful for further breeding. It may also be possible to induce haploids from interspecific hybrids between cultivated *H. brasiliensis* and other *Hevea* species to produce chromosome substitution, addition or translocation lines for the transfer of desired traits into the breeding pool.

4.3 *Populus*

Various poplar (*Populus* spp.) subspecies are widely used for production of timber, pulp and paper and reforestation purposes. In addition, hybrids among several species are used for energy production in short-term forestry cultivation based on vegetative propagation through cuttings. Almost all poplar species are dioecious with separate male and female plants, which prevents selfing and maintains highly heterozygous populations. Haploid induction by means of anther culture has been very successful for poplar species, where plant regeneration has been reported for at least seven *Populus* species (Table 1), since Sato (1974) reported the first successful cultures. A detailed review of methods and results for haploid induction in poplars with protocols and media composition has been given by Wu and Nagarajan (1990). Most successful anther cultures of poplars induce cell proliferation of callus from both microspores and somatic tissues and regenerated plants are generally a mixture of haploid, aneuploids, diploids and tetraploids. Initially, haploid plants have a tendency to undergo gradual chromosome doubling to become aneuploid, diploid or polyploid (Wu and Nagarajan 1990), but some of the non-haploid regenerants may be derived from somatic tissue of the culture. Stoehr and Zsuffa (1990b) found that the four diploid plants among the ten analysed were somatic regenerants based on analysis of an isoenzyme locus heterozygous in the donor tree. Baldursson et al. (1993) induced direct embryogenesis and plant formation without somatic callus in some *Populus* species on media without auxins. Also, the plants regenerated via direct embryogenesis were a mixture of haploids and diploids, and seven diploid regenerants of *P. trichocarpa* were all shown to have a microspore origin by isozyme analysis (Baldursson et al. 1993). Systems for direct embryogenesis via anther culture, however, have low response, and better procedures for analysis of origin of regenerants with genetic markers may solve the problem with somatic regenerants in the future.

Plants regenerated from poplar anther culture show wide genetic segregation for morphological traits (Wu and Nagarajan 1990), while segregation for traits of breeding value still needs further research. Stoehr and Zsuffa (1990b) studied the variation among 14 stable haploid clones derived from anther culture of a single donor tree of *Populus maximowiczii*. Among eight traits studied (flushing date, growth cessation, relative shoot growth rate, leaf width, petiole length, leaf teeth number, internode length and relative wood density) only genetic segregation for leaf teeth number and relative wood density were found to be significant. The low number of significant tests may be due to the small number of closely related clones in the study. However, the genetic component of relative wood density could explain 24.8% of total variation in the experiment, which would indicate good progress with selection for the trait in this segregating population derived from a single tree. Such strong genetic differences among anther-derived poplar clones have been further documented by Kiss et al. (2001) and Mofidabadi et al. (2001).

Poplars completely homozygous after chromosome doubling of a haploid may be used in the future in studies of combining ability to identify superior inbreds for hybrid seed production based on cloned parentals or as base clones for synthetic populations. The study of combining ability for major traits and subsequent seed production from selected DHs is a long-term investment for more traditional forestry with a long life cycle. Hybrid seed production based on cloned homozygous well-combining parentals for energy production in short-term forestry, however, may be a more short-term investment. Poplars, in addition, are interesting model species for genomics and molecular genetics in forestry, because of their relatively short life cycles, ease of clonal propagation and small genome. DH clones of different origin may be useful for construction of mapping populations based on homozygous parents for gene mapping and cloning.

4.4 *Citrus*

Haploid induction and application in *Citrus* species have been previously reviewed in detail by Germanà (1997) including the media and methods. *Citrus* species are widely grown worldwide for fruit production. Since the crop is vegetatively propagated through grafting or by means of nucellar embryos in seeds, the clones are generally highly heterozygous and slow to breed by traditional means. Anther cultures have been used successfully to produce microspore-derived plants in several species (Table 1), since Hidaka et al. (1979) succeeded for the first time to generate plants from the root stock *Poncirus trifoliata*, while embryos, callus and leaf-like structures were obtained from several other *Citrus* species (Germanà 1997). Anther culture methods are still highly dependent on genotypes. Alternatively, pollination with triploid pollen either in vivo (Oiyama and Kobayashi 1993) or in vitro (Germanà and Chiancone 2001) to produce parthenogenetic haploids has been successful for some genotypes of *Citrus clementina*. Such approaches may be useful for genotypes that are not responsive to anther culture.

Germanà and Chiancone (2001) pointed out a number of possible applications of such haploid or homozygous diploid plants for *Citrus* breeding, including: mutant induction and selection, fusion of haploid and diploid protoplasts to obtain seedless triploids, and fusion of haploid protoplasts with protoplasts from incompatible species to transfer new genetic traits for breeding. Pure breeding clones may also be used for studies of combining ability to identify heterotic groups for more efficient hybrid breeding in the future.

5 Examples in Gymnosperms

So far, among gymnosperm species, the only haploid induction has been in *Larix* (von Aderkas et al. 1990; von Aderkas and Bonga 1993). This is based on in vitro culture of female megagametophytes and the plant regeneration efficiency is high enough to be of interest for plant improvement. Overall, it may be said that haploid induction and use for woody plant improvement are still very much in their infancy. It should be remembered, however, that most of these woody species pose very special limitations to scientific work. Most of the species flower only once each year under natural conditions and flower-producing trees of many woody species are too big and voluminous to be cultured under controlled conditions in glasshouses or growth chambers to obtain regular flower production. The long life cycle of most of the material also makes studies of growth performance of haploid or chromosome doubled haploid offspring a very long-term investment. In light of these very special conditions, it is rather remarkable that with several of the woody species results have been obtained, which already indicate the potential of this technology for genetic improvement. For a true breakthrough in haploid induction and use also among woody plants, a better understanding of the basic principles of plant development from gametophytes is urgently needed to reduce the resources for establishment of well-functioning haploid induction systems in new species.

6 Conclusions

Haploid and doubled haploid plant production has been developed to an extent where it is efficient enough for plant improvement purposes for some herbaceous angiosperm species. In general, such plants may be obtained either androgenetically from the male gametophytes or gynogenetically from cells of the female gametophytes. For woody plants, development of efficient haploid plant formation systems has been reported for a few species. For angiosperms, such plants may be derived from microspores based on in vitro culture of anthers or isolated spores or from macrospores through selection of seeds with haploid embryos or in vitro culture of the female gametophyte. For gymnosperms the female megagametophyte may also be cultured in vitro for regeneration of haploid plants.

Reproducible DH production from gymnosperms is still limited to a few species, notably *Larix* based on plant regeneration from in vitro-cultured megagametophytes. For angiosperm woody species, methods of haploid induction have been reported for *Coffea arabica* and *Coffea canephora* based on anther/microspore culture and polyembryonic seeds, respectively, efficient enough for special applications in plant breeding. Many haploid and DH plants have been produced in the rubber tree *Hevea brasiliensis* based on anther culture technique and used to study basic genetics and applications. At least seven species of *Populus* have been reported to be responsive to anther culture and DH has been used for basic studies in breeding and genetics of *Populus maximowiczii*. In *Citrus*, haploid plants have been induced in several different species either through anther culture or through pollination with pollen from triploids to induce seeds with haploid embryos.

In spite of the limitations for development of the technology with woody species, their often long life cycle, limited season of flowering and voluminous plant sizes, it is remarkable that results from several of the species have already indicated the potential of chromosome doubled haploids, provided the technology can be fully developed.

References

Andersen SB, Madsen S, Roulund N, Halberg N, Olesen A (1997) Haploidy in ryegrass. In: Jain SM, Sopory SK, Veilleux RE (eds) In vitro haploid production in higher plants, vol 4. Kluwer, Dordrecht, pp 133–147

Anonymous (1975) Induction of haploid poplar plants from anther culture in vitro. Sci Sin 18:771–777

Baldursson S, Ahuja MR (1996a) Cytogenetics and potential of haploidy in forest tree genetics and improvement. In: Jain MS, Sopory SK, Veilleux RE (eds) In vitro haploid production in higher plants, vol 1. Kluwer, Dordrecht, pp 49–66

Baldursson S, Ahuja MR (1996b) Haploidy in forest trees. In: Jain MS, Sopory SK, Veilleux RE (eds) In vitro haploid production in higher plants, vol 3. Kluwer, Dordrecht, pp 297–336

Baldursson S, Nørgard JV, Krogstrup P, Andersen SA (1993) Microspore embryogenesis in anther culture of three species of *Populus*, and regeneration of dihaploid plants of *Populus trichocarpa*. Can J For Res 23:1821–1825
Bhatnagar SP, Singh MN (1984) Organogenesis in the cultured female gametophyte of *Ephedra foliata*. J Exp Bot 35:268–278
Bueno MA, Gomez A, Boscaiu M, Manzanera JA, Vicente O (1997) Stress-induced formation of haploid plants through anther culture in cork oak (*Quercus suber*). Physiol Plant 9:335–341
Carneiro MF (1993) Induction of doubled haploids on *Coffea arabica* cultivars via anther or isolated microspore culture. In: Proc 15th Int Scientific Colloq on Coffee, Montpellier, France, 6–11 June, Association Scientifique du Cafe, vol 2, 902 pp
Cersosimo A, Crespan M, Paludetti G, Altamura AA (1990) Embryogenesis, organogenesis and plant regeneration from anther culture in *Vitis*. Acta Hortic 280:307–314
Chavez VM, Norstog K (1992) In vitro morphogenesis of *Ceratozamia hildae* and *C. mexicana* from megagametophytes and zygotic embryos. Plant Cell Tissue Organ Cult 30:93–98
Chen Z (1985) A study on induction of plants from *Citrus* pollen. Fruit Var J 39:44–50
Chen Z (1990) Rubber (*Hevea brasiliensis* Muell. Arg.). In vitro production of haploids. In: Bajaj YPS (ed) Biotechnology in agriculture and forestry, vol 12. Haploids in crop improvement I. Springer, Berlin Heidelberg New York, pp 215–236
Chen Z, Chen F, Chien C, Wang C, Chang S, Hsu HY, Ou H, He Y, Lu T (1978) Induction of pollen plants of *Hevea brasiliensis* Muell. Arg. (in Chinese with English summary). Acta Genet Sin 5:99–107
Chen Z, Wang H, Liao H (1980) The induction of *Citrus* pollen plants in artificial media. Acta Genet Sin 7:189–192
Chen Z, Qian C, Qin M, Xu X, Xiao Y (1982) Recent advances in anther culture of *Hevea brasiliensis* (Muell.-Arg.). Theor Appl Genet 62:103–108
Chen Z, Li W, Zhang L, Xu X, Zhang S (1987) Production of haploid plantlets in cultures of unpollinated ovule of *Hevea brasiliensis* Muell. Arg. In: Ahuja MR (ed) Somatic cell genetics of woody plants. Kluwer, Dordrecht, pp 39–42
Couturon E (1982) Obtention d'haploídes spontanes de *Coffea canephora* pierre par l'utilisation du greffage d'embryons. Café Cacao Thé 26:155–160
Favre-Duchartre M (1956) Contribution à l'étude de la reproduction chez le *Ginkgo biloba*. Rev Cytol Biol Veg 17(1–2):214
Fei KW, Xue GR (1981) Induction of haploid plantlets by anther culture in vitro in apple cv „Delicious" (in Chinese with English abstract). Chin Gar Sci 4:41–44
Geraci G, Starrantino A (1990) Attempts to regenerate haploid plants from in vitro cultures of *Citrus* anthers. Acta Hortic 280:315–320
Germanà MA (1997) Haploidy in *Citrus*. In: Jain MS, Sopory SK, Veilleux RE (eds) In vitro haploid production in higher plants, vol 5. Kluwer, Dordrecht, pp 195–217
Germanà MA, Chiancone B (2001) Gynogenetic haploids of *Citrus* after in vitro pollination with triploid pollen grains. Plant Cell Tissue Organ Cult 66:59–66
Germanà MA, Wang YY, Barbagally MG, Iannolino G, Crescimanno FG (1994) Recovery of haploid and diploid plantlets from anther culture of *Citrus clementina* Hort ex Tan. and *Citrus reticulata* Blanco. J Hortic Sci 69:473–480
Guha S, Maheshwari SC (1966) Cell division and differentiation of embryos in the pollen grain of *Datura* in vitro. Nature 212:97–98
Herrera JC, Moreno LG, Acuña JR, de Peña M, Osorio D (2002) Colchicine-induced microspore embryogenesis in coffee. Plant Cell Tissue Organ Cult 71:89–92
Hidaka T, Yamada Y, Shichijo T (1979) In vitro differentiation of haploid plants by anther culture in *Poncirus trifoliata* (L.) Raf. Jpn J Breed 29:248–254
Ho RH, Raj Y (1985) Haploid plant production through anther culture in poplars. For Ecol Manage 13:133–142
Hu Han, Xi Z, Ouyang J, Wang X (1982) Production of aneuploid and heteroploid of pollen-derived plants. Plant tissue culture. Maruzen, Tokyo, pp 421–424

Huhtinen O (1978) Callus and plantlet regeneration from anther cultures of *Betula pendula* Roth. In: Proc 4th Int Congr Plant Cell Tissue Culture, Calgary, Abstract, p 169

Höfer M (1995) In-vitro androgenesis in apple. Gartenbauwissenschaft 60:12–15

Isakov YN, Butorina AK, Muraya LS (1981) Discovery of spontaneous haploids in *Pinus silvestris* and the prospects of their using in forest genetics and selection. Genetika 17:701–707

Jörgensen J (1988) Embryogenesis in *Quercus petraea* and *Fagus silvatica*. J Plant Physiol 132:638–640

Jörgensen J (1991) Androgenesis in *Quercus petrea*, *Fagus silvatica* and *Aesculus hippocastanum* (abstract). In: Ahuja RM (ed) Woody plant biotechnology. Plenum Press, New York, pp 353–354

Kadota M, Han DS, Niimi Y (2002) Plant regeneration from anther-derived embryos of apple and pear. HortScience 37:962–965

Kiss J, Kondrak M, Torjek O, Kiss E, Gyulai G, Mazik-Tokei K, Heszky LE (2001) Morphological and RAPD analysis of poplar trees of anther culture origin. Euphytica 118:213–221

Konar RN, Singh MN (1979) Production of plantlets from the female gametophytes of *Ephedra foliata* Boiss. Z Pflanzenphysiol 95:87–90

LaRue CD (1948) Regeneration in the megagametophyte of *Zamia floridana*. Bull Torrey Bot Club 75:597–603

LaRue CD (1954) Studies on growth and regeneration in gametophytes and sporophytes of gymnosperms. Brookhaven Symp Biol 6:187–208

Lashermes P, Couturon E, Charrier A (1994) Doubled haploids of *Coffea canephora* – development, fertility and agronomic characteristics. Euphytica 74:149–157

Laurain D, Trémouillaux-Guiller J, Chénieux JC (1993) Embryogenesis from microspores of *Ginkgo biloba* L., a medicinal woody species. Plant Cell Rep 12:501–505

Lespinasse Y, Godicheau M (1980) Création et description d'une plante haploíde de pommier (*Malus pumila* Mill). Ann Amélior Plant 30:39–44

Li W, Cao ZY (1993) Origin of triploid plants from anther culture of *Vitis vinifera* var Grenache. Vitis 32:191–196

Lianfang F (1990) Litchi (*Litchi chienesis* Sonn.). In vitro production of haploid plants. In: Bajaj YPS (ed) Biotechnology in agriculture and forestry, vol 12. Haploids in crop improvement I. Springer, Berlin Heidelberg New York, pp 264–274

Marinkovic N, Radojevic L (1992) The influence of bud length, age of the tree and culture media on androgenesis induction in *Aesculus–Carnea* Hayne anther culture. Plant Cell Tissue Organ Cult 31:51–59

Mofidabadi AJ, Jorabchi A, Shahrzad S, Mahmodi F (2001) New genotypes development of *Populus euphratica* OLIV. Using gametoclonal variation. Silvae Genet 50:275–279

Neuenschwander B, Baumann TW (1995) Increased frequency of dividing microspores and improved maintenance of multicellular microspores of *Coffea arabica* in medium with coconut milk. Plant Cell Tissue Organ Cult 40:49–54

Nitsch JP, Nitsch C (1969) Haploid plants from pollen grains. Science 163:83–87

Norstog K (1965) Induction of apogamy in megagametophytes of *Zamia integrifolia*. Am J Bot 52:993–999

Oiyama II, Kobayashi S (1993) Haploids obtained from diploid × triploid crosses of *Citrus*. J Jpn Soc Sci 62:89–93

Radojevic L (1978) In vitro induction of androgenic plantlets in *Aesculus hippocastanum*. Protoplasma 96:369–374

Radojevic L, Djordjevic N, Tucic B (1989) In vitro induction of pollen embryos and plantlets in *Aesculus carnea* Hayne through anther culture. Plant Cell Tissue Organ Cult 17:21–26

Raghuramulu Y, Prakash NS (1996) Haploidy in coffee. In: Mohain JS, Sopory SK, Veilleux RE (eds) In vitro haploid production in higher plants, vol 3. Kluwer, Dordrecht, pp 349–363

Sato T (1974) Callus induction and organ differentiation from anther culture in vitro of poplar. J Jpn For Soc 56:55–62

Seirlis G, Mouras A, Salesses G (1979) Tentatives de culture in vitro d'anthères et de fragments d'oranges chez les *Prunus*. Ann Amélior Plant 29:145–161

Singh MN, Konar RN, Bhatnagar SP (1981) Haploid plantlet formation from female gametophytes of *Ephedra foliata* Boiss. in vitro. Ann Bot 48:215–220
Simola LK, Honkanen J (1983) Organogenesis and fine structure in megagametophyte callus lines of *Picea abies*. Physiol Plant 59:551–561
Sondahl MR, Lauritis JA (1992) Coffee. In: Hammerschlag FA, Litz RE (eds) Biotechnology in agriculture, no 8. Biotechnology of perennial crops. CAB International, Wallingford, pp 401–420
Sondahl MR, Sharp WR (1979) Research in *Coffea* spp. and applications of tissue culture methods. In: Sharp WR, Larsen PO, Paddock EF, Raghaven V (eds) Plant cell and tissue culture, principles and applications. Ohio University Press, Columbus, Ohio, pp 527–584
Sreenivasan MS, Mamachandran M, Sundar KR (1981) Frequency of polyploids in *Coffea arabica* L. In: Vishveshwara S (ed) Proc 4th Annu Symp on Plantation Crops, Mysore, India. Placrosym 4:23–28
Sreenivasan MS, Ram AS, Prakash NS (1993) Tetraploid interspecific hybrids in coffee breeding in India. In: Proc 15th Int Sci Colloq on Coffee, vol 1, ASIC, Paris, pp 226–233
Stettler RF, Bawa KS (1971) Experimental induction of haploid parthenogenesis in black cottonwood. Silv Genet 20:15–25
Stoehr MU, Zsuffa L (1990a) Induction of haploids in *Populus maximowiczii* via embryogenic callus. Plant Cell Tissue Organ Cult 23:49–58
Stoehr M, Zsuffa L (1990b) Genetic evaluation of haploid clonal lines of a single donor plant of *Populus maximowiczii*. Theor Appl Genet 80:470–474
Tsay HS, Su CY (1985) Anther culture of papaya (*Carica papaya* L.). Plant Cell Rep 4:28–30
Tulecke W (1953) A tissue derived from the pollen of *Ginkgo biloba*. Science 117:599–600
Uddin MR, Meyer MM, Jokela JJ (1988) Plantlet production from anthers of eastern cottonwood (*Populus deltoides*). Can J For Res 18:937–941
Von Aderkas P, Bonga JM (1988) Formation of haploid embryoids of *Larix decidua*: early embryogenesis. Am J Bot 75:690–700
Von Aderkas P, Bonga JM (1993) Plants from haploid tissue of *Larix decidua*. Theor Appl Genet 87:225–228
Von Aderkas P, Dawkins MD (1993) Haploid embryogenesis in trees. In: Ahuja MR (ed) Micropropagation of woody plants. Kluwer, Dordrecht, pp 58–65
Von Aderkas P, Klimaszewska K, Bonga JM (1990) Diploid and haploid embryogenesis in *Larix leptolepis*, *L. decidua* and their hybrids. Can J For Res 20:9–14
Von Aderkas P, Klimaszewska K, Owens JN, Bonga JM (1991) Comparison of larch embryogeny in vivo and in vitro. In: Ahuja MR (ed) Woody plant biotechnology. Plenum Press, New York, pp 139–155
Wang C, Chu Z, Sun C (1975) The induction of *Populus* pollen-plants. Acta Bot Sin 18:56–62
Wu JY (1981) Induction of haploid plants from anther culture of crab apple (in Chinese with English abstract). J Northeast Agric Coll 3:105–108
Wu K, Nagarajan P (1990) Poplars (*Populus* spp). In: Bajaj YPS (ed) Biotechnology in agriculture and forestry, vol 12. Haploids in crop improvement I. Springer, Berlin Heidelberg New York, pp 237–249
Zhang YX, Lespinasse Y (1991) Pollination with gamma-irradiated pollen and development of fruits, seeds and parthenogenetic plants in apple. Euphytica 54:101–109
Zou C, Li P (1981) Induction of pollen plants of grape (*Vitis vinifera* L.). Acta Bot Sin 23:79–81

II.5 Haploids in the Improvement of Miscellaneous Crop Species (Cucurbitaceae, Liliaceae, Asparagaceae, Chenopodiaceae, Araceae and Umbelliferae)

ANIKO GÉMES JUHÁSZ[1] and MARIJANA JAKŠE[2]

1 Introduction

Over the last 20 years, one of the most intensive fields of research in plant biotechnology has been the widespread application of the in vitro haploid methods based on artificial sporophytic development from gametes. In vitro haploid methods involve the use of artificially induced stress (e.g. cold/heat, chemical factors) to influence the natural developmental and differentiation processes of male and female gametes. As a result of these stress factors, the development of the gametes is diverted from the gametophytic path of development to the sporophytic path, leading to the formation of androgenetic or gynogenetic embryos or morphogenic callus (Keller and Korzun 1996; Sopory and Munshi 1996). Doubled haploid plants originating from gametes carry the genetic information of only one set of chromosomes, so they can be regarded as genetically homozygotic.

The sporophytic developmental pathway starting from immature male gametes is known as in vitro androgenesis, while that starting from female gametes is referred to as in vitro gynogenesis. The androgenetic or gynogenetic responsiveness of a given genotype is determined by factors such as genotype, donor plant growth conditions and culture conditions.

The majority of the plant regenerants arising from in vitro andro- or gynogenesis will be haploid with regard to their genetic background, but plants with other ploidy levels may also develop spontaneously. In addition to haploid plants the most frequently found forms are spontaneous diploids. In general these also originate from a single micro- or macrospore, but their genome is doubled, e.g. in the course of spontaneous chromosome duplication (endoreduplication or nuclear fusion) in the early phase of in vitro culture. The plants formed in this way can be regarded as doubled haploids. Due to their sterility, haploid plants are of no value in plant breeding, but if their genome is doubled by means of in vivo or in vitro diploidization, valuable induced doubled haploids can be obtained.

[1] Vegetable Crops Research Institute, Budapest, 1775 Pf 95, Hungary
[2] Agronomy Department, Biotechnical Faculty, University in Ljubljana, Jamnikarjeva 101, 1000 Ljubljana, Slovenia

Biotechnology in Agriculture and Forestry, Vol. 56
Haploids in Crop Improvement II (ed. by C.E. Palmer, W.A. Keller, and K.J. Kasha)

In this chapter, we discuss the success of haploid induction of Cucurbitaceae, Liliaceae, Asparagaceae Chenopodiaceae, Araceae and Umbelliferae species.

2 Haploid Induction in Cucurbitaceae Species

Since the first reports published by Dumas de Vaulx (1979), who induced the first haploid plants of the Cucurbitaceae by interspecific crossing between *C. melo* L. and *C. ficifolius* A. Rich, most research has focused on in situ haploid induction using irradiated pollen, but as yet very few results have been achieved.

2.1 In Situ Induced Parthenogenesis

The stimulation of haploid parthenogenesis through pollination with irradiated pollen was successfully applied to haploid production in melon (Sauton and Dumas de Vaulx 1988a,b; Sauton 1989; Cuny 1992; Cuny et al. 1993; Ficcadenti et al. 1995; Lotfi et al. 2003), watermelon (Gürsöz et al. 1991; Sari et al. 1994), cucumber (Troung-Andre 1988; Niemirowicz-Szczytt and Dumas de Vaulx 1989; Sauton 1989; Przyborowski 1994; Caglar and Abak 1999; Lotfi et al. 1999) and squash (Kurtar et al. 2002). Sauton (1989) reported that the pollination season, the irradiation doses of gamma rays and the maternal genotypes significantly affected the frequency of haploid embryo induction. The haploid production rate reported by Sauton (1989) was 0.3 gynogenic embryos per 100 seeds. Przyborowsky (1994) published a better result (1.34 gynogenic embryos per 100 seeds), with 51% of the embryos being capable of further development. When Caglar and Abak (1999) tested 27 cucumber genotypes, the highest frequency of haploids (an average 3.8 embryos/fruit) was induced by low doses of irradiation (200 or 300 Gy). This team also tested the effect of various colchicine concentrations and the treatment times on the doubling process of haploids and found that 0.5% colchicine for 4 h was the most effective, leading to a doubling of the chromosome number in 60% of the treated plants on average (Caglar and Abak 1997).

2.2 In Vitro Haploid Method

The first production of haploids from unfertilized ovules of summer squash (*Cucurbita pepo* L.) was reported by Dumas de Vaulx and Chambonnet (1986). Shail and Robinson (1987) and Metwally et al. (1998a,b) also studied in vitro haploid induction and obtained haploid plants both from anther culture and unpollinated ovules of *Cucurbita pepo*.

In the case of courgettes and cucumber an effective method was established for the development of haploid plants through in vitro gynogenesis (Gémes Juhász et al. 1996, 2002a).

2.2.1 Factors Affecting Gynogenesis

In the case of *Cucurbita pepo*, ovaries harvested 1 day before anthesis gave the largest number of embryos (Dumas de Vaulx and Chambonnet 1986; Metwally et al. 1998a), while in cucumber the most embryos were found in ovaries harvested 6 h before anthesis (Gémes Juhász et al. 2002a).

According to histological studies on the female gametophyte, the best developmental stage for haploid induction seems to be the cellularization stage of embryo sac formation, when the nuclei are already in position, the membranes have sometimes developed, and the cells are fairly uniform in shape and structure. The most responsive ovaries (ovules) had nearly mature or fully mature embryo sacs. Haploid induction from megaspores or the early stages of the embryo sac can be precluded (Gémes Juhász et al. 2002a).

The effect of cold pretreatment and of heat treatment during the induction of ovaries was also studied. Metwally et al. (1998a) found that haploid induction of summer squash ovaries was more effective without cold pretreatment.

In contrast, for cucumber ovaries, the use of heat treatment (at 28 °C or even better at 35 °C) during the induction phase in the presence of growth regulator (thidiazuron) increased the rate of haploid embryo formation and plant regeneration. Correlation analysis of the effect of the temperature and treatment duration on induction and regeneration demonstrated a close linear correlation between the induction temperature and both embryo induction and plant regeneration. It was found that within the 24–35 °C temperature range a unit (1 °C) rise in temperature had a specific effect on both embryo induction and plant regeneration (Gémes Juhász et al. 2002a).

Various authors have used different culture media. Metwally et al. (1998a) applied MS (Murashige and Skoog 1962) medium supplemented with 1–5 mg/l 2,4-dichlorophenoxyacetic acid (2,4-D) and 30 g/l sucrose during both induction and regeneration. In the case of ovaries of cucumber and courgette, basal medium (CBM) supplemented with 0.02 mg/l TDZ and 4% sucrose was used for initiation. After the induction period the material was transferred to regeneration media (CBM) containing 0.05 mg/l NAA, 0.2 mg/l BA and 3% sucrose (Gémes Juhász et al. 1996, 2002a).

Cytological studies on the regenerants showed that the frequency of haploid regenerants among the gynogenic plants was highly variable. Good yields of haploids have been reported for cucumber and courgette (Gémes Juhász et al. 2003), while Metwally et al. (1998a) reported that only one third of squash regenerants were haploid.

In cucumber the highest plant regeneration rate was 7.1 plants per 100 cultured ovaries (Gémes Juhász et al. 2002a) and in summer squash 11 plants per 100 cultured ovules (Metwally et al. 1998a).

2.2.2 Studying of Chromosome Doubling

Chromosome doubling procedures in the case of cucumber and courgette are based on in vitro colchicine treatments of the young regenerants. Haploid plantlets (checked by flow cytometry) were cultured on medium containing 0.02% colchicine for 2–4 days. Genome doubling was screened also by flow cytometry. Although the 4-day colchicine treatment significantly increased the number of doubled haploids compared with 2-day treatment, only 24.1% of the 58 treated haploid plants were found to have been diploidized; 10.3% mixoploids, which contained both haploid and diploid cells, indicates that genome duplication was not complete (Gémes Juhász 2003).

2.2.3 Factors Affecting Androgenesis

Androgenesis induction through in vitro anther culture has been studied in cucumber (Lazarte and Sasser 1982), muskmelon and cucumber (Dryanovska and Ilieva 1983) and squash (Shail and Robinson 1987). All of them obtained haploid callus which yielded haploid plants at a very low frequency.

In the case of *Cucumis melo* Metwally et al. (1998b) reported a more effective method when cold pre-treatment (4 °C for 4 days) male flower (with mid- or late uninucleate stage of microspores) were put subsequently on the MS (Murashige and Skoog 1962) medium supplemented with 5 mg/l 2,4-D and 150 g/l sucrose.

3 Haploid Induction in Liliaceae Species

3.1 *Allium* Species

In the genus *Allium*, only gynogenesis has been used effectively to produce haploid plants, since efforts in the anther culture of onion failed (Campion et al. 1984; Keller 1990).

In recent years gynogenically derived haploid plants have been successfully produced in onion and shallot, and some attempts have been made in leek (Smith et al. 1991; Schum et al. 1993). Most publications have focused on onion gynogenesis, since this is the most important species within the genus. The procedure for obtaining haploid onion plants has been considerably improved since the first reports of gynogenesis in onion were published (Campion and Azzimonti 1988; Muren 1989; Keller 1990). The improvements were based on the choice of starting material, growth conditions, the medium composition and genotype.

3.1.1 Choice of Starting Material and Their Growth Conditions

Gynogenesis can be achieved by culturing three different organs: unpollinated ovules, ovaries or whole flower buds. Ovule culture was tested in early studies (Campion and Alloni 1990; Keller 1990; Campion et al. 1992; Bohanec et al. 1995) but is now considered the most laborious and the least efficient. As it has been shown that ovary or flower cultures yield more embryos with less work, no reports of ovule culture have been reported since 1995.

The simplest way of inducing gynogenic haploids in onion is from flower bud culture, which has been used in many recent studies (Cohat 1994; Gémes Juhász and Martinovich 1995; Geoffriau et al. 1997a; Martinez et al. 1997, 2000; Javornik et al. 1998; Bohanec and Jakše 1999; Michalik et al. 2000). The flower buds can either remain on the induction medium until the regeneration of embryos (60–120 days) or be transferred to fresh regeneration medium after several weeks in culture.

The size of the flowers used for inoculation varies, but in general it has been concluded that small (young) buds (2.8–3.0 mm long) produce significantly fewer embryos than older (3.5–4.5 mm long) ones, though there is considerable genotype specificity (Michalik et al. 2001). The ovaries should be excised from the flowers 3–5 days before anthesis (Muren 1989) or, according to Michalik et al. (2000), 2–3 days before anthesis.

The conditions for donor plant growth, in particular the use of low temperature, are often of great importance for the success of haploid induction procedures in many plant species. Puddephat et al. (1999) reported a ten-fold increase in yield of embryos when flower buds were harvested from donor plants raised in growth chambers at 15 °C, compared to 10 °C or to ambient conditions (24 °C) in a glasshouse. This result was confirmed in studies carried out by Michalik et al. (2001), who obtained the highest induction frequencies when flower buds were collected from plants that were grown in a growth chamber with a constant temperature of 14 °C.

As noted by Smith et al. (1991) and Cohat (1994), in the case of shallots anthers do not dehisce within the culture vessel, so selfing (formation of zygotic embryos) is not possible. The only disadvantage of whole flower bud culture over ovary culture is the possible growth of basal callus, which may be formed from the septal nectaries region, so there is an increased possibility of somatic regeneration from the callus. Gynogenic embryos arise from the top of the ovary and are mainly seen as loop structures. They are clearly distinguishable from somatic regenerants which on very rare occasions proliferate at the flower base (Gémes Juhász and Martinovich 1998; Bohanec and Jakše 1999).

3.1.2 Medium Composition

The best medium composition for the induction and regeneration stages was also extensively studied. The three most often used combinations of macro- and microelements are B5 (Gamborg et al. 1968), BDS (Dunstan and Short

1977) and MS (Murashige and Skoog 1962). The available studies do not compare media, but it seems that the three basal media have a similar effect on culture development and yield. The majority of haploid induction media for onion include 10% sucrose, which might well be one of the most important factors. Other substances, in particular several plant growth regulators, have been tested in various combinations for optimal concentration, duration of treatment and combination effects. Muren (1989) applied 2,4-D at 2 mg/l and benzylaminopurine (BAP) at 2 mg/l in the culture medium and this combination has since been approved by several other researchers and has become the standard composition of growth regulators used in onion gynogenic regeneration. In most studies, agar was the gelling agent. An increase in embryo yield was recorded by Jakše et al. (1996) when gellan-gum was used instead, but a higher proportion of hyperhydric regenerants was noted. Gellan-gum has also been reported to promote somatic regeneration from onion flower buds or ovaries (Luthar and Bohanec 1999). Often the efficiency of various medium compositions is difficult to compare, since one of the most decisive factors is genotype, which was different in most published studies.

3.1.3 Genotypic Effects

The genetic composition of the donor plants was found to be the most important factor for the success of gynogenesis. In the first experiments, yields were low, ranging from 0–3% for different genotypes. Later studies focused on more variable genetic material from different world regions. In a 3-year study, Geoffriau et al. (1997a) analysed 18 onion cultivars and populations from eastern, northern and southern Europe and four from the USA. The best yield for one cultivar was 17.4% in an optimal year. In a similar study, Bohanec and Jakše (1999) analysed 39 accessions from Europe, North America and Japan. Two European and three Japanese accessions produced no embryos, and the highest gynogenic yield was obtained from North American cultivars and inbred lines. Two inbred lines and one F_1 hybrid produced up to 22.6 embryos per 100 cultured flowers. Very high variability was found within cultivars, and even within inbred lines. When single plants were induced to flower in two consecutive years, the variation in gynogenic yield within plants between seasons was much lower than that recorded between individuals of the same line (Bohanec et al. 2003). Michalik et al. (2000) scored 11 Polish onion cultivars and 19 breeding lines for gynogenetic potential. The majority of the tested genotypes produced a low embryo yield, except for breeding line '601A' which had 10.0% embryo yield. Javornik et al. (1998) cultured flowers from selfed plants of three doubled haploids, in order to generate a second cycle of haploid plants. Only one line produced a very high yield, with a mean number of 118.3 haploid embryos per 100 cultured flowers, while the other two lines yielded only 2.3 and 0.3% haploid embryos.

The results show that genes coding for low or high gynogenic potential are present in gynogenic regenerants. This and more recent results (Bohanec et al. 2003) have confirmed that the genetic variability in gynogenesis is much higher than that resulting from culture conditions. It has been demonstrated that crossing of responsive and non-responsive onion lines resulted in increased gynogenic ability in the hybrid progeny (Bohanec et al. 1999).

3.1.4 Doubling Procedures

Unlike many other horticultural plants, spontaneous diploidization in onion is less than 10%, and according to Jakše and Bohanec (2001) the haploid ploidy level is very stable.

Comparative studies in two-step ovary cultures of onion proved that the application of 0.1 or 0.2 mg/l thidiazuron (TDZ) in the induction phase led to an increase in the ratio of spontaneous diploids and the genome analysis revealed the plants included haploids, diploids and mixoploids in the ratio 63.8% haploid, 31.4% diploid and 4.8% mixoploid. The mixoploid plants were haploid and diploid genome mosaics (Gémes Juhász 2003).

The factor that prevents doubling in later plant developmental stages is the inaccessibility of the apical meristem, with the consequence that chromosome doubling in field-grown onion plants is not possible. As a result, all chromosome doubling procedures in onion are based on in vitro treatments on explants. The majority of approaches use sliced basal parts from the shoots of in vitro elongated or micropropagated plantlets, which are treated with chromosome doubling agents (colchicine, oryzalin or trifluralin) of variable concentration and duration of exposure (Campion et al. 1995b; Geoffriau et al. 1997b; Bohanec and Jakše 1997; Nowak 2001). The major limitation of this approach is the prolonged in vitro treatment of the regenerants, causing possible losses due to their variable response to the plant growth regulators applied to stimulate shoot regrowth and other factors. In two micropropagated lines, Campion et al. (1995b) reported that 3-day exposure of shoot apices to 25 mM colchicine induced 46% doubled haploids. Similarly, Geoffriau et al. (1997b) used three micropropagated haploid clones and reported that 2.5 mM colchicine or 50 μM oryzalin resulted in 65.7 or 57.1% doubled haploids, respectively. A more recent figure was given by Nowak (2001), who reported that 72% of 419 plants did not regenerate after chromosome doubling treatments, while the remaining treated plantlets also showed extensive thickening and hyperhydration. An alternative approach, based on treatment of embryos immediately after regeneration, has been proposed (Jakše and Bohanec 2001). This approach shortens the in vitro stage by about 2–3 months, is less laborious and shows similar efficiency. A possible drawback of treating of embryos might be that the ploidy level of the embryos is not analysed prior to treatment; therefore, spontaneous diploids are treated together with haploid embryos. In these experiments, the embryos were treated with amiprophosmethyl (APM) at 10 or 50 μM, or oryzalin (10 or 50 μM) in liquid or on solid media. In general, APM was efficient and less toxic than oryzalin. Recent results (Jakše et al. 2003) showed that the highest proportion of doubled haploid plants was achieved in liquid medium at an APM concentration of 50 μM (25.3 and 36.7% after 1 or 2 days' exposure, respectively).

The low toxicity of the doubling agents is of crucial importance, since the survival rate of regenerants is only about 50% even if they are not submitted to doubling treatment (Campion et al. 1992; Puddephat et al. 1999; Martinez et al. 2000; Michalik et al. 2000). Doubling treatments cause additional losses, diminishing the number of surviving regenerants to 20–30% of the number induced.

The results of diploidization are mainly measured after 2–3 months, when the explants or embryos generate a few young leaves. When the ploidy level was measured by flow cytometry, mixoploid plants were frequently observed. Apart from 'n' and '2n' plants, mixoploid 'n+2n' and '2n+4n' plants have been detected even in control plants, where no doubling agents were used. It is not clear so far whether mixoploid regenerants develop into partially fertile diploid plants or remain haploid, as suggested by Geoffriau et al. (1997b).

3.2 *Lilium* Species

Gu and Cheng (1983) and Prakash and Giles (1986) reported successful gynogenesis in ovary culture of lily. While Gu and Cheng (1983) obtained haploid callus, Prakash and Giles (1986) showed direct embryogenesis in lily.

Haploid culture of tulip (*Tulipa gesneriana* L.) through microspore embryogenesis was first published by van den Bulk et al. (1994). Custers et al. (1997) reported that high temperature pretreatment (32 °C) of the bulbs and using early unicellular microspores had a positive effect on the embryo production. The improved protocol proved to be effective in various tulip cultivars.

4 Haploid Induction in Asparagaceae

In asparagus, pure lines cannot be obtained through conventional methods, because the cultivated *Asparagus officinalis* is a dioecious species with male and female plants. Nevertheless hybridization is easily obtained with such plants. The first step to improve rather non-uniform (heterozygous) populations was obtained through the production of hybrid varieties with few heterozygous selected parents, which resulted in double hybrids from four different parents or in clonal hybrids from two, in vitro-propagated parents. Varieties of this type have been developed in France and the USA (since 1974).

4.1 Anther Culture

Later improvements were based on producing haploids through in vitro anther culture, which leads to a production of female and male homozygous plants. Sex expression in asparagus is inherited as though controlled by a single-gene factor dominant for maleness (Ellison 1986) which allowed creation of "supermale" hybrids. Compared to female plants, male plants have better earliness, have better yield (more spears per plant) and longer life, which are important and long-known characteristics (Sneep 1953; Ellison et al. 1960). Since asparagus is a perennial crop, advantages for growers are also that male plants produce no seeds to make seedling weeds in the field that can compete with the storage roots for photosynthate, and the advantage for seed companies is that growers cannot save their own seeds and ruin the genetic quality of the cultivar (Elisson 1986).

The development of anther culture production was reviewed by Doré (1990), where it is pointed out that, among the different factors that affect androgenesis, the selection of the genotype is the most important since the aptness to develop calli versus embryoids of androgenetic origin, the capacity of a callus to regenerate shoots, and the ability to establish a crown through microcutting are all genotype-dependent characteristics. The androgenetic embryo yield varied from 0.1–40% according to genotypes; on average 2.4 embryos per 100 anthers were obtained (Falavigna et al. 1985).

A research group in northern Italy has been working with an asparagus breeding program based on in vitro culture of anthers since 1974. In the period 1976–86 the anthers were cultured in vitro following the procedures described by Doré (1974); subsequently, a more efficient method was applied (Qiao and Falavigna 1990), where a white and solid embryonic tissue, similar to that of the crown, was selected and transferred into "rooting' medium with the addition of 0.2 g/l NAA, 0.1 g/l kinetin and 0.5 g/l ancymidol, which improved the rooting and shortened the regeneration and acclimatization of anther-derived plants. Studies demonstrated that different gelling agents also did not significantly improve embryo yield (Qiao and Falavigna 1990). The main problems encountered by the Italian group during the experimental period of asparagus androgenesis were (Falavigna et al. 1999):

- Insufficient embryo yield for several asparagus genotypes
- Low percentage of in vitro plant regeneration from embryos
- Possible regeneration of heterozygous males (somatic origin)
- Inability to distinguish in vitro diploid from polyploid genotypes on the basis of morphological traits
- 41% of regenerated clones were lost because they were not diploid
- Long (several years) evaluation of clones in the field
- Difficult micropropagation of some doubled haploid clones parents of good hybrids

The physiological status of the donor plant must also be considered. Feng and Wolyn (1991) discovered that anthers from winter-grown, greenhouse flowers produced less total and embryogenic calli than those from field-grown flowers. They also studied different incubation temperatures and reported that the initial incubation of anthers at high temperatures (29–35 °C) induced a higher proportion of embryogenic callus (Feng and Wolyn 1991; Wolyn and Feng 1993), but embryo formation was asynchronous and abnormal. Anthers with microspores at the late-uninucleate stage had a three times higher frequency of forming embryogenic calli than that from anthers with microspores at all other stages of development.

Low (2 %) sucrose concentration in the induction medium induced embryo formation. The number of bipolar embryos decreased when the sucrose concentration increased. At the high (4–6 %) sucrose levels, abnormal embryo structures and callus production predominated (Ziauddin et al. 1996).

4.2 Microspore Culture

In 1994, the earliest reports of asparagus microspore culture were published (Feng and Wolyn 1994; Zhang et al. 1994), where the yields of calli and haploid or doubled haploid asparagus plants were low. A cold pretreatment of asparagus flower buds for 7 days induced symmetric rather than asymmetric mitosis-I in microspores and significantly improved the culture response (Peng et al. 1997). Peng and Wolyn (1999) established that the microculture response was also genotype dependent and that microspores in culture were very sensitive to damage by mechanical isolation, so they used shed microspores. They improved the frequency of regeneration, but all plants were regenerated through a callus phase, rather than direct embryogenesis. Early embryogenic structures were observed, but died or dedifferentiated to calli rather than developing into functional embryos. Further research is required to successfully develop mature embryos from asparagus microspores.

5 Haploid Induction in Chenopodiaceae

Most research in the family of Chenopodiaceae has focused on sugar beet haploid production. Classical techniques such as natural polyembryony and crosses between diploid and tetraploid lines or with wild species have yielded very low numbers of haploids. In vitro androgenesis has been attempted numerous times but with very limited success (van Geyt et al. 1987).

Gynogenesis has proven to be the most successful method for haploid production (Bossoutrot and Hosemans 1985; van Geyt et al. 1987; Doctrinal et al. 1989; Lux et al. 1990; Ferrant and Bouharmont 1994; Hansen et al. 1994, 1995, 1998; Gürel et al. 2000) and the only approach to produce haploids of male sterile lines. Ferrant and Bouharmont (1994) induced gynogenesis either from unopened flowers of male-fertile or from opened flowers of male-sterile mother plants. They concluded that the highest number of haploid plants could be obtained when using male-fertile donors.

5.1 Ovary Donors and Their Growth Conditions

Certain physical treatments (e.g. low or high temperature) applied to the donor plants or in vitro culture may have a strong influence on embryo induction. In *Beta vulgaris* L., 4–5 days of cold pretreatment of flower buds at 4 °C increased the embryo yield from cultured ovules (Lux et al. 1990; Gürel et al. 2000).

5.2 Effect of Stage and Origin of Ovules

The stage of female gametophyte (embryo sac) at the time of culture is very important. In the case of sugar beet, the ovaries or ovules are cultured at the mature embryo sac stage. Results of Ferrant and Bouharmont (1994) suggest that this maturity stage of the gametophyte is reached a few days before anthesis. This result is in contrast to results by Bossoutrot and Hosemans (1985) who showed that in vitro culture of sugar beet ovules was only feasible with mature embryo sacs from open flowers.

In sugar beet, the position of the ovules on the donor plants had a marked effect on their gynogenic response. The ovules from the apex of the shoots were 80 % less responsive than those from the lateral branches, and the lower branches provided the most productive ovules (D'Halluin and Keimer 1986). In this plant, even the orientation of the ovule on the medium affected response. The ovules were placed with their funiculus in contact with the medium showed best response (Pedersen and Keimer 1996).

Sugar beet gynogenic regenerants mostly originate from the egg cell (Goska et al. 1990; Ferrant and Bouharmont 1994; Pedersen and Keimer 1996) and rarely from synergids (Goska et al. 1990).

5.3 Medium Composition

The composition of the culture medium is of special importance for successful gynogenesis; in particular the content of cytokinins is crucial. Embryo yield increases with increased BAP content in the medium. The content of auxins is of minor importance (Lux et al. 1990).

The addition of charcoal to the culture medium increased the frequency of embryo formation (D'Hallum and Keimer 1986; Gürel et al. 2000) whereas $AgNO_3$ decreased or completely inhibited the haploid induction (Gürel et al. 2000).

5.4 Doubling Procedure

Different antimitotic agents were tested (amiprophosmethyl, colchicine, pronamide, trifluralin, oryzalin) on sugarbeet double haploid production. In ovule culture of sugarbeet and fodder beet, amiprophosmethyl (APM) proved to be the best doubling agent of the four chemical agents (amiprophosmethyl, pronamide, trifluralin, oryzalin) investigated, resulting in a mean of 4.7 doubled haploid plants per 100 ovules (Hansen et al. 1998).

Gürel et al. (2000) described another method of treating young haploid shoots with colchicine or trifluralin where colchicine was more effective (25.3 %) than trifluralin (18.2 %), but the difference was not significant. The uptake of both agents was better when used in liquid rather than agarose-solidified or agar-solidified media. The treatment duration of 48 h proved to be more effective than 12 h, but it was not different from 24 or 36 h.

6 Haploid Induction in Araceae

In *Spathiphyllum wallisii* gynogenic plants have been produced from the excised ovules. Eeckhaut et al. (2001) showed that the use of imidazole fungicides IMA (imazalil), PRO (prochloraz) or TRI (triflumizole) during ovary culture was not critical, although they enhanced the swelling of ovules during ovary culture and made the isolation and transfer of ovules to media easier. After 6 weeks on the TDZ-containing initiation medium, the white enlarged ovules were transferred to zeatin-containing medium where embryo clusters were formed. AFLP analysis showed that few of the plants originated from haploid cells, while the majority proved to be somaclonal regenerants.

7 Haploid Induction in Umbelliferae

In the Umbelliferae family only few reports connected with haploid production have been found. The first two haploid plants of carrot were obtained by irradiated pollen-induced parthenogenesis (Rode et al. 1987).

Andersen et al. (1990) reported a successful protocol of carrot (*Dacus carota* L.) androgenesis. They concluded that preculture treatment (1–2 days at 7 °C) of carrot umbels enhances androgenesis in anther culture. They cultured carrot anthers when microspores were at the mid-uninucleate stage of microsporogenesis. High sucrose concentration (10 %) during the initial phase and lower sucrose level (2 %) in the post-induction stage resulted in the best anther response (calli and embryo production) and later the best embryo development (Andersen et al. 1990). The presence of an appropriate concentration of growth regulators in the medium plays an important role in callus or embryo induction and embryo development. In carrot, calluses were formed on media containing 0.01–1 mg/l 2,4-D and 0–1 mg/l kinetin, while embryoids were formed on media supplemented with 1 mg/l 2,4-D (Hu et al. 1993). In celery, callus developed from tetrad and early uninucleate microspores on B5 medium supplemented with 2,4-D and adventitious embryos regenerated from calluses on BA- and NAA-containing medium or hormone-free medium (Dohya et al. 1997).

8 Conclusion

The protocols for anther or ovary/ovule culture are rather similar. The most important difference probably concerns the developmental stage of the gametophyte. While androgenic induction generally requires late stage of uninucleate microspores, the ovaries or ovules containing nearly mature embryo sacs seem to be the most suitable for gynogenesis (San and Demarly 1984; Bhojwani and Thomas 2001). The yield of both techniques depends strongly upon the genotype of the donor plants and can be enhanced by physical factors such as thermal pretreatment (Sopory and Munshi 1996; Gémes Juhász et al. 2002a) or slight irradiation (Sangwan and Sangwan 1986; Cuny et al. 1993) before in vitro culture of gametophytes.

At present, the number of species responsive to androgenesis is over ten times higher than those responsive to gynogenesis; however, there are families such as Chenopodiaceae and Liliaceae where ovary culture is quite efficient, or in some species only gynogenesis offers the possibility for haploid production, e.g. Cucurbitaceae, where androgenic response was not successful.

The successful utilization of haploids in research depends not only on the production of large numbers of haploids, but also on the efficient doubling of chromosomes to reach homozygosity and genetic stability (Rao and Suprasanna 1996). Several reports (Gémes Juhász 2003; Jakše et al. 2003) have indicated that the chromosome duplication mainly of gynogenesis-originated plants is not resolved, so great efforts are needed for this work in the future.

References

Andersen SB, Christiansen I, Farestveit B (1990) Carrot (*Dacus carota* L.): in vitro production of haploids and field trials. In: Bajaj YPS (ed) Haploids in crop improvement. Biotechnol Agric For 12:393–402

Bohanec B, Jakše M (1997) Characteristics of onion haploid induction procedure. In: Proc 1st Congr of the Genetics Society of Slovenia, 2–5 Sept, Ljubljana, Slovenia, pp 47–49

Bohanec B, Jakše M (1999) Variation in gynogenic response among long-day onion (*Allium cepa* L.) accessions. Plant Cell Rep 18:737–742

Bohanec B, Jakše M, Ihan A, Javornik B (1995) Studies of gynogenesis in onion (*Allium cepa* L.): induction procedures and genetic analysis of regenerants. Plant Sci 104:215–224

Bohanec B, Jakše M, Havey MJ (1999) Effects of genotype on onion gynogenesis and attempts at genome doubling at embryo stage – a progress report. In: Proc Gametic Embryogenesis in Monocots, COST-824 Worksh, 10–13 June, pp 37–38

Bohanec B, Jakše M, Havey MJ (2003) Genetic analyses of gynogenetic haploid induction in onion. J Am Soc Hortic Sci 128(4):571–574

Bossoutrot D, Hosemans D (1985) Gynogenesis in *Beta vulgaris*: from in vitro culture of unpollinated ovules to the production of doubled haploid plants in soil. Plant Cell Rep 4:300–303

Bhojwani SS, Thomas TD (2001) In vitro gynogenesis. In: Bhojwani SS, Soh WY (ed) Current trends in the embryology of angiosperms. Kluwer, Dordrecht, pp 489–507

Caglar G, Abak K (1997) In vitro application of haploid cucumber plants. Cucurbit Genet Coop Rep 20:21–23

Caglar G, Abak K (1999) Progress in the production of haploid embryos, plants and doubled haploids in cucumber (*Cucumis sativus* L.) by gamma irradiated pollen in Turkey. Acta Hortic 492:317–322

Cai DT, Chen DT, Zhu H, Jin Y (1988) In vitro production of haploid plantlets from the unfertilized ovaries and anthers of Hubei Photosynthetic Genic Male Sterile Rice (HPGMR). Acta Biol Exp Sin 21:401–407

Campion B, Alloni C (1990) Induction of haploid plants in onion (*Allium cepa* L.) by in vitro culture of unpollinated ovules. Plant Cell Tissue Organ Cult 20:1–6

Campion B, Azzimonti MT (1988) Evolution of ploidy level in haploid plants of onion (*Allium cepa* L.) obtained through in vitro gynogenesis. In: Proc 4th Eucarpia Allium Symp, Wellesbourne, UK, 6–9 Sept, pp 85–89

Campion B, Falavigna A, Soressi GP, Schiavi M (1984) Efforts for in vitro androgenesis in onion (*Allium cepa* L.). In: Proc 3rd Eucarpia Allium Symp, Wageningen, The Netherlands, 4–6 Sept, p 110

Campion B, Azzimonti MT, Vicini E, Schiavi M, Falavigna A (1992) Advances in haploid plant induction in onion (*Allium cepa* L.) through in vitro gynogenesis. Plant Sci 86:97–104

Campion B, Bohanec B, Javornik B (1995a) Gynogenic lines of onion (*Allium cepa* L.). Evidence of their homozygosity. Theor Appl Genet 91(4):598–602

Campion B, Perri E, Azzimonti MT, Vicini E, Schiavi M (1995b) Spontaneous and induced chromosome doubling in gynogenic lines of onion (*Allium cepa* L.). Plant Breed 114:243–246

Cohat J (1994) Obtention chez l'échalote (*Allium cepa* L. var. *agregatum*) de plantes haploides gynogénétiques par culture in vitro de boutons floraux. Agronomie 14:299–304

Cuny F (1992) Processus d'induction d'embryons haploides par du pollen irradié chez melon (*Cucumis melo* L). Responses du pollen a l'irradiation gamma. These de Docteour, Université d'Avignon et des Pays de Vaucluse, pp 139–151

Cuny F, Grotte M, Dumas de Vaulx R, Rieau A (1993) Effects of gamma irradiation of pollen on parthenogenetic haploid production in muskmelon (*Cucumis melo* L). Environ Exp Bot 33:301–312

Custers JBM, Ennik E, Eikelboom W, Dons JJM, van Lookeren MM (1997) Embryogenesis from isolated microspores of tulip; towards developing F1 hybrid varieties. Acta Hortic 430:259–266

D'Halluin K, Keimer B (1986) Production of haploid sugarbeets (*Beta vulgaris* L.) by ovule culture. In: Horn W, Jenson CJ, Odenbach W, Schielder O (ed) Genetic manipulation in plant breeding. De Gruyter, Berlin, pp 307–309

Doctrinal M, Sangwan RS, Sangwan-Norreel BS (1989) In vitro gynogenesis in *Beta vulgaris* L. Effect of plant growth regulators, temperature, genotypes and season. Plant Cell Tissue Organ Cult 17(1):1–12

Dohya N, Matsubara S, Murakami K (1997) Callus formation and regeneration of adventitious embryos from celery microspores by anther and isolated microspore cultures. J Jpn Soc Hortic Sci 65:747–752

Doré C (1974) Production des plantes homozygotes mâles et femelles à partir d'anthères d'asperge cultivée in vitro. CR Acad Sci 17(278):2135–2138
Doré C (1990) Asparagus anther culture and field trials of dihaploids and F1 hybrids. In: Bajaj YPS (ed) Biotechnology in agriculture and forestry, vol 12. Haploids in crop improvement I. Springer, Berlin Heidelberg New York, pp 322–345
Dryanovska OA, Ilieva IN (1983) In vitro anther and ovule cultures in muskmelon (*Cucumis melo* L.) C. Acad Bulg Sci 36:1107–1110
Dumas de Vaulx R (1979) Obtention de plantes haploides chez le melon (*Cucumis melo* L.) après pollinisation par *Cucumis ficifolius* A. Rich. CR Acad Sci Paris D 289:875–878
Dumas de Vaulx R, Chambonnet D (1986) Obtention of embryos and plants from in vitro culture of unfertilized ovules of *Cucurbita pepo*. In: Gruyter W (ed) Genetic manipulation in plant breeding. De Gruyter, Berlin, pp 295–297
Dunstan DI, Short KC (1977) Improved growth of tissue cultures of onion, *Allium cepa*. Physiol Plant 41:70–72
Eeckhaut T, Werbrouck S, Dendauw J, van Bockstaele E, Debergh P (2001) Induction of homozygous *Spathiphyllum wallisii* genotypes through gynogenesis. Plant Cell Tissue Organ Cult 67:181–189
Ellison JH (1986) Asparagus breeding. In: Basset MJ (ed) Breeding vegetable crops. AVI, Westport, Connecticut, pp 521–569
Ellison JH, Scheer DF, Wagner JJ (1960) Asparagus yield as related to plant vigor, earliness and sex. Proc Am Soc Hortic Sci 75:411–415
Falavigna A, Casali PE (2002) Practical aspects of a breeding program of asparagus based on in vitro anther culture. In: Uragami A (ed) Proc 10th Int Asparagus Symp. Acta Hortic 589:201–210
Falavigna A, Tacconi MG, Casali PE (1985) Asparagus breeding through anther in vitro culture: eight years of activity in Italy. In: Lougheed EC, Tiessen H (eds) Proc 6th Int Asparagus Symp. Eucarpia, University of Guelph, Guelph, pp 31–40
Falavigna A, Casali PE, Battaglia A (1999) Achievement of asparagus breeding in Italy. In: Benson B (ed) Proc 9th Int Asparagus Symp. Acta Hortic 479:67–74
Feng XR, Wolyn DJ (1991) High frequency production of haploid embryos in asparagus anther culture. Plant Cell Rep 10:574–578
Feng XR, Wolyn DJ (1994) Recovery of haploid plants from asparagus microspore culture. Can J Bot 72:296–300
Ferrant V, Bouharmont J (1994) Origin of gynogenic embryos of *Beta vulgaris* L. Sexual Plant Rep 7(1):12–16
Ficcadenti N, Veronese P, Sestili S, Crino P, Lucretti S, Schiavi M, Saccardo F (1995) Influence of genotype on the induction of haploidy in *Cucumis melo* L. by using irradiated pollen. J Genet Breed 49:359–364
Gamborg OL, Miller RA, Ojima K (1968) Nutrient requirements of suspension cultures of soybean root cells. Exp Cell Res 50:157–158
Gémes Juhász A (2003) Utilisation of in vitro androgenetic and gynogenetic methods for the development of haploid/doubled haploid plants of vegetable species. PhD Thesis, Szent István University, Budapest, pp 56–112
Gémes Juhász A, Martinovich L (1995) In vitro gynogenesis induction in Hungarian lines of onion (*Allium cepa* L.). Bull Veg Crops Res Inst 27:34–37
Gémes Juhász A, Venczel G, Balogh P (1996) Haploid plant induction in zucchini (*Cucurbita pepo* L. convar. *giromontiina* Duch) and in cucumber (*Cucumis sativus* L.) lines through in vitro gynogenesis. Acta Hortic 447:623–625
Gémes Juhász A, Martinovich L (1998) The effect of TDZ on gynogenesis induction and plant regeneration in onion. Bull Veg Crops Res Inst 28:39–47
Gémes Juhász A, Balogh P, Ferenczy A, Kristóf Z (2002a) Effect of optimal stage of female gametophyte and heat treatment on in vitro gynogenesis induction in cucumber (*Cucumis sativus* L.). Plant Cell Rep 21:105–111

Gémes Juhász A, Gajdos L, Venczel G, Sagi Zs, Zatyko L, Vagi P, Kristof Z (2002b) Production of doubled haploid breeding lines of pepper, eggplant, cucumber, zucchini and onion species. In: Proc Int Conf on Vegetables, 11–14 Nov, Bangalore, India, p 151

Geoffriau E, Kahane R, Rancillac M (1997a) Variation of gynogenesis ability in onion (*Allium cepa* L.). Euphytica 94:37–44

Geoffriau E, Kahane R, Bellamy C, Rancillac M (1997b) Ploidy stability and in vitro chromosome doubling in gynogenetic clones of onion (*Allium cepa* L). Plant Sci 122:201–208

Goska M, Jassem B, Jazdzewska E (1990) Embryo development from sugarbeet ovules in in vitro culture. In: Proc 53rd Winter Conf Int Inst Sugarbeet Research, Brussels, pp 145–154

Gürel S, Gürel E, Kaya Z (2000) Doubled haploid plant production from unpollinated ovules of sugar beet (*Beta vulgaris* L.). Plant Cell Rep 19:1155–1159

Gürsöz N, Abak K, Pitrat M, Rode JC, Dumas de Vaulx R (1991) Obtention of haploid plants induced by irradiated pollen in watermelon (*Citrillus lanatus* L.). Cucurbit Genet Coop 4:109–110

Gu ZP, Cheng KC (1983) In vitro induction of haploid plantlets from unpollinated ovaries of lily and embryological observations. Acta Bot Sin 25:24–28

Hansen AL, Plever C, Pedersen HC, Keimer B, Andersen SB (1994) Efficient in vitro chromosome doubling during *Beta vulgaris* ovule culture. Plant Breed 112:89–95

Hansen AL, Gertz A, Joersbo M, Andersen SB (1995) Short duration colchicine treatment for in vitro chromosome doubling during ovule culture of *Beta vulgaris*. Plant Breed 114:515–519

Hansen AL, Gertz A, Joersbo M, Andersen SB (1998) Antimicrotubule herbicides for in vitro chromosome doubling in *Beta vulgaris* L. ovule culture. Euphytica 101:231–237

Hu KL, Matsubara S, Murakami K (1993) Haploid plant production by anther culture in carrot (*Dacus carota* L.). J Jpn Soc Hort Sci 62(3):561–565

Jakše M, Bohanec B (2001) Studies of alternative approaches for genome doubling in onion. In: Bohanec B (ed) Proc COST Final Meet, Biotechnological Approaches for Utilisation of Gametic Cells, Bled, Slovenia 1-5 July 2000, pp 101–104

Jakše M, Bohanec B, Ihan A (1996) Effect of media components on the gynogenic regeneration of onion (*Allium cepa* L.) cultivars and analysis of regenerants. Plant Cell Rep 15:939–948

Jakše M, Havey MJ, Bohanec B (2003) Chromosome doubling procedures of onion (*Allium cepa* L.) gynogenic embryos. Plant Cell Rep 21:905–910

Javornik B, Bohanec B, Campion B (1998) Second cycle gynogenesis in onion *Allium cepa* L., and genetic analysis of the plants. Plant Breed 117:275–278

Keller J (1990) Culture of unpollinated ovules, ovaries and flower buds in some species of the genus *Allium* and haploid induction via gynogenesis in onion (*Allium cepa* L.). Euphytica 47:241–247

Keller J, Korzun L (1996) Ovary and ovule culture for haploid production. In: Jain SM, Sopory SK, Veilleux RE (eds) In vitro haploid production in higher plants, vol 1. Kluwer, Dordrecht, pp 217–235

Kurtar ES, Sari N, Abak K (2002) Obtention of haploid embryos and plants through irradiated pollen technique in squash (*Cucurbita pepo* L.). Euphytica 127:335–344

Lazarte JE, Sasser CC (1982) Asexual embryogenesis and plantlet development in anther culture of *Cucumis sativus* L. HortScience 17:88

Lotfi M, Kashi A, Onsinejad R (1999) Induction of parthenogenetic embryos by irradiated pollen in cucumber. Acta Hortic 492:323–326

Lotfi M, Alan AR, Henning MJ, Jahn MM, Earle ED (2003) Production of haploid and doubled haploid plants of melon (*Cucumis melo* L.) for use in breeding for multiple virus resistance. Plant Cell Rep 21:1121–1128

Luthar Z, Bohanec B (1999) Induction of direct somatic organogenesis in onion (*Allium cepa* L.) using a two-step flower or ovary culture. Plant Cell Rep 18:797–802

Lux H, Herrmann L, Wetzel C (1990) Production of haploid sugarbeet (*Beta vulgaris* L.) by culturing unpollinated ovules. Plant Breed 104:177–183

Martinez LE, Agüero CB, Galmarini CR (1997) Obtention of haploid plants by ovaries and ovules culture in onion (*Allium cepa* L.). Acta Hortic 433:447–453

Martinez LE, Agüero CB, López ME, Galmarini CR (2000) Improvement of in vitro gynogenesis induction in onion (*Allium cepa* L.) using polyamines. Plant Sci 156:221–226

Metwally E, Moustafa SA, El-Sawy BI, Harun SA, Shalaby TA (1998a) Production of haploid plants from in vitro culture of unpollinated ovules of *Cucurbita pepo*. Plant Cell Tissue Organ Cult 52(3):117–121

Metwally E, Moustafa SA, El-Sawy BI, Shalaby TA (1998b) Haploid plantlets derived by anther culture of *Cucurbita pepo*. Plant Cell Tissue Organ Cult 52(3):171–176

Michalik B, Adamus A, Nowak E (2000) Gynogenesis in Polish onion cultivars. J Plant Physiol 156:211–216

Michalik B, Adamus A, Samek L, Nowak E (2001) Gynogenesis in Polish onion cultivars: effect of temperature during donor plant growth. In: Bohanec B (ed) Biotechnological approaches for utilisation of gametic cells. Directorate-General for Research, European Commission, Brussels, pp 91–94

Murashige T, Skoog F (1962) Revised medium for rapid growth bioassays with tobacco tissue. Physiol Plant 15:473–497

Muren RC (1989) Haploid plant induction from unpollinated ovaries in onion. HortScience 24(5):833–834

Niemirowicz-Szczytt K, Dumas de Vaulx R (1989) Preliminary data on haploid cucumber (*Cucumis sativus* L.) induction. Cucurbit Genet Coop 12:24–25

Nowak E (2001) Gynogenic onion plants – studies of regeneration and diploidization. In: Bohanec B (ed) Proc COST 824 Final Meet, Biotechnological Approaches for Utilization of Gametic Cells, Bled, Slovenia, 1–5 July 2000, pp 95–99

Pedersen HC, Keimer B (1996) Haploidy in sugarbeet (*Beta vulgaris* L.). In: Jain SM, Sopory SK, Veilleux RE (eds) In vitro haploid production in higher plants, vol 3. Important selected plants. Kluwer, Dordrecht, pp 17–36

Peng M, Wolyn DJ (1999) Improved callus formation and plant regeneration for shed microspore culture in asparagus (*Asparagus officinalis* L.). Plant Cell Rep 18:954–958

Peng MS, Ziauddin A, Wolyn DJ (1997) Development of asparagus microspores in vivo and in vitro is influenced by gametogenic stage and cold treatment. In Vitro Cell Dev Biol Plant 33(4):263–268

Prakash J, Giles KL (1986) Production of doubled haploids in oriental lilies. In: Horn W, Jensen CJ, Odenbach W, Schieder O (eds) Genetic manipulation in plant breeding. De Gruyter, Berlin, pp 335–337

Przyborowski J (1994) Main factors affecting cucumber (*Cucumis sativus* L.) haploid embryo development and haploid plant characteristics. Plant Breed 112:70–75

Puddephat IJ, Robinson HT, Smith BM, Lynn J (1999) Influence of stock plant pretreatment on gynogenic embryo induction from flower buds of onion. Plant Cell Tissue Organ Cult 57(2):145–148

Qiao YM, Falavigna A (1990) An improved in vitro anther culture method for obtaining doubled haploid clones of asparagus. Acta Hortic 271:145–150

Rao PS, Suprasanna P (1996) Methods to double haploid chromosome numbers. In: Jain SM, Sopory SK, Veilleux (eds) In vitro haploid production in higher plants, vol 1. Kluwer, Dordrecht, pp 317–339

Rode JC, Dumas de Vaulx R (1987) Obtention de plantes haploides de carotte (*Dacus carota* L.) issues de parthenogenese induite in situ par du pollen irradie et culture in vitro des graines immatures. CR Acad Sci III 305:225–229

San LH, Demarly Y (1984) Gynogenesis in vitro and biometric studies of doubled haploids obtained by three techniques in *Hordeum vulgare* L. In: Lange W, Zeven AC, Hogenboom NG (eds) Proc 10th Congr Efficiency in Plant Breeding, Eucarpia, Wageningen, 19–24 June 1983, p 347

Sangwan RS, Sangwan BS (1986) Effets des rayons gamma sur l'embryogénese somatic et l'androgénese chez devers tissues végétaux cultivés in vitro. In: Nuclear techniques and in vitro culture for plant improvement. IAEA, Vienna, pp 181–185

Sari N, Abak K, Pitrat M, Rode JC, Dumas De Vaulx R (1994) Induction of parthenogenetic haploid embryos after pollination by irradiated pollen in watermelon. HortScience 29:1189–1190

Sauton A (1989) Haploid gynogenesis in *Cucumis sativus* induced by irradiated pollen. Rep Cucurbit Genet Coop 12:22–23

Sauton A, Dumas de Vaulx R (1988a) Doubled haploid production in melon (*Cucumis melo* L.). In: Proc Eucarpia Meetg on Cucurbitaceae, Avignon-Monfavet, France, pp 119–128

Sauton A, Dumas de Vaulx R (1988b) Effect of season and genotype on gynogenic haploid production in muskmelon, *Cucumis melo* L. Sci Hortic 35:71–75

Schum A, Mattiesch L, Timmann EM, Hofmann K (1993) Regeneration of dihaploids via gynogenesis in *Allium porrum* L. Gartenbauwissenschaft 58(5):227–232

Shail JW, Robinson RW (1987) Anther and ovule culture of *Cucurbita*. Cucurbit Genet Coop 10:92

Smith BM, Godwin RM, Harwey E, Werner CP (1991) Gynogenesis from whole flower buds in bulb onion (*Allium cepa* L.) and leeks (*Allium porrum* L.). J Genet Breed 45:353–358

Sneep J (1953) The significance of andromonoecy for the breeding of *Asparagus officinalis* L. Euphytica 2:89–95

Sopory SK, Munshi M (1996) Anther culture. In: Jain SM, Sopory SK, Veilleux RE (eds) In vitro haploid production in higher plants, vol 1. Kluwer, Dordrecht, pp 145–176

Troung-Andre I (1988) In vitro haploid plants derived from pollination by irradiated pollen of cucumber. In: Proc Eucarpia Meetg on Cucurbitaceae, Avignon-Monfavet, France, pp 143–144

Van den Bulk RW, de Vries van Hulten HPJ, Custers JBM, Dons JJM (1994) Induction of embryogenesis in isolated microspores of tulip. Plant Sci 104:101–111

Van Geyt JPC, Spechmann GJ, D'Hallium K (1987) In vitro induction of haploid plants from unpollinated ovules and ovaries of the sugar beet (*Beta vulgaris* L.). Theor Appl Genet 73:920–925

Wolyn DJ, Feng XR (1993) Genotype, temperature, and sampling date affect embryogenesis in asparagus anther culture. HortScience 28(3):216–217

Zhang CJ, Wang HL, Ma Y, Kang YQ (1994) Regeneration of haploid plants from isolated microspores of asparagus (*Asparagus officinalis* L.). Plant Cell Rep 13:637–640

Ziauddin A, Feng XR, Wolyn DJ (1996) Advances in asparagus anther culture. In: Nichols M, Swain D (eds) Proc 8th Int Asparagus Symp. Acta Hortic 415:231–235

II.6 Haploids in the Improvement of Linaceae and Asteraceae

KARIN NICHTERLEIN[1, 3], and RENATE HORN[2]

1 Linaceae

1.1 Introduction

The family Linaceae includes herbs and shrubs. Its genus *Linum* comprises almost 200 species, which are spread over the temperate and warm temperate region of the northern hemisphere, mostly in Europe and Asia, but also in America. Linseed (syn. flax, *Linum usitatissimum*) is the only important crop species of the Linaceae. It is one of the oldest cultivated plants, having been grown for about 8,000 years. The crop is adapted to warm and cool temperate climates. It can be used as an oilseed and as a fibre crop. In addition, the seed is traditionally used in therapeutics because of its laxative properties. The oilseed types (linseed or oil flax) compared to fibre types (flax) have shorter and thicker stems with more branches, whereas flax produces fewer capsules and smaller seeds than linseed. The taller flax type is usually cultivated in cool, temperate regions and used for textiles (Lühs and Friedt 1994). The annual global production of flax fibre and tow is about 647,000 Mt, with China, Spain, France and the Russian Federation as the major producers (FAO 2003). The production of fibre in the EU-15 countries dropped between 25 and 52% from 2000 to 2001 (EC 2002).

The shorter linseed types are mainly grown in the warmer climates. Shorter fibres from the oilseed types are used in the manufacture of fine papers and for industrial fibre products such as the interior panelling of some cars (Berglund 2002). The seed is processed into a variety of diverse products from breads to linseed oil for environmentally friendly paints and concrete preservation to linoleum flooring. Linseed contains oil with a high alpha-linolenic acid (50%) essential for humans but with short shelf life due to autoxidation, a high percentage of dietary fibre, both soluble and insoluble, and the highest content of lignans of all grain crops. It is considered as a healthy food additive. The rapid drying property of the oil to form a natural

[1] Joint FAO/IAEA Division of Nuclear Techniques in Food and Agriculture, Vienna, Austria
[2] Department of Genetics and Biochemistry, Clemson University, 100 Jordan Hall, Box 340324, Clemson, South Carolina 29634-0324, USA
[3] Present address: FAO, Regional Office for Europe, Viale delle Terme di Caracalla, Rome, Italy

Biotechnology in Agriculture and Forestry, Vol. 56
Haploids in Crop Improvement II (ed. by C.E. Palmer, W.A. Keller, and K.J. Kasha)

plastic-like film explains its traditional use as industrial oil for paints and varnishes. Although the oil is not suitable for frying at high temperatures, linseed oil is used for low-temperature stir-frying in rural linseed-producing areas of China (Berglund 2002). The meal left after the extraction of oil is used in animal feed for cattle and horses. The world production of linseed is about 2.28 million Mt. Major producers of linseed are Canada (704,000 Mt), China (560,000 Mt), the USA (399,000 Mt) and India (220,000 Mt). Production in the EU-15 dropped over the last decade from about 325,000 to 78,500 Mt due to policy changes. Argentina in 1992, still a major producer (342,900 Mt), abandoned linseed production. Production increased by a factor of five in the USA and more than doubled in Canada over the last decade (FAO 2003).

Linseed is regarded as a self-pollinating crop with about 4% cross-fertilization. The pedigree method and bulk breeding method have been mostly used to develop improved breeding lines and new varieties (Bergmann and Friedt 1997). Major breeding objectives are the improvement of yield (seed or fibre) and oil content or fibre quality, and resistance to lodging and diseases, the latter causing considerable yield losses. Good progress has been made in the development of varieties with resistance to rust and moderate resistance to *Fusarium*; however, maintaining the resistance to rust and wilt continues to be the objective in development of new varieties (Anonymous 2002). Mutation induction and interspecific hybridizations have been used as tools in linseed breeding to create new variation in the fatty acid composition (Nichterlein et al. 1989). The induction of low-linolenic acid content mutants after ethyl methanesulfonate (EMS) treatment and their recombination lead to the development and release of so-called Solin varieties with levels of linolenic acid under 2%, similar to traditional sunflower oil (Dribnenki and Green 1995; Dribnenki et al. 1996, 1999; Prairie Registration Recommending Committee for Grains 2002). Linseed oil from such varieties is more stable and more suitable for cooking at higher temperatures and long-term storage. Linseed was one of the first species to be genetically modified by using recombinant DNA technology, and the first transgenic herbicide-tolerant variety was produced but has recently been deregistered and is not commercially available anymore (Anonymous 2002).

Traditional breeding methods are time-consuming and at least 10–12 years are needed from the initial cross or induced mutation to the release of a new variety (Steiss et al. 1998). The development of low linolenic acid Solin varieties involved the induction of two different fatty acid mutants and their recombination (Green and Marshall 1984). In this case, it took even more than two decades from the mutation induction to the official registration of the first low linolenic variety 'Linola 947' (Dribnenki and Green 1995). The identification and rapid fixation of rarely segregating desired gene combinations in linseed breeding can be done through doubled-haploid (DH) techniques. Doubled haploids have been produced in linseed using in vivo methods by selection of polyembryonic seedlings or through in vitro techniques (Bergmann and Friedt 1997). In this chapter, the recent progress of various techniques for the development of doubled haploids in linseed and their use in research and breeding is reviewed.

1.2 Haploids Through Twinning Genotypes and Their Use in Genetic Research and Mapping

The phenomenon of haploids from polyembryonic seeds in linseed and the development of genotypes with frequencies of haploid–diploid twin seedlings reaching more than 30% were reviewed by Bergmann and Friedt (1997). The twinning character of such a high twinning genotype, 'RA91', provides a useful tool for the development of doubled haploids in genetic research and molecular genetics. It has been used to develop a recombinant DH population for studying wilt resistance in linseed and identifying markers tightly linked to wilt resistance (Spielmeyer et al. 1998b). Flax wilt caused by the soil fungal pathogen *Fusarium oxysporum* f.sp. *lini* is considered as one of the most devastating diseases of linseed. One mapping parent was the wilt-resistant DH plant 'CRZY8/RA91' with Solin quality, developed from the wilt-resistant Solin line CRZY8 and high twinning genotype RA91. This wilt-resistant Solin-type plant was used in a cross with the Australian variety 'Glenelg', genetically related to CRZY8, but highly wilt-susceptible. The haploid components of the resulting F_2 twin seedlings were colchicinized and a population of 143 homozygous recombinant lines was developed. The segregation of wilt resistance in the homozygous DH lines was studied in the glasshouse using highly *Fusarium*-infested soil and under field conditions (hot spot for wilt). The segregation ratio in the DH lines suggests the involvement of two independent major genes with additive effects in wilt resistance of the cross, and probably also minor genes for wilt resistance. The elimination of heterozygosity and genetic variation within the individual recombinant lines assisted in clarifying the inheritance of *Fusarium* wilt resistance. The estimation of the disease response was more precise on the basis of each DH line when compared with the accuracy of an individual estimate in F_2 progeny. The use of the DH population also confirmed that a glasshouse screening method of DH lines was a reliable indicator of field resistance to *Fusarium* wilt (Spielmeyer et al. 1998b). This study provided the basis for the molecular analysis of the resistance mechanism as the DH population was used to develop an AFLP (amplified fragment length polymorphism) linkage map of linseed with 213 marker loci (Spielmeyer et al. 1998a). Two quantitative trait loci were identified on independent linkage groups with major effects on resistance to *Fusarium* wilt. Although a very useful tool in genetic research and molecular studies, today, DH production through twinning genotypes may not be the most suitable DH technique for variety development. In breeding, it requires the use of a twinning genotype in a cross, and later the removal of the twinning trait through selection. Recent improvements in the development of doubled haploids from anther culture would favour this technique in linseed improvement.

1.3 Doubled Haploids Through Anther and Microspore Culture and Use in Crop Improvement

In order to increase the efficiency of haploid plants independent from the hybridization with twinning lines, anther and microspore culture systems have been developed for linseed. Early studies on haploids from anthers or isolated microspores of linseed in China and Europe were reviewed by Bergmann and Friedt (1997). A detailed protocol of the various steps involved in anther culture of linseed from donor plant growth, collection of buds, preselection of anthers and the in vitro protocol has recently been compiled (Nichterlein 2003).

A characteristic of androgenesis in linseed anther and microspore culture is the predominance of the organogenic pathway in which plant regeneration is achieved through the intermediate callus phase (Nichterlein et al. 1991; Nichterlein and Friedt 1993). The controlled growth conditions of donor plants at 14/8 °C day/night in a 16-h photoperiod have proven to be beneficial for anther response (Nichterlein et al. 1991) and have been used frequently in further studies on optimization of the protocol and for production of doubled haploids in linseed (Friedt et al. 1995; Chen et al. 1998a–c, 1999; Chen and Dribnenki 2002). Donor plants from glasshouse cultivation at 12–22 °C were used for anther culture of eight linseed varieties, but callus induction was low and failed to form shoots (Kurt and Evans 1998). Anthers from field-grown plants of linseed and fibre flax varieties have also been used, with reduced anther response (Pretova and Obert 2000). For the development of doubled haploids through anther and microspore culture, anthers with the majority of microspores in the late uninucleate stage (Nichterlein et al. 1991; Nichterlein and Friedt 1993) or mid-uninucleate stage were selected (Chen et al. 1998c). Incubation of anthers for 1 day at 35 °C before culturing at 25 °C had no effect on genotype 'M3696/NorLin' F_1, but significantly increased the regeneration efficiency of the genotype 'AC McDuff/AC Emerson' F_1 (Chen et al. 1998c) and was therefore used in further experiments with a wider range of genotypes (Chen et al. 1998b, 1999).

A solid anther induction medium, A22, consisting of a modified Murashige and Skoog (MS) medium (10% 165 mg/l ammonium nitrate), with glutamine (750 mg/l), 1-naphthaleneacetic acid (NAA) (1 mg/l), 6-benzylaminopurine (BA) (2 mg/l), 6% sucrose and 0.4% agarose was first recommended for anther culture (Nichterlein et al. 1991). The induction medium has been optimized by the addition of 10 mg/l thiamine HCl and changing of the growth regulators to 2,4-dichlorophenoxyacetic acid (2,4-D) (2 mg/l) and BA (1 mg/l). The latter increased the overall efficiency of regeneration from 14 to 43% of anthers cultured from McDuff × McEmerson F_1 (Chen et al. 1998b). Increasing the sucrose concentration from 6 to 15% resulted in a reduction of the overall efficiency of regeneration due to inhibition of callus formation and regeneration from somatic anther tissues. The overall efficiency of DH production was not significantly changed, but the frequency of microspore-derived plants increased from 67 to 94%. In subsequent experiments with various F_1 hybrids, 9% sucrose in the induction medium was used, resulting in 85–90% microspore-derived plants as determined by molecular analysis. Replacing sucrose by lactose resulted in an increase in callus induction, but the effect on shoot regeneration seemed to be genotype dependent. Only one of three genotypes tested showed significantly improved shoot production from anthers cultured in lactose (Chen and Dribnenki 2002).

N6 basal medium is used for regeneration of anther calli. It is supplemented with glutamine (375 mg/l), asparagine (250 mg/l) and serine (125 mg/l), the plant growth regulator zeatin (1 mg/l), maltose (30 g/l) and 0.4% gelrite (Nichterlein et al. 1991). The replacement of zeatin by various concentrations of thidiazuron was tested but had a negative effect on shoot regeneration (Chen et al. 1998c). Shoot elongation could be improved by subsequently transferring cultures to the same modified N6 medium but without growth regulators (Nichterlein et al. 1991). Recently, the effect of various sucrose concentrations (0–5 to 10–30 g/l) on elongation of anther-derived shoots from 'AC McDuff/96–150' was examined using MS medium, supplemented with 0.2 mg/l indoleacetic acid (IAA), 1 mg/l silver nitrate and 0.4% phytagel. After 4 weeks of culture the percentage of transferable shoots (at least 6 cm long) was the highest (88%) on the medium containing 10 g/l sucrose, compared to 48% of the 5 g/l sucrose medium and 8% of the 30 g/l sucrose medium. The protocol with reduced sucrose content of the shoot elongation medium has been successfully used for a wide range of genotypes of diverse origin (Chen et al. 2003).

Genotypic effects on anther response and shoot regeneration were observed and both processes seem to be under independent genetic control (Nichterlein et al. 1991; Nichterlein and Friedt 1993; Chen et al. 1999; Chen and Dribnenki 2002). The European linseed variety 'Atalante' and its hybrids showed good callus formation and regeneration. The overall response of anthers (percentage anthers producing anther-derived lines) of Atalante hybrids was 1.8–4.0 (Friedt et al. 1995). In a screening of 44 Canadian breeding materials, the overall response varied from 0–10%. Advanced breeding lines with 5–10% overall anther response could be identified with low cadmium types, high oil types, high linolenic types and in certain hybrids with AC McDuff. For breeding purposes, hybrids from high responding genotypes and genotypes with desirable agronomic traits were crossed for DH production. The efficiency of anther culture of F_1 hybrids could be improved by combining genotypes showing high callus formation with genotypes showing good plant generation; often F_1 hybrids showed better response than their respective homozygous parents (Chen et al. 1999). For the use of anther culture in breeding of low linolenic genotypes, one Solin breeding population and one Solin genotype have been identified with a high shoot regeneration capacity of anthers (Chen and Dribnenki 2002).

There has been a continuous improvement of the anther culture technique for linseed. At the beginning, the frequency of spontaneous doubled haploids in regenerated anther culture derived plants from an Atalante hybrid was 36%, with 50% of somatic origin as shown by segregation of their progenies (Friedt et al. 1995). With the modified induction protocol and AC McDuff hybrids the frequency of microspore-derived plants was about 55%, of which 38% showed spontaneous doubling of chromosomes as found using PCR-based markers (Chen et al. 1999). The rate of microspore-derived plants could be increased to more than 90% with a frequency of spontaneous doubling of 78% through the further optimization of the medium composition (Chen et al. 1998b).

The isolation and culture of microspores would avoid the regeneration of heterozygous plants from diploid non-gametic tissue, which are of no value to breeders. A protocol for microspore culture of linseed was developed and DH plants could be produced from donor plants grown under the same conditions as for anther culture. Microspores were cultured in modified NLN 82 medium containing 0.1 mg/l benzyladenine, 0.3 mg/l indole-3-acetic acid, 0.5 mg/l naphthalene-1-acetic acid and 13% sucrose. Calli produced were transferred to the regeneration medium described for anther culture (Nichterlein and Friedt 1993). In a study comparing anther culture and microspore culture of linseed, it was concluded that the response of anther cultures is higher than of microspore culture with regard to callus formation (Steiss et al. 1998). However, the calli from microspore cultures generally showed a better differentiation into shoots (Steiss et al. 1998). With the recent improvements of the anther culture protocol increasing the percentage of microspore-derived plants with spontaneous doubling (Chen et al. 1998b; Chen and Dribnenki 2002), the efficiency of developing doubled haploids in linseed through anther culture seems to be higher than through the currently available microspore culture system (Steiss et al. 1998).

Anther culture has been used for basic genetic studies and applied crop improvement programmes. Anther culture was tested as a tool to study the inheritance of rust resistance and to develop rust-resistant varieties. Flax rust caused by the fungus *Melampsora lini* is a major and constant threat to linseed production because it can survive locally and has the ability to produce new races that attack hitherto resistant varieties. Maintaining the resistance to rust therefore continues as an objective in linseed improvement programmes (Anonymous 2002). The inheritance pattern of two resistance genes and six molecular markers in two DH populations of linseed was investigated after confirming the microspore-derived origin by molecular techniques. It was found that the inheritance of the two independent rust resistance genes K^1 and M^3 and of three molecular markers was according to the expected Mendelian pattern in the DH populations of AC McDuff/AC Emerson F_1 and AC Emerson/McGregor F_1 (Chen et al. 2001). Therefore, it can be concluded that microspore-derived linseed populations should be as effective for resistance breeding to rust as traditional breeding populations. However, three molecular marker loci showed segregation distortion in favour of alleles of the parent AC McDuff. AC McDuff was more responsive to the used anther culture protocol (Chen et al. 2001). Similar segregation distortions were observed in barley and rice DH populations, suggesting that genes affecting survival in anther or microspore culture favour the more responsive parent of a hybrid (Zivy et al. 1992; Xu et al. 1997; Manninen 2000). A comparison of 43 DH lines with 29 F_5 pedigree lines developed from three different F_1 hybrids of French and Russian varieties and a primitive form, *L. usitatissimum candidum*, revealed that these pedigree lines were more productive than the DH lines. DH lines matured 4 days earlier, which can be advantageous in environments requiring early maturity (Steiss et al. 1998). Experiments for the field evaluation of DH lines developed in a low linolenic linseed breeding programme are in progress (Chen and Dribnenki 2002).

1.4 Conclusions and Future Prospects of Doubled Haploids in Linseed Improvement

The protocols for DH production in linseed have been improved over the last few years and their first applications in genetic studies and breeding have occurred. DH lines have been produced through hybridization with twinning genotypes, anther or microspore culture and these are useful tools in genetic studies. Well-responding genotypes have been identified for the development of doubled haploids through anther culture and these can be used as parents in breeding programmes that incorporate doubled haploids. However, further research on the improvement of the anther culture protocol is desirable in order to improve the response of more recalcitrant genotypes and to further increase the efficiency of DH production to make its routine use in linseed improvement more economic. Molecular techniques could help to identify and map the genes controlling the in vitro response of linseed.

Although no release of a DH-derived linseed or flax variety has been reported yet, breeders have incorporated anther culture techniques in linseed and flax improvement programmes, for high and low linolenic acid types (Chen et al. 1999; Chen and Dribnenki 2002). Through application of improved culture protocols, a number of DH lines from a wide range of different genotypes and F_1 hybrids have been developed and evaluated in a field trial. A first field evaluation comparing linseed lines developed through DH techniques with conventionally developed pedigree lines does not allow us to draw final conclusions about the value of the technique for breeding due to the very small population sizes used in the study (Steiss et al. 1998). Additional evaluations of the yield potential and agronomic characteristics of DH lines are needed using larger DH and pedigree populations.

The DH technique seems to be suitable for breeding of linseed resistant to rust, despite the observation of segregation distortion favouring the more anther culture responsive genotype. It can be expected that the production of DH lines will complement the traditional breeding methods of linseed in the future, provided breeders have easy access to the necessary resources such as climate chambers and tissue culture laboratories.

2 Asteraceae

2.1 Introduction

The Asteraceae (Compositae) family represents the largest of all plant families with over 20,000 species (Cronquist 1988), which are mostly herbs, a few shrubs and less commonly trees. The family comprises a considerable number of economically important crops, e.g. sunflower, safflower and nigerseed as oilseeds, lettuce, Jerusalem artichoke, endive, chicory as vegetables, chamomile and *Artemisia* species as medicinal plants, the insecticide-producing *Pyrethrum* and many ornamentals. However, the majority of the more than 40 crops, medical and horticultural species throughout most tribes in this family have relatively low economic value and restricted use. Lettuce (*Lactuca sativa*) and sunflower (*Helianthus annuus*) are the most economically important species in Asteraceae. The worldwide production of lettuce increased by more than 50% from 12.42 million Mt in 1992 to 18.75 million Mt in 2002 (FAO 2003). Major producers of lettuce are China (8.0 million Mt), the USA (4.35 million Mt), Spain (914,900 Mt) and Italy (845,593 Mt). The major breeding objectives of lettuce are the improvement of yield and resistance especially to downy mildew caused by the obligate oomycete *Bremia lactucae*. For sunflower seeds, world production is about 23.85 million Mt (FAO 2003). Major producers of sunflower seeds are Argentina (3.84 million Mt), the Russian Federation (3.6 million Mt), Ukraine (3.27 million Mt) and China (1.9 million Mt). Production in the EU-15 dropped from 4.09 million Mt in 1992 to 2.80 million Mt in 2002, with France (1.52 million Mt) and Spain (757,200 Mt) still remaining the major producers. Sunflower seeds are used to produce oil for human consumption or industrial purposes, and oil cake, rich in proteins, for feeding cattle. Worldwide, cultivated sunflower is the fourth most important oil crop behind soybean, palm oil and rapeseed. The major breeding objectives of sunflower are the improvement of seed yield, oil content and oil quality (conventional and high oleic) as well as resistance breeding with special emphasis on *Sclerotinia sclerotiorum*, *Plasmopara halstedii* and *Phomopsis helianthi* for moderate mild climates and drought tolerance and *Orobranche* resistance for dry hot climates. Drought leads to heavy losses in sunflower, for example in southern parts of Europe, especially in countries like Spain where due to water stress production per hectare is more than two times lower than the French average. Improvement of drought tolerance and water use efficiency by exploiting the genetic diversity within the genus *Helianthus* is one of the strategies to reduce yield losses. The genus *Helianthus* consists of 49 species (Schilling and Heiser 1981), offering a broad possibility to introduce new traits via interspecific hybridization (Korell et al. 1996a,b).

The Compositae family lags behind several other families in molecular analyses of individual species. With two exceptions, *Lactuca sativa* (lettuce) and *Helianthus annuus* (sunflower), little molecular research on genome organization or on genetic engineering of species in this family has been published. For genome research, sunflower has become a model species among the Asteraceae. In the genus *Helianthus*, genetic maps based on RFLP (restriction fragment length polymorphism), RAPD (random amplified polymorphic DNA), AFLP and also SSR (simple sequence repeat) markers have been developed for *H. annuus* (Berry et al. 1995; Gentzbittel et al. 1995, 1999; Leon et al. 1995; Flores-Berrios et al. 2000; Tang et al. 2002; Yu et al. 2003) and *H. anomalus* (Rieseberg et al. 1993). Genetic loci for quantitative traits such as oil content (Leon et al. 1995), days to flowering (Leon et al. 2001) and monogenic traits such as resistance to diseases like downy mildew (Mouzeyar et al. 1995; Brahm et al. 2000) and rust (Lawson et al. 1998) were localized on the sunflower genome maps. Apart from the genus *Helianthus*, detailed or partial genetic maps have been constructed in three other genera: *Lactuca* (Kessli et al. 1994; Witsenboer et al. 1997; Waycott et al. 1999; Jeuken et al. 2001), *Cichorium* (de Simon et al. 1997) and *Microseris* (Vanhouten et al. 1994; Bachmann and Hombergen 1997). Genetic maps and identified markers closely linked to agronomically important traits now facilitate crop improvement by marker-assisted selection. The use of molecular markers can help to avoid time- and cost-intensive resistance tests and allows selection in early stages of plant development.

The application of haploid techniques as anther culture or microspore culture can also considerably accelerate breeding programmes by providing homozygous DH lines after a single in vitro culture step and avoiding the time-consuming process of developing inbred lines by selfing over several generations and years. Application of haploid techniques and molecular and biochemical characterization of doubled haploids in the genus *Helianthus* have been previously reviewed by Friedt et al. (1997). This chapter here will focus on improvements in anther and microspore culture since then and will also address the new possibilities by the use of parthenogenesis in the genus *Helianthus* that had not been exploited before and may represent an interesting alternative for sunflower.

2.2 Production of Doubled Haploids by Anther and Microspore Culture in the Genus *Helianthus*

Establishment of haploid techniques in the genus *Helianthus* has proven to be challenging, as sunflower has been recalcitrant to all in vitro culture techniques (Friedt 1992; Yang et al. 1995; Hahne 2001). As wild species of the genus *Helianthus* were considered to have a better regeneration capacity, haploid techniques were approached not only in the cultivated sunflower but also in wild species of the genus *Helianthus* and interspecific hybrids between cultivated sunflower and annual or perennial wild species.

The early research reports on anther culture in the genus *Helianthus* showed the difficulties in the development of haploid techniques for sunflower. Different compositions of culture media used in these attempts to establish anther and microspore culture in sunflower have been extensively reviewed by Friedt et al. (1997).

With anther culture, regenerated plants can develop from anther walls (somatic origin) or from the microspores (androgenic origin). Histological studies of Zhong et al. (1995) suggested that most of their anther-derived embryos in sunflower were of somatic origin, representing regenerants from the anther walls. Even microspore preparations needed additional gradient purification to avoid development of plants with somatic origin (Coumans and Zhong 1995). These investigations clearly demonstrated the necessity to prove the androgenic origin of plants regenerated by haploid techniques in sunflower. The possibilities of using biochemical and molecular methods to identify the origin of the regenerated plants from anther and microspore culture have been discussed by Friedt et al. (1997).

Different pretreatments, media compositions, culture conditions and additives have been further tested to optimize anther and microspore culture in sunflower, wild species of the genus *Helianthus*, as well as in interspecific hybrids (Table 1). In cultivated sunflower, Badigannavar and Kuruvinashetti (1996) investigated the effect of cold pretreatment of anthers, dark incubation and different concentration of 2,4-D, BA and NAA on callus formation. Using the genotypes 'KBSH-1', 'BSH-1' and 'Morden', cold pretreated anthers with uninucleate microspores resulted in maximum callus induction. Uninucleate microspores of anthers divided symmetrically to give rise to multicellular proembryo-like bodies. The combination of 2,4-D (2 mg/l) and BA (1 mg/l) or NAA (1 mg/l) and BA (1 mg/l) gave the best results. Occasionally, shoot bud formation was observed for KBSH-1 and BSH-1. Saji and Sujatha (1998) achieved as high as 100% callus formation and embryo formation at a maximum frequency of 44%, plating anthers of KBSH-1, MSFH-8, Jwalamukhi and Morden. Although complete embryos were formed, the frequency of their conversion to whole plantlets was low (14.3%). Therefore, the embryogenic pathway was bypassed to obtain multiple shoots by transferring embryogenic calli with developing embryos to MS medium supplemented with 0.5 mg/l BA. When donor capitula were subjected to a cold treatment (4 °C) for 4 days prior to anther plating, 27% of the embryos regenerated to plantlets. This represented a four-fold increase as compared to the non-treated control (7%). Agar and sucrose concentration and the carbohydrate source proved to be the determining factors for maximum callus formation. Furthermore, the genotype revealed a strong effect on the frequency of embryo formation from anther-derived callus. Rapid whole plantlet recovery starting within 7 weeks after anther inoculation was reported (Saji and Sujatha 1998). Cytological studies revealed a frequency of 8.3% haploids in rooted shoots.

Vasic et al. (2000) placed anthers of six sunflower varieties on six different culture media to develop DH lines. Organogenesis and somatic embryogenesis were estimated as percentage of anthers with calli or somatic embryos. Sil-

Table 1. Overview of the production of doubled haploids by anther and microspore culture or parthenogenesis in the genus *Helianthus* since 1996

Doubled-haploid production by anther and microspore culture	*Authors*
Cold pretreatment, dark incubation and phytohormone variation	Badigannavar and Kuruvinashetti (1996)
Optimization of anther culture from interspecific hybrids of *H. annuus* with *H. resinosus* and *H. tuberosus* with regard to induction of callus formation and direct embryogenesis; characterization of regenerants by isoenzymes	Nurhidayah et al. (1996)
Effect of agar- and sucrose concentration as well as carbohydrate source on callusing and embryo formation as well as culture conditions	Saji and Sujatha (1998)
Screening of perennial and annual wild species of *Helianthus* for androgenesis using different nutrient media	Nenova et al. (2000)
Organogenesis and somatic embryogenesis in anther culture and influence of silver nitrate	Vasic et al. (2000)
Doubled-haploid production by induced parthenogenesis	
Anther culture of six interspecific hybrids and induced parthenogenesis of eight wild species by γ-irradiated pollen	Todorova et al. (1997a)
Induction of parthenogenesis by γ-irradiation (three doses) of pollen from four donors on four recipient hybrids	Todorova et al. (1997b)
Testing effect of pollen donor in 88 combinations (10 donors, pollen mixtures and 2 mother genotypes)	Todorova and Ivanov (1999)

ver nitrate was found to have a positive effect on organogenesis and somatic embryogenesis.

The most promising results with regard to anther culture were obtained by Nurhidayah et al. (1996) using interspecific hybrids of cultivated sunflower with the wild species *H. tuberosus*, *H. laetiflorus* and *H. resinosus*. Regeneration occurred after callus formation via direct embryogenesis (Fig. 1), which yielded up to 2,000 regenerated plants. Under optimized conditions, 92.7% of the anthers were embryogenic with an average of 8.5 embryoids per anther (Nurhidayah et al. 1996). Isozyme analysis of 1,200 plants regenerated from the F_1 hybrid (40/3) of the interspecific cross *H. annuus* × *H. resinosus* revealed that the plantlets could be distinguished into four groups. For two of the groups the androgenic origin could be verified by their isozyme segregation pattern. Additional grouping of the regenerants due to morphological traits indicated that the majority of plants were of androgenic origin. Todorova et al. (1997a) also used six interspecific hybrids between the cultivated sunflower and the wild species *H. decapetalus*, *H. hirsutus*, *H. rigidus*, *H. mollis*, *H. eggertii* and *H. laevigatus* for anther culture. Plant regeneration and shoot multiplication were observed in F_1 hybrids with *H. eggertii*, *H. mollis*, *H. decapetalus* and *H. rigidus*. Nenova et al. (2000) studied 20 perennial and

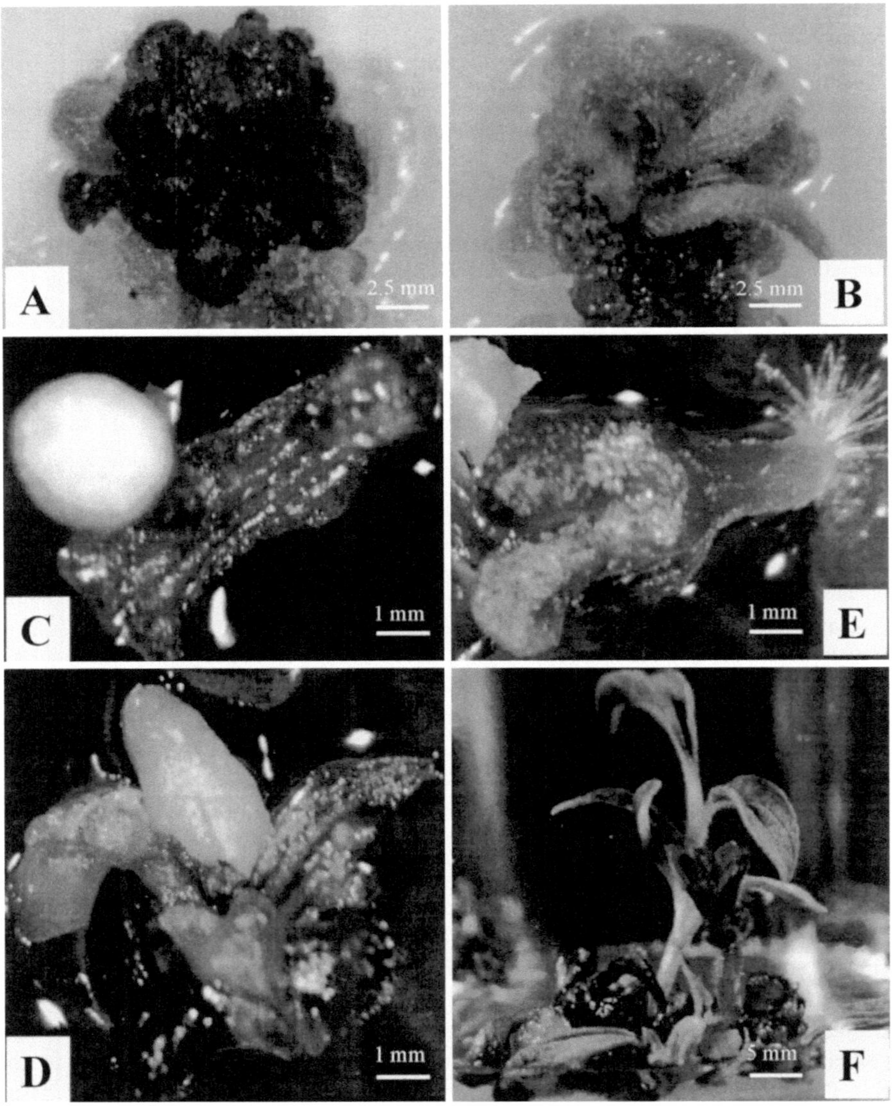

Fig. 1A–C. Anther culture of interspecific hybrids between *H. annuus* and *H. tuberosus* (11/51) or *H. resinosus* (40/3) under continuous light. **A** Shoot induction on callus of genotype 11/51 on MS-I3 6 weeks after plating the anthers, 30 °C. **B** Development of a shoot from callus of genotype 11/51 on MS-I3 8 weeks after plating anthers, 30 °C. **C** Globular embryo formed on an anther of genotype 40/3 after 15 days, 35 °C, **D** Multiple embryos induced on an anther of genotype 40/3 after 15 days, 30 °C. **E** Embryo at the cotyledonary stage formed on an anther of genotype 40/3 after 20 days, 30 °C. **F** Shoot development from embryos of genotype 40/3 on MS-R3 after 6 weeks, 30 °C. (Pictures taken from Nurhidayha et al. 1996)

annual *Helianthus* species and were able to regenerate shoots from anther culture of *H. mollis*, *H. salicifolius* and *H. smithii*. Differences between the species regarding anther culture could be observed and the ploidy level of the regenerants was determined by flow cytometry.

2.3 Doubled-Haploid Production of Sunflower by Induced Parthenogenesis

Although large efforts have been made to establish DH production for sunflower in recent years, none of the improvements has made anther or microspore culture in sunflower efficient enough to be applicable in plant breeding programmes. Induced parthenogenesis might represent an alternative for haploid production in the genus *Helianthus*. Todorova et al. (1997a) tried anther culture as well as induced parthenogenesis with interspecific hybrids and wild species, respectively. Although only seven plants of the variant *H. laetiflorus* × Rf147 (pollen treated with 600 Gy gamma rays) could be transferred to the greenhouse, the method via the gynogenic pathway seemed to be promising. Todorova et al. (1997b) further improved the method of induced parthenogenesis by irradiated pollen for the application in cultivated sunflower. Pollen was irradiated with doses of 300, 600 and 900 Gy. From nearly 2,300 embryos, 296 haploid plantlets were obtained using commercial hybrids as female recipients in combination with four restorer lines as pollinators. Most of the haploid plantlets (second–third leaf stage) spontaneously doubled during the next 20 days as could be demonstrated by flow-cytometric measurements. The dose of 600 Gy proved the most suitable one for this purpose. The efficiency of the method was dependent on the genotype of both the pollen donor and the recipient. Therefore, the authors studied 88 additional combinations involving 10 pollinators and two female genotypes (Todorova and Ivanov 1999) to better identify performing pollen donors and to further characterize the interactions between donors, recipient and applied dose of gamma radiation. In total, 705 embryos were cultivated, giving rise to 205 plants that reached maturity. After selfing, 104 plants produced seeds and 65 agronomically useful DH lines were selected based on fertility restoration, branching and downy mildew resistance. The line Z-8-A irradiated with 900 Gy proved to be superior as pollen donor compared to the lines Rf 937 and Rf 147 used in previous studies (Todorova et al. 1997b; Todorova and Ivanov 1999). Although strong interactions among pollen donors and female genotypes were observed, the reports indicate the possibility of a broad application of producing doubled haploids in sunflower.

2.4 Conclusions and Future Prospects of Doubled-Haploid Production in Asteraceae

Sunflower has proven to be recalcitrant in all applications based on its regeneration capacity in tissue culture, which has so far considerably hampered progress in sunflower breeding programmes. Wild species of the genus *Helianthus* and derived interspecific hybrids have been a valuable source for the introductions of agronomically important genes (Korell et al. 1996a,b) and have also led to some progress in tissue culture research (Weber et al. 2000; Hahne 2001). Molecular approaches to map the traits controlling organogenesis might help to identify genotypes with higher regeneration capacities in the near future (Berrios et al. 2000). General improvement in tissue culture of sunflower addressing the precocious flowering (Ivanov et al. 1998), the heritability of in vitro regenerability (Nestares et al. 2002) and genetic analysis of organogenesis (Sarrafi et al. 1996; Berrios et al. 1999; Hewezi et al. 2003) might also accelerate improvements in the establishment of DH techniques in sunflower.

Although the Asteraceae family represents the largest of all plant families, there have been only very few reports of application of haploid techniques on other genera such as *Cichorium* (Theiler and Hunter 1995; Varotto et al. 1996; Theiler et al. 1997), *Hierachium* (Bicknell and Borst 1996), *Guizotia* (Adda et al. 1994) and *Carthamus* (Prasad et al. 1990, 1991). This is mainly due to the limited economical importance of these species. The results are summarized in Table 2. However, none of the reports represents a breakthrough that might help to improve haploid techniques in the genus *Helianthus*.

Table 2. Overview of haploid techniques in other genera apart from *Helianthus* in Asteraceae

Anther culture and microspore culture in other genera	*Authors*
Influence of genotype and cold pretreatment on anther culture in safflower (*Carthamus tinctorius*)	Prasad et al. (1990)
Influence of culture medium and cytological studies of anther-derived plants in safflower	Prasad et al. (1991)
Anther culture of nine niger (*Guizotia abyssinica*) genotypes – androclonal variation in regenerated plants	Adda et al. (1994)
Regeneration of dihaploid chicory (*Cichorium intybus*) via microspore culture	Theiler and Hunter (1995)
Anther culture of chicory with different media	Varotto et al. (1996)
Callusing of anthers from *Hierachium pilosella* – dependence on the stage of microsporogenesis	Bicknell and Borst (1996)
RAPD characterization of microspore-derived plants of chicory	Theiler et al. (1997)
Doubled-haploid production via gynogenic pathway	
In situ gynogenetic haploid plants of chicory after intergeneric hybridization with *Cicerbita alpina*	Dore et al. (1996)

In summary, the reports to date indicate that progress has been made in the application of haploid techniques in sunflower although all methods based on in vitro culture of anthers or microspores display a strong genotype dependency and have only been established for specific genotypes (i.e. interspecific hybrids). Haploid production via the gynogenic pathway by induced parthenogenesis employing irradiated pollen represents a promising way to bypass in vitro culture problems of the androgenic pathway. However, further research is necessary to optimize haploid production for use in sunflower breeding programmes.

References

Adda S, Reddy TP, Kishor PBK (1994) Androclonal variation in niger (*Guizotia abyssinca* Cass). Euphytica 79:59–64

Anonymous (2002) Growing flax – production, management and diagnostic guide, 4th edn. Flax Council of Canada and Saskatchewan Flax Development Commission, Winnipeg and Saskatoon

Bachmann K, Hombergen EJ (1997) From the phenotype via QTL to virtual phenotype in *Microseris* (*Asteraceae*): predictions from multilocus marker genotypes. New Phytol 137:9–18

Badigannavar AM, Kuruvinashetti MS (1996) Callus induction and shoot bud formation from cultured anthers in sunflower (*Helianthus annuus* L.). Helia 19:39–46

Berglund DR (2002) Flax: new uses and demands. In: Janick J, Whipkey A (eds) Trends in new crops and new uses. ASHS Press, Alexandria, pp 358–360

Bergmann R, Friedt W (1997) 1. Haploidy and related biotechnological methods in linseed (*Linum usitatissimum* L.) In: Jain MJ, Sopory SK, Veilleux RE (eds) Current plant science and biotechnology in agriculture, vol 5. Oil, ornamental and miscellaneous plants. Kluwer, Dordrecht, pp 1–16

Berrios EF, Gentzbittel L, Alibert G, Griveau Y, Berville A, Sarrafi A (1999) Genetic control of in vitro-organogenesis in recombinant inbred lines of sunflower (*Helianthus annuus* L.). Plant Breed 118:359–361

Berrios EF, Gentzbittel L, Kayal H, Alibert G, Sarrafi A (2000) AFLP mapping of QTLs for in vitro organogenesis traits using recombinant inbred lines in sunflower (*Helianthus annuus* L.). Theor Appl Genet 101:1299–1306

Berry ST, Leon AJ, Hanfrey CC, Challis P, Burkholz A, Barnes SR, Rufener GK, Lee M, Caligari PDS (1995) Molecular marker analysis of *Helianthus annuus* L. 2. Construction of an RFLP linkage map for cultivated sunflower. Theor Appl Genet 91:195–199

Bicknell RA, Borst NK (1996) Isolation of reduced genotypes of *Hieracium pilosella* using anther culture. Plant Cell Tissue Organ 45:37–41

Brahm L, Röcher T, Friedt W (2000) PCR-based markers facilitating marker assisted selection in sunflower for resistance to downy mildew. Crop Sci 40:676–682

Chen Y, Dribnenki P (2002) Effect of genotype and medium composition on flax *Linum usitatissimum* L. anther culture. Plant Cell Rep 21:204–207

Chen Y, Hausner G, Kenaschuk E, Procunier D, Dribnenki P, Penner G (1998a) Identification of microspore-derived plants in anther culture of flax (*Linum usitatissimum* L.) using molecular markers. Plant Cell Rep 18:44–48

Chen Y, Kenaschuk E, Dribnenki P (1998b) High frequency of plant regeneration from anther culture in flax, *Linum usitatissimum* L. Plant Breed 117:463–467

Chen Y, Kenaschuk, EO, Procunier D (1998c) Plant regeneration from anther culture in Canadian cultivars of flax (*Linum usitatissimum* L.). Euphytica 102:183–189

Chen Y, Kenaschuk E, Dribnenki P (1999) Response of flax genotypes to doubled haploid production. Plant Cell Tissue Organ Cult 57:195–198

Chen Y, Kenaschuk E, Dribnenki P (2001) Inheritance of rust resistance genes and molecular markers in microspore-derived populations of flax. Plant Breed 120: 82–84

Chen Y, Lin S, Duguid S, Dribnenki P, Kenaschuk E (2003) Effect of sucrose concentration on elongation of shoots from flax anther culture. Plant Cell Tissue Organ Cult 72:181–183

Cronquist A (1988) The evolution and classification of flowering plants, 2nd edn. Allen Press, Kansas

Coumans M, Zhong DN (1995) Doubled haploid sunflower (*Helianthus annuus*) plant production by androgenesis – fact or artifact. 2. In vitro isolated microspore culture. Plant Cell Tissue Organ Cult 41:203–209

De Simon M, Morgante M, Lucchin M, Parrini P, Marcocco A (1997) A first linkage map of *Cichorium intybus* L. using a one-way pseudo-testcross and PCR-derived markers. Mol Breed 3:415–425

Dore C, Pringent J, Desperez B (1996) In situ gynogenetic haploid plants of chicory (*Cichorium intybus* L.) after intergeneric hybridization with *Cicerbita alpina* Walbr. Plant Cell Rep 15:758–761

Dribnenki JCP, Green AG (1995) Linola^{TM} '947' low linolenic acid flax. Can J Plant Sci 75:201–202

Dribnenki JCP, Green AG, Atlin GN (1996) Linola^{TM} 989 low linolenic flax. Can J Plant Sci 76:329–331

Dribnenki JCP, McEachern SF, Green AG, Kenaschuk EO, Rashid KY (1999) Linola^{TM} '1084' low-linolenic acid flax. http://www.provenseed.com/proven/linola/linolaprint.html (24.1.2002)

EC (2002) Area, yield and production of fibre flax and hemp. Directorate-General for Agriculture, European Commission. http://europa.eu.int/comm/agriculture/agrista/2000/table-en/en41112.pdf (8.10.2002)

FAO (2003) FAOSTAT agriculture data. http://apps.fao.org/page/collections?subset=agriculture (4.5.2003)

Flores-Berrios E, Gentzbittel L, Kayyal H, Alibert G, Sarrafi A (2000) AFLP mapping of QTLs for in vitro organogenesis traits using recombinant inbred lines in sunflower (*Helianthus annuus* L.). Theor Appl Genet 101:1299–1306

Friedt W (1992) Present state and future prospects of biotechnology in sunflower breeding. In: Seiler G (ed) Field crops research, vol 30. Elsevier, Amsterdam, pp 425–442

Friedt W, Bickert C, Schaub H (1995) In vitro breeding of high-linolenic, doubled-haploid lines of linseed (*Linum usitatissimum* L.) via androgenesis. Plant Breed 114:322–326

Friedt W, Nurhidayah T, Röcher T, Köhler H, Bergmann R, Horn R (1997) Haploid production and application of molecular methods in sunflower (*Helianthus annuus* L.) In: Jain SM, Sopory SK, Veilleux RE (eds) In vitro haploid production in higher plants, vol 5. Oil, ornamental and miscellaneous plants. Kluwer, Dordrecht, pp 17–35

Gentzbittel L, Vear F, Zhang YX, Bervillé A, Nicolas P (1995) Development of a consensus linkage RFLP map of cultivated sunflower (*Helianthus annuus* L.). Theor Appl Genet 90:1079–1086

Gentzbittel L, Mestries E, Mouzeyar S, Mazeyrat F, Badaoui S, Vear F, Tourvieille de Labrouhe D, Nicolas P (1999) A composite map of expressed sequences and phenotypic traits of the sunflower (*Helianthus annuus* L.) genome. Theor Appl Genet 99:218–234

Green AG, Marshall DR (1984) Isolation of induced mutants in linseed (*Linum usitatissimum*) having reduced linolenic acid content. Euphytica 33:321–328

Hahne G (2001) Sunflower. In: Hui YH, Khatchtourians GG, McHughen A, Nip WK, Scorza R (eds) Handbook of transgenic plants. Dekker, New York, pp 813–833

Hewezi T, Jardinaud F, Alibert G, Kallerhoff J (2003) A new approach for efficient regeneration of a recalcitrant genotype of sunflower (*Helianthus annuus*) by organogenesis induction on split embryonic axes. Plant Cell Tissue Organ Cult 73:81–86

Ivanov P, Encheva J, Ivanova I (1998) A protocol to avoid precocious flowering of sunflower plantlets in vitro. Plant Breed 117:582–584

Jeuken M, van Wijk R, Peleman J, Lindhout P (2001) An integrated interspecific AFLP map of lettuce (*Lactuca*) based on two *L. sativa* × *L. saligna* F_2-populations. Theor Appl Genet 103:638–647

Kessli RV, Paran I, Michelmore RW (1994) Analysis of a detailed genetic-linkage map of *Lactuca sativa* (lettuce) constructed from RFLP and RAPD markers. Genetics 136:1435–1446

Korell M, Brahm L, Horn R, Friedt W (1996a) Interspecific and intergeneric hybridization in sunflower breeding. I. General breeding aspects. Plant Breed Abstr 66:925–931

Korell M, Brahm L, Friedt W, Horn R (1996b) Interspecific and intergeneric hybridization in sunflower breeding. II. Specific uses of wild germplasms. Plant Breed Abstr 66:1081–1091

Kurt O, Evans GM (1998) Anther culture potential of linseed (*Linum usitatissimum* L.): effects of genotypes and pretreatment on callus formation and differentiation. J Agric For 22:553–560

Lawson WR, Goulter KC, Henry RJ, Kong GA, Kochmann JK (1998) Marker-assisted selection for two rust resistance genes in sunflower. Mol Breed 4:227–234

Leon AJ, Lee M, Rufener GK, Berry ST, Movers RP (1995) Use of RFLP markers for genetic linkage analysis of oil percentage in sunflower seeds. Crop Sci 35:558–564

Leon AJ, Lee M, Andrade FH (2001) Quantitative trait loci for growing degree days to flowering and photoperiod response in sunflower (*Helianthus annuus* L.). Theor Appl Genet 102:497–503

Lühs W, Friedt W (1994) The major oil crops. In: Murphy DJ (ed) Designer oil crops. VCH, Weinheim, pp 1–130

Manninen OM (2000) Associations between anther-culture response and molecular markers on chromosome 2H, 3H, and 4H of barley (*Hordeum vulgare* L). Theor Appl Genet 100:57–62

Mouzeyar S, Roeckel-Drevet P, Gentzbittel L, Philippon J, Tourvieille de Labrouhe D, Vear F, Nicolas P (1995) RFLP and RAPD mapping of the sunflower *Pl1* locus for resistance to *Plasmopara halstedii* race 1. Theor Appl Genet 91:733–737

Nenova N, Christov M, Ivanov P (2000) Anther culture regeneration from some wild *Helianthus* species. Helia 23:65–72

Nestares G, Zorzoli R, Mroginski L, Picardi L (2002) Heritability of in vitro regeneration capacity in sunflower. Plant Breed 121:366–368

Nichterlein K (2003) Anther culture of linseed (*Linum usitatissimum* L.). In: Maluszynski M, Kasha, KJ, Forster BP, Szarekjo I (eds) Doubled haploid production in crop plants. A manual. Kluwer, Dordrecht, pp 1–480

Nichterlein K, Friedt W (1993) Plant regeneration from isolated microspores of linseed (*Linum usitatissimum* L.). Plant Cell Rep 12:426–430

Nichterlein K, Nickel M, Umbach H, Friedt W (1989): Recent progress and prospects of biotechnology in breeding of linseed (*Linum usitatissimum*). Fat Sci Technol 91:272–275

Nichterlein K, Umbach H, Friedt, W (1991) Genotypic and exogenous factors affecting shoot regeneration from anther callus of linseed (*Linum usitatissimum* L.). Euphytica 58:157–164

Nurhidayah T, Horn R, Röcher T, Friedt W (1996) High regeneration rates in anther culture of interspecific sunflower hybrids. Plant Cell Rep 16:167–173

Prairie Registration Recommending Committee for Grains (2002) PRRCG report. Oilseeds subcommittee. New lines recommended. Flax. 13. Meristem land and science. http://www.meristem.com/prrcg/PRRCG.pdf

Prasad BR, Khadeer MA, Seeta P, Anwar SY (1990) Influence of genotype and cold pretreatment on anther culture response in safflower, *Carthamus tinctorius* L. Indian J Exp Biol 28:924–927

Prasad BR, Khadeer MA, Seeta P, Anwar SY (1991) In vitro induction of androgenic haploids in safflower (*Carthamus tinctorius* L.). Plant Cell Rep 10:48–51

Pretova A, Obert B (2000) Progress in flax embryogenesis. In: Bohanec B (ed) Proc COST 824 Final Meeting, Biotechnological Approaches for Utilization of Gametic Cells, Bled, Slovenia 1–5 July. Directorate General for Research, European Commission, pp 165–169

Rieseberg L, Coi H, Chan R, Spore C (1993) Genomic map of a diploid hybrid species. Heredity 70:285–293

Saji KV, Sujatha M (1998) Embryogenesis and plant regeneration in anther culture of sunflower (*Helianthus annuus* L.). Euphytica 103:1–7

Sarrafi A, Roustan JP, Fallot J, Alibert G (1996) Genetic analysis of organogenesis in the cotyledons of sunflower (*Helianthus annuus* L.). Theor Appl Genet 92:225–229

Schilling EE, Heiser CB (1981) Intrageneric classification of *Helianthus* (*Compositae*). Taxon 30:393–403

Spielmeyer W, Green AG, Bittisnich D, Mendham N, Lagudah ES (1998a) Identification of quantitative trait loci contributing to *Fusarium* wilt resistance on an AFLP linkage map of flax (*Linum usitatissimum*). Theor Appl Genet 97:633–641

Spielmeyer W, Lagudah ES, Mendham N, Green AG (1998b) Inheritance of resistance to flax wilt (*Fusarium oxysporum* f.sp. Schlecht) in a doubled haploid population of *Linum usitatissimum* L. Euphytica 101:287–291

Steiss R, Schuster A, Friedt W (1998) Development of linseed for industrial purposes via pedigree-selection and haploid-technique. Ind Crop Prod 7:303–309

Tang S, Yu JK, Slabaugh MB, Shintami DK, Knapp SJ (2002) Simple sequence repeat map of the sunflower genome. Theor Appl Genet 105:1124–1136

Theiler R, Hunter CS (1995) Regeneration of dihaploid chicory (*Cichorium intybus* L. *varfoliosum hegi*) via microspore culture. Plant Breed 114:18–23

Theiler R, Lüscher M, Hunter CS (1997) Differentiation between diploid and doubled haploid chicory (*Cichorium intybus* L.) by RAPD-PCR analysis. Acta Hortic 447:367–370

Todorova M, Ivanov P (1999) Induced parthenogenesis in sunflower: effect of pollen donor. Helia 22:49–56

Todorova M, Nenova N, Ivanov P, Christov M (1997a) Plant regeneration through anther culture and induced parthenogenesis in the genus *Helianthus*. Biotechnol Equip 11:A27–A30

Todorova M, Ivanov P, Shindrova P, Christov M, Ivanov I (1997b) Doubled haploid production of sunflower (*Helianthus annuus* L.) through irradiated pollen-induced parthenogenesis. Euphytica 97:249–254

Vanhouten W, Vanraamsdonk L, Bachmann K (1994) Intraspecific evolution of *Microseris pygmaea* (*Asteraceae*, *Lactuceae*) analysed by cosegregation of phenotypic characters (QTLs) and molecular markers (RAPDs). Plant Syst Evol 190:49–67

Vasic D, Skoric D, Jocic S (2000) Anther culture of sunflower cultivars. In: Proc 15th Int Sunflower Conf, Toulouse, France 12–15 June, L52–L55

Varotto S, Luccin M, Parrini P (1996) Anther culture in *Cichorium intybus* L. Italus-Hortus 3:12–15

Waycott W, Fort SB, Ryder EJ, Michelmore RW (1999) Mapping morphological genes relative to molecular markers in lettuce (*Lactuca sativa* L.). Heredity 82:245–251

Witsenboer H, Vogel J, Michelmore RW (1997) Identification, genetic localization, and allelic diversity of selectively amplified microsatellite polymorphic loci in lettuce and wild relatives (*Lactuca* spp.). Genome 40:923–936

Weber S, Horn R, Friedt W (2000) High regeneration potential in vitro of sunflower (*Helianthus annuus* L.). Plant Cell Rep 116:271–280

Xu Y, Zhu L, Xiao J, Huang N, McCouch SR (1997) Chromosomal regions associated with segregation distortion of molecular markers in F_2, backcross, doubled haploid, and recombinant inbred populations in rice (*Oryza sativa* L.). Mol Gen Genet 253:535–545

Yang HY, Yan H, Zhou C (1995) In vitro production of haploids in *Helianthus*. In: Bajaj YPS (ed) Biotechnology in agriculture and forestry, vol 10. Legumes and oilseed crops. Springer, Berlin Heidelberg New York, pp 472–486

Yu JK, Tang S, Slabaugh MB, Heesacker A, Cole G, Herring M, Soper J, Han F, Chu WC, Webb DM, Thompson L, Edwards KJ, Berry S, Leon AJ, Grondona M, Olungu C, Maes N, Knapp SL (2003) Towards a saturated molecular genetic linkage map for cultivated sunflower. Crop Sci 43:367–387

Zhong D, Michauxferriere N, Coumans M (1995) Assay for doubled haploid sunflower (*Helianthus annuus*) plant production by androgenesis. Fact or artifact. 1. In vitro anther culture. Plant Cell Tissue Organ Cult 41:91–97

Zivy M, Devaux P, Blaisonneaux J, Jean RTH (1992) Segregation distortion and linkage studies in microspore-derived double haploid lines of *Hordeum vulgare* L. Theor Appl Genet 83:919–924

II.7 Challenges and Limitations to the Use of Haploidy in Crop Improvement

C.E. DON PALMER and W.A. KELLER[1]

1 General Aspects

There have been a number of recent reviews on the uses of doubled haploids in plant improvement (Kao 1996; Kush and Virmani 1996; Pelletier 1998; Bohanec 2002). Consequently this chapter will concentrate on the limitations and potentials of haploids and doubled haploids in plant improvement while emphasizing the recent literature. There is an impressive list of plant species in which haploids and doubled haploids have been produced (Maluszynski et al. 1996, 2003). However, most of the published protocols are developed by genotype selection, manipulation of donor plant growth conditions and modifications to culture media composition and culture conditions.

There have been attempts at understanding the fundamental basis for gametophyte to sporophyte transition and morphogenesis, but our knowledge of how the process is regulated is still fragmentary (Kyo et al. 2003). What are the signals, genes and gene products involved in this process? For this technology to be widely used in crop improvement, greater insight into the molecular and biochemical basis for induction of gametophyte embryogenesis must be achieved.

There are a number of emerging crop species of medicinal, herbal and neutraceutical value where genetic improvement would benefit from doubled haploid technology. While an empirical approach to haploidy is still useful, a fundamental understanding of the process would be invaluable for the development of genotypes with reliable and diverse characteristics.

Significant advances have been made in the use of doubled haploids in the improvement of some species, e.g. members of the Brassicaceae, Poaceae and Solanaceae. However, there has been little improvement in the recovery of doubled haploids from members of the Fabaceae. Soybean, *Glycine max* (L.) Merr.; in spite of its commercial importance, anther culture remains the only route to haploidy (Hu et al. 1996; Rodrigues et al. 2004). In all cases, haploids are produced at a frequency too low to be of value in a soybean breeding program.

[1] NRC – Plant Biotechnology Institute, 110 Gymnasium Place, Saskatoon, Saskatchewan, S7N 0W9, Canada

Biotechnology in Agriculture and Forestry, Vol. 56
Haploids in Crop Improvement II (ed. by C.E. Palmer, W.A. Keller, and K.J. Kasha)

Similarly, in woody species, the use of doubled haploids in breeding has been slow (see Chap. II.4, this Vol.). The wide application of haploid technology to the improvement of woody species would be a valuable addition to other biotechnological techniques aimed at improvement of these species (Campbell et al. 2003). In tree species, because of their robust structure, one major limitation to haploid production is the difficulty in providing controlled donor plant growth conditions known to be important for success in many other species (Palmer and Keller 1999). As a consequence, most research has to rely on pretreatment of isolated flower buds or inflorescences to compensate for these conditions. The main challenges to the application of doubled haploids for improvement of these species are likely to be (1) improved efficiency of embryo production and maturation (Bueno and Manzanera 2003; Bueno et al. 2003) and (2) direct embryogenesis from microspores without a callus phase to reduce gametoclonal variation and improve genetic stability of the embryos (Deutsch et al. 2004). Nevertheless, gametoclonal variation may prove useful for the isolation of novel genotypes.

2 Androgenesis

This topic has been extensively reviewed recently (Chupeau et al. 1998) and is the main route to haploid production. It is clear that a variety of stress factors are involved in the induction of microspore embryogenesis of competent cells (see Chap. I.3, this Vol.) (Henry 1998; Touraev et al. 2001). Such stresses probably inhibit gametophytic development and initiate sporophyte development by mechanisms still unknown. High-temperature stress is frequently used for induction of microspore embryogenesis and the mechanism of induction probably involves heat shock proteins (HSPs) (Segui-Simarro et al. 2003) and heat shock transcription factors may be involved (Lohmann et al. 2004). The importance of heat stress to embryo induction is complicated by observations that antimicrotubule agents can induce embryogenesis by a mechanism not involving HSPs (Zhao et al. 2003). Furthermore, pH changes have been shown to induce embryogenesis without the need for temperature stress (Barinova et al. 2004). Consequently, there are still gaps in our knowledge of the induction and development of embryos from microspores. It appears that microspores at the uninucleate or early binucleate stage of development are competent to respond to the inductive stress signals. With an understanding of the molecular and biochemical basis for this stage restriction, culture conditions may be manipulated to allow embryogenesis over a wider range of stages, and recovery of haploids from a larger number of species, even the currently recalcitrant ones.

In a few cases, notably in *Brassica*, microspore embryogenesis approximates the zygotic counterpart (see Chap. II.2, this Vol.). However, the environment in which microspore embryos develop is very different from that of the zygotic embryo in terms of nutrient availability and transport, hormones and osmotic conditions. Yet there is little comparative physiological and biochemical information on these systems. This is required if microspore-derived embryos are to be truly used as representatives of the zygotic embryo. With isolated microspore culture there is still a need for optimization of culture conditions for synchronized cell division, embryo development and maturation to ensure recovery of normal embryos at high frequency. In *Brassica*, microspore culture cell division can be as high as 90%, while embryo recovery can be only 5% of the cultured cells (Telmer et al. 1995). This suggests that culture conditions are limiting embryo development.

In the case of cereals, there is little emphasis on isolated microspore embryo induction, development and maturation and recovery on a large scale (Ryan et al. 1999; Zheng 2003). Development of a protocol applicable to cereals in a genotype-independent manner would facilitate the use of these embryos in studies of storage product metabolism, and for direct use as artificial seeds.

A major challenge to the widespread adaptation of microspore-derived haploids in crop improvement is to overcome recalcitrance and genotypic differences within species. This would allow easy production of large numbers of embryos without genotypic restrictions. Molecular markers have been associated with embryogenic response of microspores (Zhang et al. 2003) and are useful for selecting responsive genotypes.

Some chemicals have been shown to increase the frequency of microspore embryogenesis in responding genotypes and to induce embryogenesis in non-responding ones (Zheng et al. 2001). If this response proves applicable to a wide range of species and if the mode of action can be unraveled, this may prove very useful in overcoming recalcitrance. The process of microspore embryogenesis is still a black box, but undoubtedly the genes that control zygotic and somatic embryogenesis are expressed during this process. Unlike in vitro somatic embryogenesis which is induced by plant hormones, especially auxins (Mordhorst et al. 1997), microspore embryogenesis appears independent of exogenous hormones.

In *Arabidopsis* the somatic embryogenesis receptor kinase 1 gene (AtSERKI) expression marks cells competent for embryogenesis and overexpression of AtSERKI lead to enhanced somatic embryogenesis (Hecht et al. 2001; Nolan et al. 2003). Other genes such as LEAFY COTYLEDON (LEC1, LEC2) and BABY BOOM (BBM) are involved in somatic embryogenesis (Lotan et al. 1998; Boutilier et al. 2002, see also Chap. I.5, this Vol.). Ectopic expression of these genes in developing microspores may provide some clues for the basis of recalcitrance. In addition, a thorough investigation of gene expression profiles of microspores cultured under inductive and non-inductive conditions should provide useful information on genes and gene interactions related to embryo induction and development.

With the identification of gene products critical to the induction process it may be possible to develop protocols for high-frequency embryo induction without species or genotypic restrictions.

3 Gynogenesis

In many members of the Liliaceae, Cucurbitaceae and Chenopodiaceae, gynogenesis is the main route to doubled haploid production and can be relatively efficient (see Chap. II.5, this Vol.; Sita 1996; Bohanec 2002; Alan et al. 2003). Where there is male sterility resulting in lack of pollen development or unresponsive microspores, gynogenetic haploids may be the only means of homozygous line development. The main limitations to overcome are: genotype dependence, low frequency of haploids, low rate of induced chromosome doubling and low fertility in some species (Alan et al. 2003). The low embryogenic frequency may be overcome by inducing somatic embryogenesis on the developing haploid embryo as such cells are usually totipotent (Ikeda-Iwai et al. 2002). Even where androgenesis is successful there may be value in the gynogenetic approach as novel genetic recombinations may be uncovered and there is absence of albinism among the regenerants (Sita 1996; Eders and Chalyk 2002).

4 Parthenogenesis and Wide Hybridization Crosses

The chromosome elimination method of haploid production is extensively used in monocots (Mujeeb-Kazi and Riera-Lizarazu 1996) but has not been explored in dicots. In species where androgenesis and gynogenesis are impractical this method may be an alternative. The limitations are that the mode of chromosome elimination is not known, embryo yield depends on the genotype of the pollinating parent, embryo rescue and in vitro culture are required and an efficient system for chromosome doubling. In spite of these limitations haploid embryos can be produced at frequencies justifying incorporation into a breeding program, and the agronomic performance of doubled haploid wheat lines from maize × wheat was similar to single seed descent populations (Guzy-Wrobelska and Szarejko 2003), and this method is preferred to anther culture for doubled haploid production (Ma et al. 1999). Where parthenogenetic haploids occur without the need for embryo rescue this simplifies the procedure, and the selection of superior pollinating species can augment the frequency of haploid recovery, as is the case with potato (Peloquin et al. 1996; Straadt and Rasmussen 2003).

Thomas et al. (2003) have provided a list of 200 varieties that were developed by the use of doubled haploid technology. They have also outlined a number of considerations to be taken into account when using this technology in a plant breeding program.

In general, whatever the route to doubled haploid production for effective contribution to a breeding program, the method should be simple, cost-effective, genotype-independent and efficient. The doubled haploids should be genetically stable and chromosome doubling agents should not induce genetic variability (see Pelletier 1998). The full potential of haploid embryogenesis will only be realized when there is a better understanding of what triggers this process and controls the fate of the haploid cell. Recent insights into the genetic control of fertilization-independent endosperm and seed development and apomictic embryo development should aid our understanding of parthenogenetic embryo development (Ohad et al. 1999; Estrada-Llana et al. 2002; Köhler et al. 2003). Perhaps a re-examination of taxa where the gametophyte represents the dominant stage of the life cycle will help us to better understand this process (Mable and Otto 1998; Wang et al. 2000).

5 Emerging Uses of Doubled Haploids

Haploidy is likely to play a major role in the future in breeding programs utilizing induced mutation as this allows rapid selection and fixation of traits (Lee et al. 2003; Szarejko 2003; Vagera et al. 2004). It will also prove valuable in the development of chromosome addition lines (Wang et al. 1993; Henry 1998; Kaneko et al. 2003).

One use that has received considerable attention recently is the development of doubled haploid populations for gene mapping and identification of quantitative trait loci (QTL) (Kush and Virmani 1996; Forster and Thomas 2003). The advantages of these populations compared to other populations have been discussed by Forster and Thomas (2003). The use of this technology for the rapid production of the recombinant inbred lines, which are valuable in gene mapping and QTL detection, is limited only by the relatively few species in which doubled haploids can be recovered efficiently. There are several reports where doubled haploid populations are used for such purposes (Yang et al. 2003; Gupta et al. 2004; Snowdon and Friedt 2004).

5.1 Transformation

A highly synchronized microspore or haploid cell culture where embryogenesis occurs at high frequency is an ideal target for transformation as transgenics can be recovered rapidly and the inserted trait fixed in the homozygous state by chromosome doubling. Such transformation can be effected by *Agrobacterium*, particle bombardment, electroporation or micro-injection (Dormann et al. 1995; Heberle-Bors 1998; Henry 1998; Touraev et al. 2001; Shim and Kasha 2003). For technical reasons the microspore system is more suitable for transformation than other haploid systems such as gynogenesis or parthenogenesis. The use of this system is limited by the susceptibility of the target cells to transformation and the efficiency of embryogenesis over a wide range of species. In a highly efficient system there would be the potential to recover a large number of individual transformants useful for selection, and direct embryogenesis should reduce the incidence of gametoclonal variation.

Germline transformation is another aspect of haploid technology where the aim is to transform microspores which then mature into viable pollen (Touraev et al. 2001). Upon pollination and fertilization the introduced gene is transferred to the next generation. This has been demonstrated in a few cases (Touraev et al. 2001; Aronen et al. 2003) and has the advantage of normal seed development without the need for in vitro culture. For this approach to transformation to have any impact on crop improvement, effective and efficient culture systems must be developed for early stage microspore culture and pollen maturation in vitro (Touraev and Heberle-Bors 1999; Touraev et al. 2001; Barinova et al. 2004).

The pollen maturation process should approximate that of in vivo maturation and germination; pollen tube growth and sperm cell formation should not be altered by the culture process. A large number of microspores must be transformed at an early enough nuclear stage to ensure male gamete transformation and effective pollination.

An effective protocol for the culture of meiocytes capable of in vitro development and microspore formation would be a useful step towards large-scale pollen culture. Before this can be achieved there needs to be a better understanding of the nutritional and hormonal aspects of microsporogenesis.

References

Alan AR, Mutschler MA, Brants A, Cobb E, Earle ED (2003) Production of gynogenic plants from hybrids of *Allium cepa* L and *A. roylei* Stearn. Plant Sci 165:1201–1211

Aronen JS, Nikkonen JO, Häggman HM (2003) The production of transgenic Scots pine (*Pinus sylvestris* L) via the application of transformed pollen in controlled crossings. Transgenic Res 12:375–378

Barinova I, Clement C, Martin L, Bailieul F, Soukupova H, Heberle-Bors I, Touraev A (2004) Regulation of developmental pathways in cultured microspores of tobacco and snapdragon by medium pH. Planta 219:141–146

Bohanec B (2002) Double haploids in onions. In: Rabinowitch HD, Currah L (eds) *Allium* crop science: recent advances. CAB International, Wallingford, pp 145–158

Boutilier K, Offringa R, Sharma VK, Kieft H, Quellet T, Zhang LM, Hattori J, Liu CM, Van Lammeren AAM, Miki BLA (2002) Ectopic expression of BABY BOOM triggers a conversion from vegetative to embryonic growth. Plant Cell 14:1737–1749

Bueno M, Manzanera JA (2003) Oak anther culture. In: Maluszynski M, Kasha KJ, Forster BP Szarejko I (eds) Doubled haploid production in crop plants: a manual. Kluwer, Dordrecht, pp 297–301

Bueno MA, Gomez A, Sepulveda F, Segui JM, Testillano PS, Manzanera JA, Risueno M-C (2003) Microspore-derived embryos from *Quercus suber* anthers mimic zygotic embryos and maintain haploidy in long-term anther culture. J Plant Physiol 160:953–960

Campbell MM, Brunner AM, Jones HM, Strauss SH (2003) Forestry's fertile crescent: the application of biotechnology to forest trees. Plant Biotech J 1:141–154

Chupeau Y, Caboche M, Henry Y (eds) (1998) Androgenesis and haploid plants. Springer, Berlin Heidelberg New York

Deutsch F, Kumlehn J, Ziegenhagen B, Fladung M (2004) Stable haploid poplar callus lines from immature pollen cultures. Physiol Plant 120:613–632

Dormann M, Wang HM, Datla N, Ferrie AMR, Keller WA, Oelck MM (1995) Transformation of freshly isolated *Brassica* microspores and regeneration of fertile homozygous plants. In: Proc 9th Int Rapeseed Congr, Cambridge, UK, 4–7 July, Groupe Consultating International de Recherche sur le Colza, pp 816–818

Eders J, Chalyk S (2002) In vitro haploid induction in maize. Theor Appl Genet 104:703–708

Estrada-Llana AA, Illanca-Mamrani W, Acosta-Garcia G, Leon-Martinez G, Becerra-Flora A, Perez-Ruiz R, Vielle-Calzada JP (2002) Beyond promiscuity: from sexuality to apomixes in flowering plants. In Vitro Cell Dev Biol Plant 38:146–151

Forster BP, Thomas WTB (2003) Doubled haploids in genetic mapping and genomics. In: Maluszynski M, Kasha KJ, Forster BP, Szarejko I (eds) Doubled haploid production in crop plants: a manual. Kluwer, Dordrecht, pp 376–390

Gupta V, Mukhopadhyay A, Arumujan N, Sodhi YS, Pental D, Pradhan AK (2004) Molecular tagging of erucic acid trait in oilseed mustard (*Brassica juncea*) by QTL mapping and single nucleotide polymorphisms in FAEI gene. Theor Appl Genet 108:743–749

Guzy-Wrobelska J, Szarejko I (2003) Molecular and agronomic evaluation of wheat doubled haploid lines obtained through maize pollination and anther culture methods. Plant Breed 122:305–313

Heberle-Bors E (1998) Experimental control of pollen development. In: Chupeau Y, Caboche M, Henry Y (eds) Androgenesis and haploid plants. Springer, Berlin Heidelberg New York, pp 38–53

Hecht V, Vielle-Calzada JP, Hartog MV, Schmidt ED, Boutilier K, Grossniklaus U, de Vries SC (2001) The *Arabidopsis* somatic embryogenesis receptor kinase 1 gene is expressed in developing ovules and embryos and enhances embryogenic competence in culture. Plant Physiol 127:803–816

Henry Y (1998) What are the next priorities for the future? In: Chupeau Y, Caboche M, Henry Y (eds) Androgenesis and haploid plants. Springer, Berlin Heidelberg New York, pp 113–119

Hu C, Yin G, Bodanese-Zanettini MH (1996) Haploid of soybean. In: Jain SM, Sopory SK, Veilleux RE (eds) Haploid production in higher plants, vol 3. Kluwer, Dordrecht, pp 377–395

Ikeda-Iwai, Satoh S, Kamada H (2002) Establishment of a reproducible tissue culture system for the induction of *Arabidopsis* somatic embryos. J Exp Bot 53:1575–1580

Kaneko Y, Bang SW, Torii-Abe J, Eduardo RB, Matsuzawa Y (2003) Chromosome pairing and haploid plants of radish derived from alien monosomic addition lines. Plant Breed 122:450–554

Kao KN (1996) Future prospects for crop improvement through anther and microspore culture. In: Jain SM, Sopory SK, Veilleux RE (eds) In vitro haploid production in higher plants. Kluwer, Dordrecht, pp 367–373

Kim M, Kim J, Yoon M, Choi D-Il, Lee M (2004) Origin of multicellular pollen and pollen embryos in cultured anthers of pepper (*Capsicum annuum*). Plant Cell Tiss Org Cult 77:63–72

Köhler C, Hennig H, Spillane C, Pieu S, Greissem W, Grossuiklaus U (2003) The polycomb-group protein MEDEA regulates seed development by controlling expression of the MADS-box gene PHERES I. Genes Dev 17:1540–1553

Kush GS, Virmani SS (1996) Haploids in plant breeding. In: Jain SM, Sopory SK, Veilleux RE (eds) In vitro haploid production in higher plants, vol 1. Kluwer, Dordrecht, pp 11–33

Kyo M, Hattori S, Yamaji N, Pechan P, Fukui H (2003) Cloning and characterization of cDNAs associated with the embryogenic dedifferentiation of tobacco immature pollen grains. Plant Sci 164:1057–1066

Lee SY, Cheong JI, Kim TS (2003) Production of doubled haploids through anther culture of hybrid plants derived from mutagenized fertilized egg cells. Plant Cell Rep 22:218–223

Lohmann C, Eggers-Schumacher G, Wunderlich M, Schoff F (2004) Two different heat shock transcription factors regulate immediate early expression of stress genes in *Arabidopsis*. Mol Gen Genet 271:11–21

Lotan T, Ohto M, Yee KM, West MAL, Lo R, Kwang RW, Yamagiski K, Fischer RL, Goldberg RB, Harada JJ (1998) *Arabidopsis* LEAFY COTYLEDON1 is sufficient to induce embryo development in vegetative cells. Cell 93:1195–1205

Ma H, Busch RH, Riera-Lizarazu O, Rines HW, Dill-Macky R (1999) Agronomic performance of lines derived from anther culture, maize pollination and single-seed descent in a spring wheat cross. Theor Appl Genet 99:432–436

Mable BK, Otto SP (1998) The evolution of life cycles with haploid and diploid phases. Bioassays 20:454–462

Maluszynski M, Szarejko I, Sigurbjorusson B (1996). Haploid and mutation techniques. In: Jain SM, Sopory SK, Veilleux RE (eds) In vitro haploid production in higher plants, vol 1. Kluwer, Dordrecht, pp 67–73

Maluszynski M, Kasha KJ, Szarejko I (2003) Published doubled haploid protocols in plant species. In: Maluszynski M, Kasha KJ, Forster BP, Szarejko I (eds) Doubled haploid production in crop plants, a manual. Kluwer, Dordrecht, pp 309–335

Mordhorst AP, Toonen MAJ, de Vries SC (1997) Plant embryogenesis. Crit Rev Plant Sci 16:535–576

Mujeeb-Kazi A, Riera-Lizaraza O (1996) Polyhaploid production of Tritaceae by sexual hybridization. In: Jain SM, Sopory SK, Veilleux RE (eds) In vitro haploid production in higher plants, vol 1. Kluwer, Dordrecht, pp 275–296

Nolan K, Irwanta RR, Rose RJ (2003) Auxin-up-regulates MtSERK1 expression in both *Medicago truncatula* root forming and embryogenic cultures. Plant Physiol 133:218–230

Ohad N, Yadegari R, Morgossian L, Hannon M, Michaeli D, Harada J, Goldberg RB, Fischer RL (1999) Mutations in FIE, a WD polycomb group gene, allow endosperm development without fertilization. Plant Cell 11:407–415

Palmer CE, Keller WA (1999) Haploidy in *Brassica*. In: Gomez-Campo F (ed) Biology of *Brassica* and coeno-species. Elsevier, Amsterdam, pp 267–286

Peloquin SJ, Babert AC, Ortiz R (1996) Nature of 'pollinator' effect in potato (*Solanum tuberosum* L.) haploid production. Ann Bot 77:539–542

Pelletier G (1998) Use of haplo-diploidisation for plant breeding. In: Chupeau Y, Caboche M, Henry Y (eds) Androgenesis and haploid plants. Springer, Berlin Heidelberg New York, pp 113–119

Rodrigues LR, Terra T de F, Bered F, Bodanese-Zanettini MH (2004) Origin of embryo-like structures in soybean anther culture investigated using SSR markers. Plant Cell Tissue Organ Cult 77:287–289

Ryan AB, Castillo AM, Valles MP, Sanz JM, Cistue L (1999) Dessicated doubled-haploid embryos obtained from microspore culture of barley cv. Igri. Plant Cell Rep 18:924–928

Segui-Simarro JM, Testillano PS, Risueno MC (2003) HSP30 and HSP90 change their expression and subcellular localization after microspore embryogenesis induction in *Brassica napus* L. J Struct Biol 142:379–391

Shim YS, Kasha KJ (2003) Barley microspore transformation protocol by biolistic gun. In: Maluszynski M, Kasha KJ, Forster BP, Szarejko I (eds) Doubled haploid production in crop plants: a manual. Kluwer, Dordrecht, pp 363–366

Sita GL (1996) Gynogenic haploids in vitro. In: Jain SM, Sopory SK, Veilleux RE (eds) In vitro haploid production in higher plants, vol 5. Kluwer, Dordrecht, pp 175–193

Snowdon RJ, Friedt W (2004) Molecular markers in *Brassica* oilseed breeding: current status and future possibilities. Plant Breed 123:1–8

Straadt IK, Rasmussen OS (2003) AFLP analysis of *Solanum phureja* DNA introgression into potato dihaploid. Plant Breed 122:352–356

Szarejko I (2003) Doubled haploid mutation production. In: Maluszynski M, Kasha KJ, Forster BP, Szarejko I (eds) Doubled haploids production in crop plants, a manual. Kluwer, Dordrecht, pp 351–361

Telmer CA, Newcomb W, Simmonds DH (1995) Cellular changes during heat shock induction and embryo development of cultured microspores of *Brassica napus* cv. Topaz. Protoplasma 185:106–112

Thomas WTB, Forster BP, Gertsson B (2003) Doubled haploids in breeding. In: Maluszynski M, Kasha KJ, Forster BP, Szarejko I (eds) Doubled haploids production in crop plants: a manual. Kluwer, Dordrecht, pp 337–349

Touraev A, Heberle-Bors E (1999) Microspore embryogenesis and in vitro pollen maturation in tobacco. In: Hall R (ed) Plant cell culture protocols, vol 3. Humana Press, Totowa, New Jersey, pp 281–291

Touraev A, Pfosser M, Heberle-Bors E (2001) The microspore: a haploid multipurpose cell. Adv Bot Res 35:53–109

Vagera J, Novotny J, Ohnoutkova L (2004) Induced androgenesis in vitro in mutated populations of barley, *Hordeum vulgare*. Plant Cell Tissue Organ Cult 77:55–61

Wang G, Ji J, Wang YB, Hu H, King IP, Snape JW (1993) The genetic characterization of a novel multi-addition doubled haploid lines derived from triticale × wheat hybrids. Theor Appl Genet 87:531–536

Wang M, van Bergen S, van Duijn B (2000) Insights into a key developmental switch and its importance for efficient plant breeding. Plant Physiol 124:523–530

Yang ZP, Gilbert J, Somers DJ, Fedak G, Procunier JD, McKenzie IH (2003) Marker assisted selection of *Fusarium* head blight resistance genes in two doubled haploid populations of wheat. Mol Breed 12:309–317

Zhang IL, Aoki S, Takahata Y (2003) RAPD markers linked to microspore embryogenics ability in *Brassica* crops. Euphytica 131:207–213

Zhao J, Newcomb W, Simmonds D (2003) Heat-shock proteins 70 kDa and 19 kDa are not required for induction of embryogenesis of *Brassica napus* L. cv Topaz microspores. Plant Cell Physiol 44:1417–1421

Zheng MY (2003) Microspore culture in wheat (*Triticum aestivum*) – doubled haploid production via induced embryogenesis. Plant Cell Tissue Organ Cult 73:213–230

Zheng MY, Liu W, Weng Y, Polle E, Konzak CF (2001) Culture of freshly isolated wheat (*Triticum aestivum* L.) microspores treated with inducer chemicals. Plant Cell Rep 20:685–690

Zuo JR, Nia QW, Frugis G, Chua NH (2002) The WUSCHEL gene promotes vegetative to embryonic transition is *Arabidopsis*. Plant J 30:349–359

Subject Index

MIX
Papier aus verantwortungsvollen Quellen
Paper from responsible sources
FSC® C105338

If you have any concerns about our products,
you can contact us on
ProductSafety@springernature.com

In case Publisher is established outside the EU,
the EU authorized representative is:
Springer Nature Customer Service Center GmbH
Europaplatz 3, 69115 Heidelberg, Germany

Printed by Libri Plureos GmbH
in Hamburg, Germany